# Miniaturized Electrochemical Devices

Evidently, electrochemical sensing has revolutionized the electroanalytical detections in the world. Since the 19th century, a huge amount of growth has been visible on various fronts, such as biosensors, energy devices, semiconductor devices, communication, embedded systems, sensors etc. However, the major research gap lies in the fact that most of the reported literatures are bulk systems; hence there are limitations for practical applications.

Research in these domains has been carried out by both academia and industry, whereby academics is the backbone whose intellectual outputs have been widely adopted by the industry and implemented for consumers at large. In order to impart portability to the electrochemical sensors for point-of-care application, the collaboration of electrochemistry, microfluidics, electronics and communication as an interdisciplinary forum is crucial. The miniaturization, automation, IoT enabling and integration are the requirements for building the mentioned research gap. The conversion of electrochemical sensing theoretical concepts to practical applications in real time via miniaturization and integration of microfluidics will enhance this domain. In this context, of lately, several research groups have developed miniaturized microdevices as electrochemical-sensing platforms. This has led to a demand of offering a reference book as a guideline for the PhD programs in electrochemistry, MEMS, electronics and communication. Undoubtedly, this will have a huge impact for R&D in industries, public-funded research institutes and academic institutions.

The book will provide a single forum to understand the current research trends and future perspectives of various electrochemical sensors and their integration in microfluidic devices, automation and point-of-care testing. For students, the book will become a motivation for them to explore these areas for their career standpoints. For the professionals, the book will become a thought-provoking stage to manoeuvre the next-generation devices/processes for commercialization.

# Miniaturized Electrochemical Devices

## Advanced Concepts, Fabrication, and Applications

Edited by
### Sanket Goel
### Khairunnisa Amreen

**CRC Press**
Taylor & Francis Group
Boca Raton  London  New York

CRC Press is an imprint of the
Taylor & Francis Group, an **Informa** business

First edition published 2023
by CRC Press
2385 NW Executive Center Drive, Suite 320, Boca Raton, FL 33431

and by CRC Press
4 Park Square, Milton Park, Abingdon, Oxon, OX14 4RN

*CRC Press is an imprint of Taylor & Francis Group, LLC*

© 2023 selection and editorial matter, Sanket Goel and Khairunnisa Amreen; individual chapters, the contributors

ISBN: 978-1-032-39271-4 (hbk)
ISBN: 978-1-032-54173-0 (pbk)
ISBN: 978-1-003-41549-7 (ebk)

DOI: 10.1201/b23359

Typeset in Times
by MPS Limited, Dehradun

# Contents

# About the Editors

**Dr. Sanket Goel** is a professor with the Department of Electrical and Electronics Engineering and dean, sponsored research and consultancy division at BITS-Pilani, Hyderabad campus. He joined the institution in 2015; during his tenure, he headed the EEE Department at BITS-Pilani (2017–2020). Prior to this, he led the R&D department and was an associate professor at the University of Petroleum & Energy Studies (UPES), Dehradun, India (2011–2015).

Dr. Goel received his BSc (H-Physics) from the Ramjas College, Delhi University; MSc (physics) from IIT Delhi; PhD (electrical engineering) from the University of Alberta, Canada on NSERC fellowship; and MBA in international business from Amity University in 1998, 2000, 2006 and 2012, respectively. He has worked with two Indian national labs: Institute of Plasma Research, Gandhinagar (2000–2001) and DEBEL-DRDO, Bangalore (2006). As an NIH fellow, Sanket performed his postdoctoral work at the Stanford University, US (2006–2008), and worked as a principal investigator at A*STAR, Singapore (2008–2011).

His current research interests are MEMS, microfluidics, nanotechnology, materials and devices for energy, biochemical and biomedical applications, science policy and innovation and entrepreneurship. As a principal investigator, Sanket has been implementing several funded projects (from DRDO, DST, ISRO, MNRE, Government of India; UNESCO; European Commission) and has been collaborating with various groups in India and abroad.

During the course of his career, Dr. Goel has won several awards, including the Fulbright-Nehru fellowship (2015), DST Young Scientist Award (2013), American Electrochemical Society's Best students paper award (2005), University of Alberta PhD thesis award (2005) etc. As of April 2023, he has more than 325publications and 26patents to his credits. He has delivered 95invited talks, guided/guiding 45PhDs, and several masters and bachelors students.

Dr. Goel is a senior member, IEEE; life member, Institute of Smart Systems and Structures; and life member, Indian Society of Electrochemical Chemistry. Currently he is an associate editor of *IEEE Transactions on NanoBioscience, IEEE Sensors Journal, IEEE Access, Applied Nanoscience*, and guest editor of a special issue on sensors: "3D Printed Microfluidic Devices". He is also serving as a visiting professor with UiT, The Arctic University of Norway.

**Dr. Khairunnisa Amreen** is working as a young scientist post-doctorate for ICMR-DHR at Department of Electrical and Electronics Engineering, BITS-Pilani, Hyderabad campus under the supervision of Prof. Sanket Goel. Before this, she was a SERB-National post-doctorate fellow with the Department of Electrical and Electronics Engineering, BITS-Pilani, Hyderabad campus.

Prior to this, she worked as an assistant professor at the Dept. of PG chemistry, St. Ann's College for women, Hyd, India. Dr. Amreen graduated from Osmania University, Telangana, India and received her M.Sc. (analytical chemistry) from Vellore Institute of technology (VIT), Vellore, TN. India. She received her doctoral degree (PhD in electrochemistry) from VIT, Vellore with specialization in electroanalytical sensing. Dr. Amreen has worked as a teaching cum research associate and as a CSIR-senior research fellow at VIT.

Her research interests are biosensing, electrochemical sensing, nanotechnology, microfluidic electrochemical sensing, 3D-printed electrodes and material synthesis. During the course of her career, she has been awarded fellowships and awards like Science and Engineering Board-National Post doctorate, CSIR-SRF, Research award from VIT in recognition for her contribution to research and publications in reputed journals and Indo-Asian Research Excellence award.

She has 50 research publications and three Indian patents filed to her credit. She has also co-authored 12 book chapters for reputed publications like IEEE, Elsevier, IOP and Science. She has presented papers at more than 15 international/national conferences. Dr. Amreen is a passionate speaker and has delivered several invited talks as a resource person. She believes in working for the benefit of the society through her research and is presently working on fabricating devices for health management.

# Contributors

**Saroj Sundar Baral**
Department of Chemical Engineering
BITS Pilani
K K Birla Goa Campus
Goa, India

**Manish Bhaiyya**
MEMS
Microfluidics and Nanoelectronics Lab
Department of Electrical and
    Electronics Engineering
Birla Institute of Technology and
    Science Pilani
Hyderabad Campus
Hyderabad, Telangana, India

**Mitul Bhalerao**
Department of Biological Sciences
Center for Genetics and Molecular
    Microbiology
Birla Institute of Technology and
    Sciences-Pilani
Hyderabad Campus
Hyderabad, Telangana, India

**Chanchal Chakraborty**
Birla Institute of Technology and
    Science (BITS) Pilani
Hyderabad Campus
Hyderabad, Telangana, India

**Jhansi Chintakindi**
Department of Chemical Engineering
BITS Pilani Hyderabad Campus
Hyderabad, Telangana, India

**Dev Choudhary**
Department of Chemical Engineering
BITS, Pilani
K K Birla Goa Campus
Goa, India

**Sri Satya Omkar Dadi**
Birla Institute of Science and
    Technology
Pilani, Rajasthan, India

**Yogeshwar Devarakonda**
Department of Biological Sciences
Center for Genetics and Molecular
    Microbiology
Institute of Eminence
Birla Institute of Technology and
    Sciences-Pilani
Hyderabad, Telangana, India

**Arnab Dutta**
Department of Chemical Engineering
Birla Institute of Technology and
    Science (BITS) Pilani
Hyderabad Campus
Hyderabad, Telangana, India

**Asmita Dileep Gaonkar**
Department of Chemistry
Birla Institute of Technology & Science
    (BITS) Pilani
K K Birla Goa Campus
Zuarinagar, Goa, India

**Sanket Goel**
Electrical and Electronics Engineering
    Department
BITS Pilani Hyderabad Campus
Hyderabad,  Telangana, India

**Navneet Gupta**
Department of Electrical and
    Electronics Engineering
Birla Institute of Technology and
    Science
Pilani, Rajasthan, India

**R. K. Gupta**
Department of Physics
Birla Institute of Technology and
 Science
Pilani (BITS-Pilani)
Rajasthan, India

**Subbalakshmi Jayanty**
Department of Chemistry
Birla Institute of Technology and
 Science
Pilani-Hyderabad Campus
Hyderabad, Telangana, India

**Ashutosh Joshi**
Department of Physics
Birla Institute of Technology and
 Science
Pilani (BITS-Pilani)
Rajasthan, India

**Surendran Jyothis**
Mechanical Engineering Department
BITS Pilani Hyderabad Campus
Hyderabad, Telangana, India

**Sayan Kanungo**
Department of Electrical & Electronics
 Engineering and Materials Center for
 Sustainable Energy & Environment
Birla Institute of Technology and
 Science-Pilani
Hyderabad, Telangana, India

**Santanu Koley**
Department of Mathematics
Birla Institute of Technology and
 Science-Pilani
Hyderabad Campus
Hyderabad, Telangana, India

**Jegatha Nambi Krishnan**
Department of Chemical Engineering
BITS Pilani
K K Birla Goa Campus
Goa, India

**Reddi Lakshmi**
Department of Biological Sciences
Center for Genetics and Molecular
 Microbiology
Birla Institute of Technology and
 Sciences-Pilani
Hyderabad Campus
Hyderabad, Telangana, India

**Dr. Afkham Mir**
Department of Chemical Engineering
BITS Pilani Hyderabad Campus
Hyderabad, Telangana, India

**Debirupa Mitra**
Department of Chemical Engineering
Birla Institute of Technology and
 Science (BITS) Pilani
Hyderabad Campus
Hyderabad, Telangana, India

**Sangeeta Jana Mukhopadhyay**
Department of Electrical and
 Communication Engineering
Dr. Sudhir Chandra Sur Institute of
 Technology and Sports Complex
JIS Group
Surermath, Kolkata, India

**Ramendra Kishor Pal**
Department of Chemical Engineering
Birla Institute of Technology and
 Science (BITS) Pilani
Hyderabad Campus
Hyderabad, Telangana, India

**Koustav Pan**
Department of Chemical Engineering
BITS Pilani
K K Birla Goa Campus
Goa, India

**Shraddha Paniya**
Department of Chemistry
Birla Institute of Technology & Science
  (BITS) Pilani
K K Birla Goa Campus
Zuarinagar, Goa, India

**Divyansh Singh Patel**
Birla Institute of Science and
  Technology
Pilani, Rajasthan, India

**Prasant Kumar Pattnaik**
MEMS
Microfluidics and Nanoelectronics Lab
Department of Electrical and
  Electronics Engineering
Birla Institute of Technology and
  Science Pilani
Hyderabad Campus
Hyderabad, Telangana, India

**Prathiksha P Prabhu**
Department of Chemical Engineering
BITS Pilani
K K Birla Goa Campus
Goa, India

**Susmita Roy**
Birla Institute of Technology and
  Science (BITS) Pilani
Hyderabad Campus
Hyderabad, Telangana, India

**Neethu R. S.**
Department of Biological Sciences
Center for Genetics and Molecular
  Microbiology
Birla Institute of Technology and
  Sciences-Pilani
Hyderabad Campus
Hyderabad, Telangana, India

**Vyom Sharma**
Indian Institute of Technology Kanpur
Kanpur, Uttar Pradesh, India

**Kumar Shivesh**
MEMS
Microfluidics and Nanoelectronics Lab
Department of Electrical and
  Electronics Engineering
Birla Institute of Technology and
  Science-Pilani
Hyderabad Campus
Hyderabad, Telangana, India

**Sarthak Singh**
Department of Chemical Engineering
BITS Pilani
K K Birla Goa Campus
Goa, India

**Ravindran Sujith**
Mechanical Engineering Department
BITS Pilani Hyderabad Campus
Hyderabad, Telangana, India

**Kirtimaan Syal**
Department of Biological Sciences
Center for Genetics and Molecular
  Microbiology
Birla Institute of Technology and
  Sciences-Pilani
Hyderabad Campus
Telangana, India

**Reva Teotia**
Department of Electrical and
   Electronics Engineering
Birla Institute of Technology and
   Science
Pilani, Rajasthan, India

**Aditya Tiwari**
Department of Electrical and
   Electronics Engineering
Birla Institute of Technology and
   Science-Pilani
Hyderabad, Telangana, India

**Kshma Trivedi**
Department of Mathematics
Birla Institute of Technology and
   Science-Pilani
Hyderabad Campus
Hyderabad, Telangana, India

**Manjuladevi V.**
Department of Physics
Birla Institute of Technology and
   Science
Pilani (BITS-Pilani)
Rajasthan, India

**Vipin Valappil**
Department of Mathematics
Birla Institute of Technology and
   Science-Pilani
Hyderabad Campus
Hyderabad, Telangana, India

**Kiran Vankayala**
Department of Chemistry
Birla Institute of Technology & Science
   (BITS) Pilani
K K Birla Goa campus
Zuarinagar, Goa, India

# 1 Electrochemical Micromachining

*Sri Satya Omkar Dadi*
Birla Institute of Science and Technology, Pilani,
Rajasthan, India

*Vyom Sharma*
Indian Institute of Technology Kanpur, Kalyanpur,
Kanpur, Uttar Pradesh, India

*Divyansh Singh Patel*
Birla Institute of Science and Technology, Pilani,
Rajasthan, India

## CONTENTS

## 1.1 INTRODUCTION

Electrochemical micromachining (ECMM) is an advanced micromachining process used for the processing of conducting as well as semiconducting materials. In this process, the material to be machined (referred to as a workpiece) is connected to an external DC process energy source (PES) and a counter-electrode (referred to as a tool) is connected to the negative terminal of the PES. In between these two electrodes, an electrically conducting fluid (referred to as an electrolyte) is supplied, which completes the electrical circuit. In the gap between these two electrodes (referred to as inter-electrode gap (IEG)), an electric field is established and material starts dissolving from the workpiece at an ionic level. If the tool material is inert to the electrolyte solution used, no material gets removed from it in the process and therefore, the tool has theoretically infinite life. This feature of ECMM makes it one of the most promising techniques for micromachining operations (McGeough 1974). The difference between ECMM and electrochemical machining (ECM) is in terms of the dimension of the tool used, the magnitude of the applied potential, the IEG, the conductivity of the electrolyte used, and the relative tool/workpiece feed rate (De Barr and Oliver 1968). The form of the tool determines the shape and dimensions of the workpiece; hence, pre-defined tool forms are utilised to create the required features on the workpiece. In ECMM, the tool dimensions are generally in the order of a few micrometres and, hence, result in a high current density on the workpiece surface beneath the tool.

According to the classical theory of ECM, positive metal ions leave the workpiece during electrolysis, and electrons leave the tool to reach the opposite ends. The quantification of the physical change occurring at the workpiece surface is done using the first and second law of electrolysis proposed by Michael Faraday in the early 19th century. E. Shiptalsky, a Russian physicist, devised ECM in 1911. In 1929, Russian engineers V. N. Gusev and L. Rozhkov improved the technique, and in 1959, the Anocut Engineering Company (U.S.) commercialised the ECM model. This technology gained popularity in the 1960s and 1970s for machining hard, intricately formed aerospace materials. It was also used in the tool manufacturing industries of Russia and Western Europe. By the end of the 1990s, the ECM was used for machining many types of materials in aerospace, textile, biomedical, and automotive industries. Electrochemical manufacturing processes can be classified into four categories as electrochemical machining, electrochemical micromachining, electrochemical polishing and electrochemical deposition, where material removed or deposited are based on Faraday's laws of electrolysis. Pulsed voltage as input in ECM resulted in localised material dissolution and led to the evolution of ECMM. The schematic of ECM is shown in Figure 1.1(a). The variation in current density causes the selective dissolving of metal from the anode's surface, therefore replicating the form of the tool on the workpiece, as shown in Figure 1.1(b). A pump supplies high-velocity electrolyte between two electrodes to flush away reaction products and disperse heat. The IEG must be maintained at a minimum (5–50 m) to reduce resistance and increase current flow across the gap, which increases material removal (MRR) (B. Bhattacharyya, Munda, and Malapati 2004).

ECM and ECMM can be differentiated by the differences in the dimensions of tool diameter, IEG, electrolyte flow velocity, tool/workpiece feed rate and applied potential. ECMM employs an intermittent voltage supply consisting of a current pulse and a relaxation phase of zero current, which enables more efficient removal of electrolysis products from the constricted IEG (Bijoy Bhattacharyya 2015b).

Following are some advantages of the ECM/ECMM process:

1. The surface of the machined workpiece is free of burrs and has a good surface finish (Ra between 0.05 μm to 0.4 μm) (B. Bhattacharyya, Munda, and Malapati 2004).

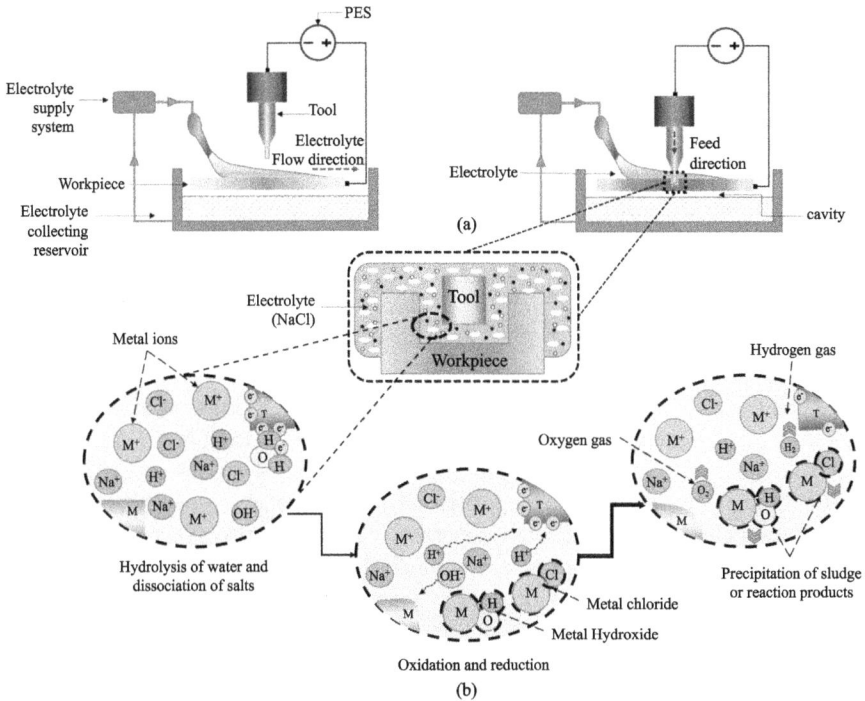

**FIGURE 1.1** (a) Schematic of ECM. (b) Anodic dissolution of metals.

2. The tool and workpiece are not in contact while machining. Thus, the surface of the workpiece is devoid of mechanical residual stresses.
3. No heat-affected zone is created and, hence, thermal stresses are absent in the machined workpiece.
4. The process performance is independent of the physical and thermal properties of the workpiece.
5. Protracted tool life.

Following are the limitations of the ECM/ECMM process:

1. The tool and workpiece materials must conduct electricity.
2. Sand inclusions and spots in the material of the workpiece make it difficult to machine the component.
3. The stray current in the machining zone of ECMM cannot be eliminated completely. They cause unwanted material removal and reduce machining accuracy and surface finish.
4. For each new micro-fabrication, new research is essential. So, for the process to be cost effective, higher production numbers are required.

Micromachining is a subtractive manufacturing process that employs mechanical microtools with defined cutting-edge geometries in the fabrication of parts or features in the micrometer range at least for some of their dimensions (Mativenga 2018). Communications, optics, computers, MEMS, micro-fluids, automotive, biomedical, aerospace, electronics, refrigeration, and air conditioning industries have been adapting micromachining rapidly due to increasing product complexity. The market is migrating towards smaller products due to availability, cost, energy, space limitations, and for better technology. Advanced engineering materials can be machined using a variety of micromachining techniques, which are chosen depending on requirements such as accuracy, surface finish, cost, and production rate.

### 1.1.1 FARADAY'S LAWS OF ELECTROLYSIS

The two fundamental laws of electrolysis stated by Michael Faraday, in the year 1834, serve as the foundation for electrodeposition and dissolution processes. The statements can be represented as (Bijoy Bhattacharyya 2015a):

1. The mass of a substance deposited on the electrode or dissolved from it during electrolysis is directly proportional to the quantity of current passed.
2. The mass of different elemental materials deposited or dissolved for a given quantity of electricity is directly proportional to the elements equivalent weight.

Faraday's laws can be mathematically represented as shown in Equation (1.1):

$$m = \frac{ItA}{zF} \qquad (1.1)$$

where,

m = mass (g),
I = current (amp),
t = time (s),
A = gram equivalent weight of anode material (g/mol),
z = valency number of anode material, and
F = Faraday's constant ($\approx$ 96,500 C/mol).

### 1.1.2 MATERIAL REMOVAL RATE

Material removal rate in ECM/ECMM can be expressed in three different forms as: $MRR_g$ for mass material removal rate in g/s $MRR_v$ for volumetric removal rate in mm$^3$/s and $MRR_I$ for liner material removal rate in m/s (Jain 2011). The following assumptions are made for simplifying the analysis:

i. Electrical conductivity (k) of the electrolyte is to remain constant in IEG.
ii. The electrodes' conductivity is very large compared to that of electrolyte.
iii. The anode dissolution occurs at one fixed valency.

iv. Across the electrodes, the effective working voltage is $(V - \Delta V)$, where $V$ is the applied voltage and $\Delta V$ is a fraction of $V$ that includes over-voltage, electrode voltage, etc.

From Equation (1.1), $MRR_g$ can be calculated as

$$MRR_g = \frac{\eta IA}{zF} \qquad (1.2)$$

where $\eta$ is current efficiency which is the ratio of actual mass of the material dissolved to theoretical mass dissolved.

$MRR_v$ can be evaluated from Equation (1.2) as

$$MRR_v = \frac{\eta IA}{zF\rho_a} \qquad (1.3)$$

where $\rho_a$ is the anode material density $(g/mm^3)$.

$MRR_I$ is evaluated as

$$MRR_I = \frac{\eta JA}{zF\rho_a} \qquad (1.4)$$

where $J$ is the current density which is expressed as

$$J = \frac{I}{A} \qquad (1.5)$$

where $A$ is the common electrode area $(mm^2)$ through which the current flows. Current density can also be expressed as

$$J = \frac{k(V - \Delta V)}{h} \qquad (1.6)$$

where $h$ is IEG. In ECM/ECMM, the tool is usually fed at a constant rate towards the workpiece. At equilibrium, $MRR_I$ (the rate of reduction in material thickness) is equal to the rate at which the tool is fed to the workpiece (feed rate). Thus, on substituting the value of current density from Equation (1.6) in Equation (1.4), the feed rate (f) can be expressed as

$$f = \frac{\eta Ak(V - \Delta V)}{zh_eF\rho_a} \qquad (1.7)$$

where $h_e$ is the IEG when the equilibrium is attained.

### 1.1.3 WORKING PRINCIPLE OF ECMM

The anodic dissolution occurs because of simultaneous oxidation and reduction phenomena when the electric current flows through the system. The motion of cations and anions present in the electrolyte solution are responsible for the charge flow and hence the material removal from the anode takes place as shown in Figure 1.1(b).

Depending on the nature of the electrolyte used, the reactions can be summarized as follows (McGeough 1974):

i. In acidic electrolytes (e.g., HCl, $H_2SO_4$)

Reactions at anode:

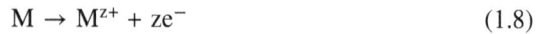

$$M \rightarrow M^{z+} + ze^- \tag{1.8}$$

where M is the metal, z is the valency of the metal ion, and $e^-$ is the charge of the electron.

Reactions at cathode:

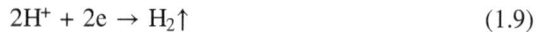

$$2H^+ + 2e \rightarrow H_2\uparrow \tag{1.9}$$

ii. In neutral electrolytes (e.g., $NaNO_3$, NaCl)

Reaction at the anode:

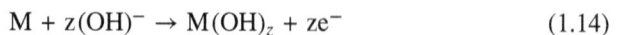

$$M \rightarrow M^{z+} + ze^- \tag{1.10}$$

or

$$M + z(OH)^- \rightarrow M(OH)_z + ze^- \tag{1.11}$$

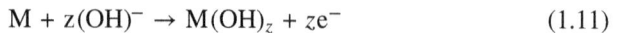

Reaction at the cathode:

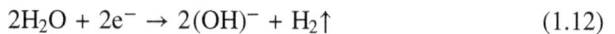

$$2H_2O + 2e^- \rightarrow 2(OH)^- + H_2\uparrow \tag{1.12}$$

The hydroxyl ions formed at the cathode and metal ions from the anode react in the bulk electrolyte to form metal hydroxide (McGeough 1974).

e.g.,

$$M^{2+} + 2(OH)^- \rightarrow M(OH)_2 \tag{1.13}$$

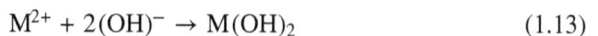

iii. In alkaline electrolytes (e.g., KOH, NaOH)

Reaction at the anode:

$$M + z(OH)^- \rightarrow M(OH)_z + ze^- \tag{1.14}$$

Reaction at the cathode:

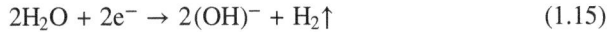

$$2H_2O + 2e^- \rightarrow 2(OH)^- + H_2\uparrow \tag{1.15}$$

The produced metal hydroxide forms as sludge and precipitates in the solution, which is filtered out from the liquid.

### 1.1.4 ECMM SETUP

The experimental setup of ECMM consists of a PES, electrolyte supply system which holds a pump and a purifier, xyz axes and computer numerical control (CNC) motion controller, workpiece table, machining chamber, and micro-tools. Figure 1.2 shows a schematic of an ECMM setup. Perspex or another non-corrosive material is used to fabricate the machining chamber. It holds the job and collects the electrolyte with sludge throughout the procedure. The purified sludge-free electrolyte is re-circulated to the machining chamber by the electrolyte supply system. The PES comprises of a DC power supply, a function generator, and a voltage signal modulation circuit. The tool and workpiece are connected to PES's output terminals from the voltage signal modulation circuit. An oscilloscope is connected in parallel to the VSM output to monitor ECMM input voltage.

### 1.1.5 CLASSIFICATION OF ECMM

ECMM can be classified into maskless ECMM and through-mask ECMM (TM-ECMM), as shown in Figure 1.3, based on the localisation effect of the material removal mechanism (Bijoy Bhattacharyya 2015c). In TM-ECMM, the material dissolution is confined to the desired area on the workpiece surface by using a photoresist pattern or a mask. In maskless ECMM, material removal is controlled

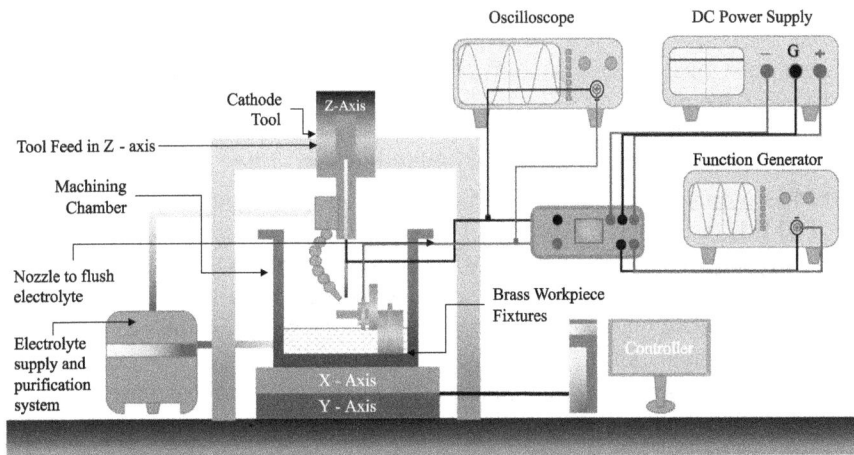

**FIGURE 1.2** Schematic showing the setup of ECMM.

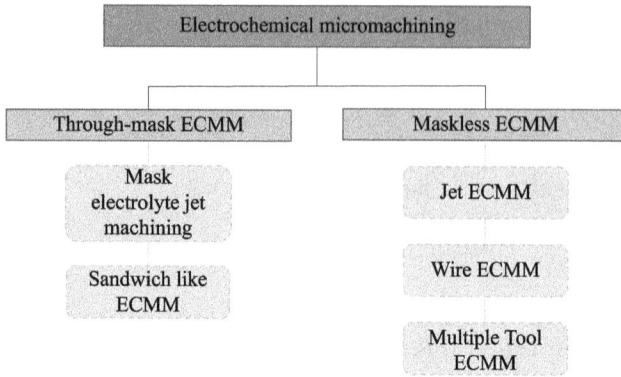

**FIGURE 1.3**   Schematic showing ECMM classification.

by several techniques that localize the material removal without employing a mask or photoresist pattern.

### 1.1.5.1   Through-Mask ECMM

TM-ECMM is an effective ECMM approach for making textures and 2D micro-features. Unmasked workpiece areas are anodically dissolved rapidly. The schematic of TM-ECMM is shown in Figure 1.4(i). This process requires a mask or photoresist and workpiece surface preparation. Isotropic material removal at exposed workpiece areas causes undercutting, where material beneath the mask dissolves. TM-ECMM mask design must consider undercutting and material dissolving for various electrolytes and metal combinations. The etch factor (EF) quantifies the localised material dissolution (Madore and Landolt 1997).

$$EF = \frac{2h}{W_1 - W} \tag{1.16}$$

where h is the micro-groove depth, $W_1$ is the micro-groove width, and W is the width of the slit in the mask.

#### 1.1.5.1.1   Mask Electrolyte Jet Machining (MJEM)

MJEM is a coalesence of TM-ECMM and jet electrochemical micromachining (Jet-ECMM), as shown in Figure 1.4(ii). TM-ECMM (selective localised dissolution at large areas) and Jet-ECMM, with its configurable liquid tool (i.e. electrolyte jet), are combined to generate micro-structures with excellent precision and consistency (Wu et al. 2020). The electrolyte jet operates as a tool and the negative terminal of the power supply is connected to its nozzle. This method produces high-aspect-ratio microfeatures like microdimples and microchannels. The conductive mask used reduces electric field intensity at the edges, which helps to minimise undercutting.

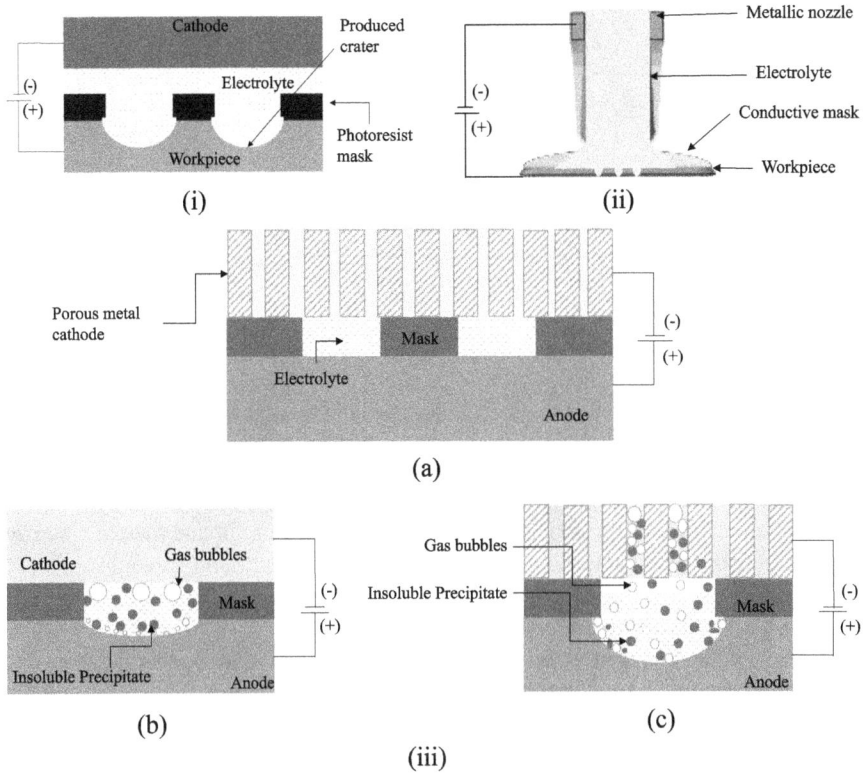

**FIGURE 1.4** Schematic of (i) Through-mask ECMM, (ii) conductive mask jet electrochemical micromachining (Chen et al. 2018). (iii) Sandwich-like ECMM: (a) Workpiece mask and porous cathode in sandwich configuration, (b) accumulation of bubbles and sludge, and (c) reaction products escaping through the porous cathode (Chen et al. 2018).

### 1.1.5.1.2  Sandwich-Like ECMM

This process has a layered sandwich-like construction of cathode, mask and anode with no gap between them as shown in Figure 1.4(iii). This configuration increases the current density, reduces overcut, and improves EF while micromachining (Zhang, Qu, and Chen 2016). A dry film mask applied to the workpiece maintains a minimal IEG during machining. In contrast to through-mask ECMM, localised current density reduces the mean diameter of micro-dimples. Using a porous cathode allows reaction products to escape, resulting in deeper micro-dimples, as shown in Figure 1.4(iii) (Zhang, Qu, and Fang 2017).

### 1.1.5.2  Maskless ECMM

Maskless ECMM uses a microtool or electrolyte jet to remove material and machine microfeatures. A 2D or 3D pattern can be made by selectively dissolving metal from the workpiece's surface. The machining parameters control anodic dissolution current density. Low IEG reduces stray currents. Passivating electrolytes like sodium nitrate

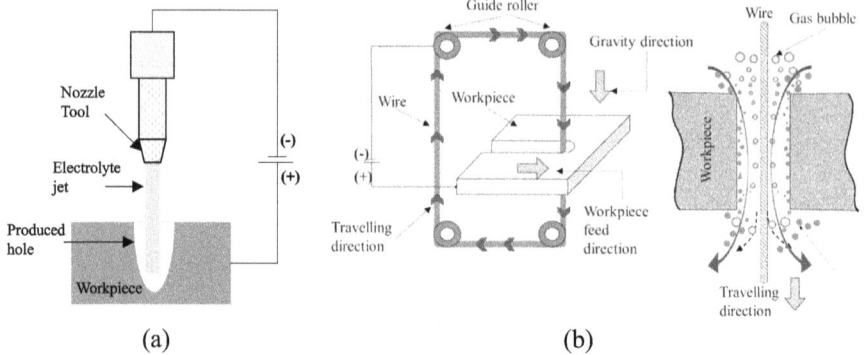

(a)                                                                                      (b)

**FIGURE 1.5**   Schematic of (a) Jet-ECMM and (b) Wire-ECMM.

reduce stray corrosion in the machining zone (Bhattacharyya 2015c). These techniques often require precise tool movement, a system to monitor and maintain narrow IEG and sensing, and controlling devices to work on closed-loop strategies.

### 1.1.5.2.1  Jet ECMM

In Jet ECMM, a stream of electrolyte jet from a nozzle, connected as cathode, acts as a tool as shown in Figure 1.5(a). A thin jet stream localises the current density and dissolves the workpiece. The jet's diameter controls material removal. Material dissolution is affected by input voltage, electrolyte type, concentration, pressure, and nozzle diameter. Moving the jet along a defined path can machine the desired profile (Natsu, Ooshiro, and Kunieda 2008). By reversing the polarity, rapid prototyping with selective deposition is also possible using an electrolyte jet (Kunieda, Katoh, and Moril, n.d.). Thin metal masks and low aspect ratio micro-features on thin metal foils can be machined effectively using Jet ECMM.

### 1.1.5.2.2  Wire-ECMM

In wire ECMM, a tool in the form of wire is used as a cathode. Microfeatures are fabricated by tracing the wire along a specified trajectory, as shown in Figure 1.5(b). This process is preferred over ECMM when a profile on the workpiece requires a trepanning action (Sharma et al. 2020). A wire can looped to travel in the machining zone for uninterrupted material removal, which also enhances mass transport and removal of reaction products in the IEG. Ultrasonic vibrations (to the workpiece/wire) can be used to achieve the same (Xu et al. 2016; Jiang et al. 2018).

### 1.1.5.2.3  Multiple-Tool ECMM

In multiple-tool ECMM, fabrication of microfeatures is relatively faster as the machining area is increased (Park and Chu 2007). Through a pre-designed array of micro-tools or wires or jets, as shown in Figure 1.6, complex 3D microfeatures can be machined. Positioning and moving tools on the workpiece in a defined path, creates desired features. Designing adjustable power supply, tools, and tool holders

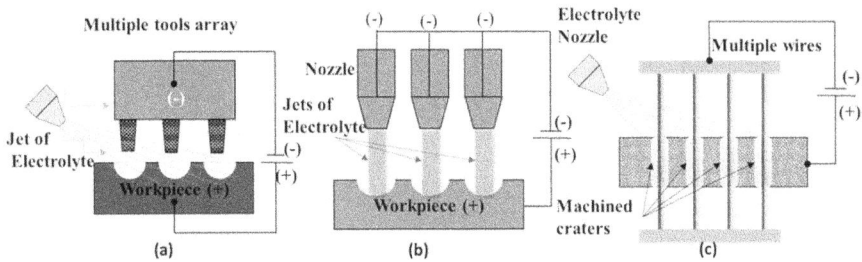

**FIGURE 1.6** Schematic of (a) multi-tool ECMM, (b) multi-jet ECMM, and (c) multi-wire ECMM.

is challenging. Various types of microfeatures can be fabricated using an array of micro-pins, micro-jets, and micro-wires, as shown in Figure 1.6.

### 1.1.6 A SHORT CHRONOLOGICAL REVIEW

Industries adopted ECM in middle of the 20th century, and since then, with active research in this domain, the process has evolved significantly. In 1989, textures were the microfeatures fabricated initially through ECMM using photoresist masks and thin metal films (West et al. 1992; Rosset and Landolt 1989). The first patent on ECMM was filed in 1992, where thin metal films were patterned using ECMM with a tool and TM-ECMM (Madhav Datta and Romankiw 1994). Materials such as stainless steel (Abouelata, Attia, and Youssef 2022), nickel (Bi, Zeng, and Qu 2020), titanium alloy (Praveena Gopinath et al. 2021), silicon (Harding et al. 2016), copper (Soundarrajan and Thanigaivelan 2021), tungsten (Gao and Qu 2019), and aluminium alloy (Bi, Zeng, and Qu 2021) were commonly used in ECMM. The chronological advancement of ECMM is shown in Figure 1.7.

Stage 1: Early works (1989–2005)
Early works were more focused on texturing, their defect elimination, and improvisation using TM-ECMM (Shenoy, Datta, and Romankiw 1996). ECMM of silicon and Jet ECMM were explored (Özdemir and Smith 1992) (M. Datta et al. 1989).
Stage 2: Process development (2006–2019)
Studies were more focused on improving productivity with the use of multiple tool electrodes (M. H. Wang and Zhu 2008), precision using tool vibration (K. Wang 2011), pulsed voltage supply (Schuster et al., n.d.), and tool geometry (Qi, Fang, and Zhu 2018), and several optimization techniques (Labib et al. 2011) (Samanta and Chakraborty 2011). Multiple variants of ECMM evolved, and their progress is depicted in Figure 1.8.
Stage 3: Recent advancements (2020–2022)
Recent works were more focused on novel tools (Singh Patel et al. 2020) and tool materials such as polymer graphite electrodes (Pradeep, Sundaram, and Pradeep Kumar 2019), hybrid techniques (Saxena, Qian, and Reynaerts 2020), performance simulation using finite element method (Tsai et al. 2020), machining on metallic glass (Hang et al. 2020), and sustainable techniques (Patel et al. 2021).

Enhancements in voltage waveforms, and sustainable techniques.

Improved productivity, precision with improved tool design.

Optimization techniques and hybrid techniques are explored. .

Developed new techniques such as wet stamping and Wire ECMM.

**2020: (Pradeep et al.)** Polymer graphite electrode for ECMM.

Pulse ECMM was developed.

**2016: (Wang et al.)** Liquid membrane Electrochemical Etching for Twin nano tip tools for ECMM.

**2011: (Wu et al.)** TMECMM with pulsed DC voltage power supply.

**2020: (Zhao et al.):** Dual-Sensing system monitoring Ion Concentration change in ECMM

Major research in the area of TMECMM.

**2008: (Jiang et al.):** Etchant Layer Confined Technique for ECMM of titanium alloy at a resolution of 0.503 microns.

**2011: (Labib et al.)** Designed Fuzzy logic control approach for ECMM to control feed and Electrolyte flow.

**2017: (He et al.)** Improvement of hydrogen bubble detachment form the tool During ECMM.

**2020: (Wu et al.)** ECMM through mask electrolyte jet machining.

**2001, 2003: (Chauvy et al.)** TMECMM of Titanium through a patterned oxide film generated through, a patterned photoresist **(2001)**, through Excimer laser irradiation on the oxide film **(2003)**, and numerical investigation on unusual cavity shapes **(2003)**.

**2009: (Wang et al.)** Multiple tool electrode micro hole array fabrication.

**2013: (Pa et al.)** Reuse of solar cell silicon wafers using magnetic assistance in ultrasonic ECMM.

**2017: (Wang et al.)** A new method Air shield ECMM is proposed for fabrication of micro dimple array.

**2020: (Meng et al.)** Dynamic liquid membrane Electrochemical modification of Carbon Nano Fiber for wire ECMM.

**1989: (Datta et al.)** Through mask ECMM with a photoresist mask, Jet and laser-Jet ECMM.

**2010:(Wang et al.)** High precision ECMM using vibrating tungsten wire electrode with pulsed voltage supply.

**2015: (Cen et al.)** Flexible mask for TMECMM on cylindrical surfaces.

**2018: (Zhao et al.)** Controlled Electrochemical nanomachining with Adjustable capacitance...

**2020: (Fan et al.)** Jet ECMM with Conductive mask for micro grooves fabrication.

**1992: (Ozdemir et al.)** New phenomenon observed during ECMM of Silicon.

**2010: (Tang et al.)** Micromachine on copper and Nickle using Electrochemical wet stamping.

**2015: (Chang et al.)** Atmospheric dual layer dielectric coating on electrodes as tools for ECMM.

**2019: (Xu et al.)** ECMM with sinusoidal signals.

**2020: (Patel et al.)** a novel Flexible tool for texturing free form surfaces.

**1996: (Shenoy et al.)** Effect of mask wall angle, Investigation of Island formation in TMECMM.

**2002: (Lee et al.)** Pulse ECMM was developed for Micro grooves.

**2005: (Kamda et al.)** Solid state ECMM.

**2021: (Gu et al.)** Bi-Directional Pulse ECMM.

| 1980-1999 | 2000-2005 | 2006-2010 | 2011-2015 | 2016-2019 | 2020-2022 |
|---|---|---|---|---|---|
| Stage – 1: Early works (preliminary studies) | | Stage – 2: Process development and refinement | | | Stage – 3: Current trend |

Progress trend

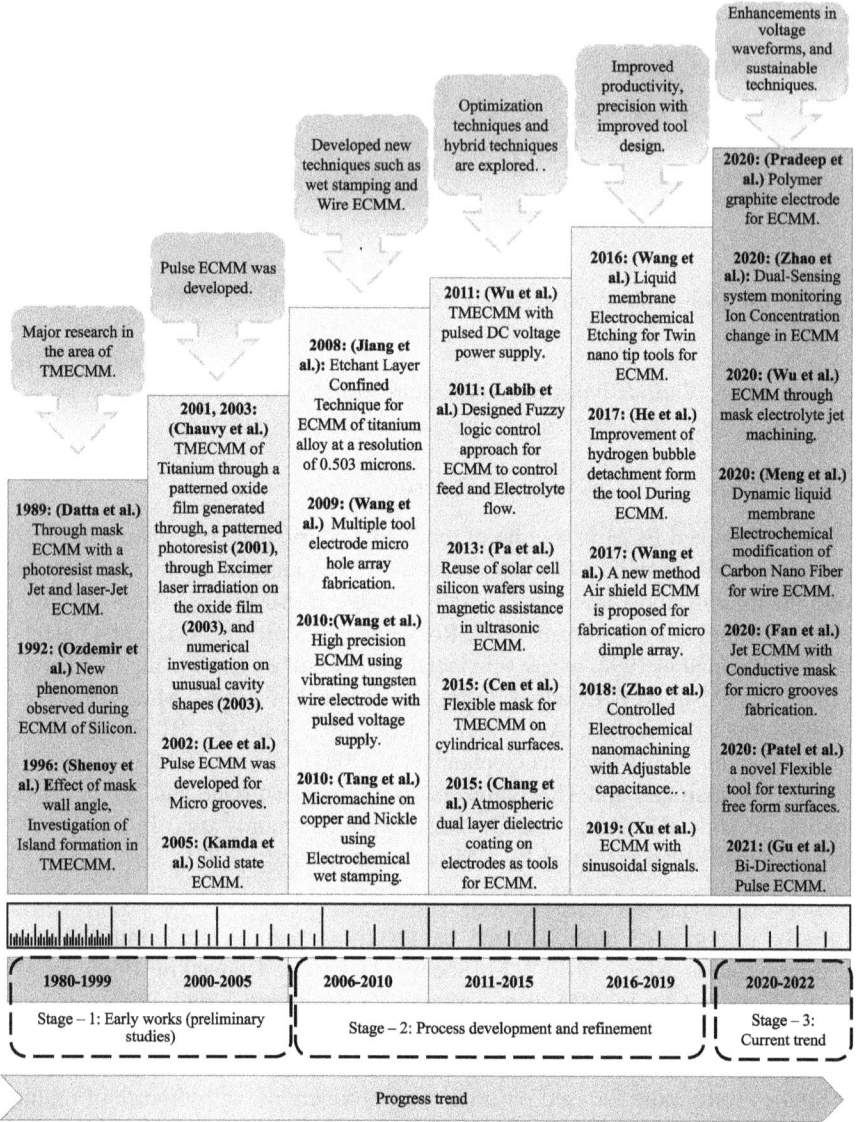

**FIGURE 1.7**   Schematic showing the chronological development of ECMM process.

## 1.2   PROCESS PARAMETERS IN ECMM

Various parameters influencing the process performance in ECMM are shown schematically in Figure 1.9. Pulse-on time and duty cycle affect pulse energy. Pulse energy increases overcut during machining. High frequency pulse power is preferable than constant DC. Pulsed DC eliminates reaction products formed during anodic dissolution, enhancing machining performance (Kumar et al. 2020). An increase in IEG increases resistance and reduces MRR. Current density affects

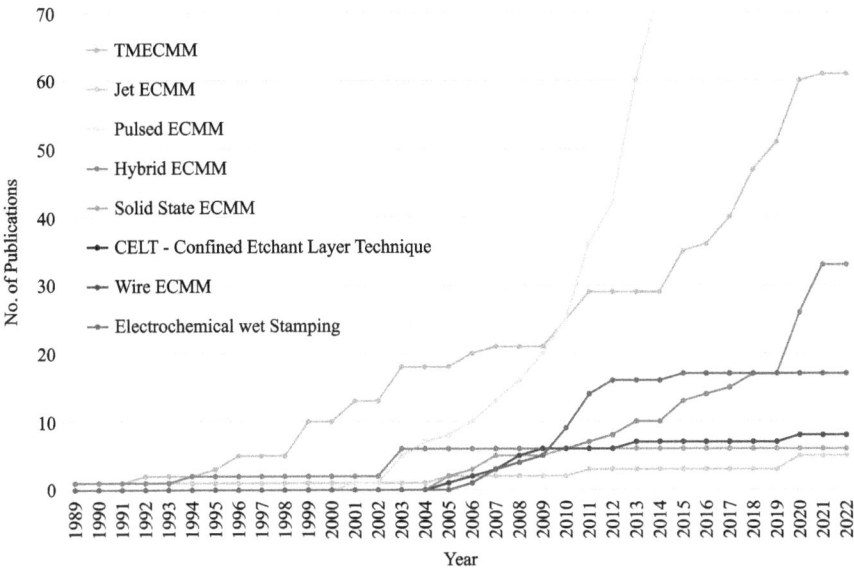

**FIGURE 1.8** Number of publications in different categories of ECMM.

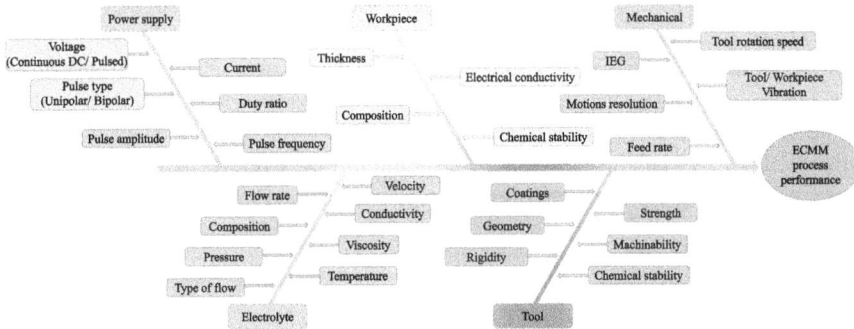

**FIGURE 1.9** Fishbone diagram showing the process parameters of ECMM.

material removal rate (since current density depends on the applied voltage, electrolyte conductivity, and the cross-sectional area being machined). Table 1.1 shows the generalized process parameters for ECMM.

## 1.3 POWER SUPPLY IN ECMM

The amount of workpiece material dissolved depends on the electric current passing between the electrodes, making the power supply crucial. Direct current power sources cause continuous material removal and high reaction products that can be eliminated using high-velocity electrolyte. Accumulating reaction products may affect the electrolyte's temperature, lowering its conductivity, forming deposits on

**TABLE 1.1**

**Process Parameters for ECMM (B. Bhattacharyya, Munda, and Malapati 2004)**

| Power Supply | Pulsed Supply |
|---|---|
| Tool size | 1 μm–999 μm |
| Inter electrode gap | 5 μm–50 μm |
| Voltage | < 10 V |
| Current | < 1 A |
| Pulse duration | 10 ns–500 ns |
| Waveform | Unipolar, Bipolar, Rectangular, Sinusoidal, Triangular |
| Frequency | 10 kHz–5 MHz |
| Surface finish | ≤ 0.5 μm |
| Electrolyte conductivity | 1 S/cm–50 S/cm |
| Duty ratio | 0.05–0.5 |

the tool, slows anodic breakdown, and affects machining accuracy and performance. Using a pulsed power source, it mitigates these and other issues, such as non-localized electric field and electrolyte boiling (Leese and Ivanov 2016). Therefore, ECMM always uses a pulsed DC power source.

On-time and off-time pulses make up pulsed DC power. Patel et al. created a novel approach for lowering pulse energy where the waveform of the pulse voltage is altered as shown in Figure 1.10 (Patel et al. 2020). Using a low-frequency triangle pulse voltage has enhanced accuracy and minimised overcut. It is generated by using a function generator's output function, a constant DC voltage, and a voltage

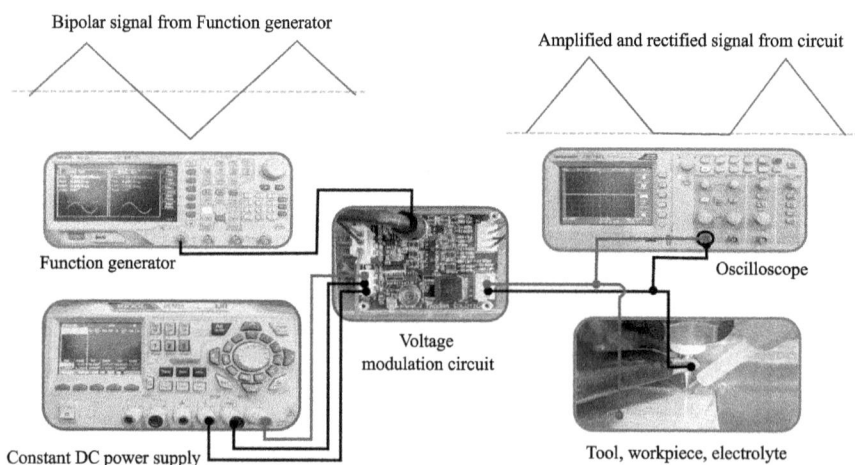

**FIGURE 1.10** Construction of a power supply used for generating rectangular, sinusoidal, and triangular voltage signals for machining. (Sharma et al. 2018).

signal modulation circuit. The circuit is powered by constant DC, Bipolar signals from the function generator are rectified and amplified by the circuit. This circuit can handle 5 MHz signals with 0 to 1 duty ratio.

## 1.4   ELECTROLYTES IN ECMM

The electrolyte in ECMM completes the circuit between the tool and workpiece, removes heat, and removes reaction products during machining. Based on the electrolyte, reaction by-products and anodic film affect machining quality (Bijoy Bhattacharyya 2015b). Hence, for ECMM of a metal, a suitable electrolyte selection is mandatory.

In ECMM, the micron tool allows current densities of 150 A/cm$^2$ in the limited machining zone. ECMM relies on the electrical conductivity of electrolytes to move ions between electrodes. Electrolyte conductivity increases with temperature. After applying the pontial, machining begins and electrolyte ion concentration increases. This increases charge carriers per unit volume, improving electrolyte conductivity. Metal ions discharged into the electrolyte create precipitates with negative ions, reducing charge carriers. Hydrogen generation replaces a considerable amount of cathode electrolyte. Precipitates and hydrogen bubbles reduce electrolyte conductivity. Hopenfield et al. postulated electrolyte conductivity, as shown in Equation (1.17) (Hopenfeld and Cole 1969).

$$k = k_0(1 + \alpha\Delta T)(1 - \alpha_v)^n \qquad (1.17)$$

where $k_o$ is the initial conductivity of the electrolyte, $\alpha$ is temperature coefficient of specific conductance, $\Delta T$ is the temperature difference, $\alpha_v$ is void fraction in the electrolyte, and n is exponent (value ranges from 1.5 to 2).

Jain et al. proposed an improved equation for calculating electrolyte conductivity by considering all three phases, as shown in Equation (1.18) (Jain and Chouksey 2017).

$$k_{eff} = k_0(1 + \alpha\Delta T)(1 - \alpha_v)^a \left(1 - b\left(1 - \frac{k_d}{k_o}\right)\alpha_s\right)^c \qquad (1.18)$$

where $\alpha_s$ is the sludge fraction, $k_d$ is the conductivity of the dispersed medium, and a, b, and c are constants. Figure 1.11(a) shows how reaction products affect conductivity, resulting in a tapered crater at the anode. Temperature and reaction products (oxygen at the anode, hydrogen gas at the cathode, and sludge) impact electrolyte conductivity (Bannard 1977). For reliable and reproducible results, the electrolyte should be regarded as a three-phase fluid whose conductivity may be calculated using Equation (1.18).

The most commonly used electrolytes in ECMM are sodium chloride and sodium nitrate. They are relatively inexpensive, and unlike acidic electrolytes, they do not corrode machinery over time. Sodium chloride is a non-passivating electrolyte with higher conductivity than sodium nitrate, ensuring faster machining and a bright surface finish while machining stainless steel. Sodium nitrate offered lower current densities, favoring passivation and a higher machining resolution at a given electrolyte

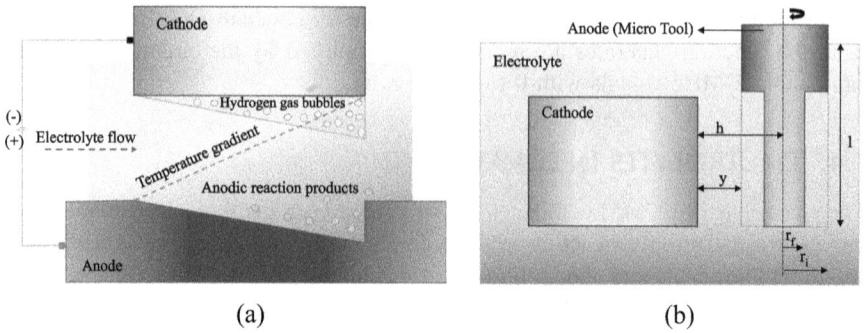

FIGURE 1.11   (a) Schematic showing the effect of reaction products and temperature on the anode shape in ECMM and (b) tool fabrication.

FIGURE 1.12   Classification of the types of electrolytes used in wire-ECMM (Sharma et al. 2020).

concentration (Leese and Ivanov 2016). The classification of electrolytes used in wire-ECMM is shown in Figure 1.12 (Sharma et al. 2020). The electrolyte selected determines the process' chemical reactions. Acidic, neutral, and basic electrolytes are categorised by pH. Less than 7 is acidic, 7 is neutral, and above 7 is basic (alkaline). In an acidic electrolyte, soluble reaction products are generated as the hydroxide ions formed at cathode react with the concentrated hydrogen ions in bulk acidic solution (Sharma et al. 2020). This minimises the sludge interference and allows reducing the IEG. Unlike acidic electrolytes, basic electrolytes yield high hydroxide ions that interact with metal ions to precipitate metal hydroxides (sludge), which may obstruct IEG. Table 1.2 shows the different electrolytes used for different metals and alloys.

## 1.5   TOOL DESIGN IN ECMM

Kamaraj et al. proposed a method to fabricate micro-tool for ECMM and is shown in Figure 1.11(b) (Kamaraj and Sundaram 2013). In Figure 1.11(b), $l$ electrode's length immersed in the electrolyte, $h$ represents the average gap between the vertical surface of cathode and the rotating surface of anode, $y$ represents the shortest distance between the anode (tool) surface and the cathode surface, and $r_f$ and $r_i$

**TABLE 1.2**

**List of Electrolytes Preferably Used for the Processing of Different Materials in ECMM (Bhattacharyya 2015)**

| Materials | Electrolyte |
|---|---|
| **Aluminum and its alloys** | Sodium chloride (NaCl), sodium nitrate (NaNO$_3$) |
| **Molybdenum, tungsten** | Sodium hydroxide (NaOH) |
| **Nickel and its alloys** | NaCl |
| **Tungsten carbide** | NaCl + NaOH + triethanolamine (C$_6$H$_{15}$NO$_3$) |
| **Titanium and its alloys** | NaCl, Sodium Bromide (NaBr), NaCl + NaNO$_3$ |
| **Gold** | Lithium chloride (LiCl), dimethyl sulfoxide (CH$_3$)$_2$SO |

represent the final and initial radius of rotating electrode. Following are the assumptions made while deriving the equation for tool diameter:

1. During machining, the overpotential ($\Delta V$), electrolyte conductivity (k), and current efficiency ($\eta$) are assumed to be constant.
2. Uniform current density on the tool is assumed.
3. Electrolyte flow and pressure effects are neglected as the electrolyte movement is only due to tool rotation.

The volume for a micro-tool is $v = \pi r^2 l$.

The rate of change of volume (MRR$_v$) can be expressed as

$$\frac{dv}{dt}(MRR_v) = 2\Pi rl\frac{dr}{dt} \tag{1.19}$$

The surface area of an anode immersed in an electrolyte is $A_c = 2\pi rl$. From Equation (1.5) current density can be written as

$$J = \frac{I}{2\pi rl} \tag{1.20}$$

from Equations (1.6) and (1.20), we get

$$I = \frac{k(V - \Delta V)}{h}(2\Pi rl) \tag{1.21}$$

A relationship between the tool radius and the machining parameters should be established to find the diameter of the micro-tool using Equations. (1.3), (1.19), and (1.21); we get the following result,

$$\frac{dr}{dt} = \frac{\eta A k(V - \Delta V)}{zhF\rho_a} \tag{1.22}$$

By integrating both sides of the Equation (1.22) from initial radius as $r_i$ to final radius $r_f$, change in tool radius can be obtained as

$$r_i - r_f = \int_0^t \frac{\eta Ak(V - \Delta V)}{zhF\rho_a} dt \tag{1.23}$$

Since a pulsed voltage source is employed, machining occurs only during the on-time ($t_{on}$) and not during the off-time. Therefore, from pulse frequency the effective machining time can be obtained. Thus,

$$r_i - r_f = \sum_0^n \int_0^{t_{on}} \frac{\eta Ak(V - \Delta V)}{zhF\rho_a} dt \tag{1.24}$$

where $n$ is the number of pulses in time period $t$, which is product of time and pulse frequency ($n = f \times t$). From Equation (1.24), we can calculate the final diameter ($D_f$) of micro-tool as follows:

$$D_f = D_i - 2\frac{\eta Ak(V - \Delta V)}{zhF\rho_a} t_{on} ft \tag{1.25}$$

where $D_i$ is the initial diameter Equation (1.25) and can be utilized for predicting the final diameter of the micro-tool for a specified parametric condition of pulsed electrochemical micromachining.

## 1.6   RECENT ADVANCEMENTS IN ECMM

The research is carried out on the improvement of accuracy and performance efficiency. The adjustment of process parameters that affect the accuracy and surface roughness of the manufactured parts is a major challenge for the ECMM. The methodologies explored for gaining better control over the process are:

1. Insulating workpiece and tool with a non-conducting coating/mask (M Datta 1998),
2. Ultrasonic vibration assistance to the workpiece/tool (Cliffon, Imai, and Mcgeough, n.d.),
3. Power supply with micro-/nano-second voltage pulses (M Datta and Lanwlt, n.d.; Bannard 1977),
4. Sludge removal through effective electrolyte supply (Jain 2002),
5. Pulsating electrolyte supply (Qu et al. 2013; Fang et al. 2014),
6. Altering the shape of voltage pulse (Patel et al. 2020).

Low overcut, excellent surface quality, improved precision, and increased machining efficiency are the broad objectives of the current researches in ECMM. Various advancements include the use of neural network (NN) to predict the inner and outer diameter of a hole drilled through SUS 304 stainless-steel foil using electrochemical micro-drilling at a given voltage, feed rate, and pulse on time (Lu et al. 2017). Sustainable electrochemical machining has come into consideration for reducing electrolyte wastage (Patel 2021) and reducing the toxicity of the used electrolytes at the time of disposal, which is achieved by using an atomized electrolyte flushing technique where a thin moving electrolyte mist film of $\approx$ 100 μm is supplied in the confined IEG (Patel 2021). Titanium alloys are extensively employed in aerospace, armaments, biomedicals, and other industries. The reaction products created during the electrochemical machining of titanium alloys are insoluble and stick to the alloy's machining surface, which causes difficulty in machining using standard wire-ECM. Hence, an inner-jet electrolyte flushing with a tube electrode is employed to circumvent such issues. In addition, the outer-jet electrolyte flushing aids in the removal of electrolysis products in the machining zone (Yang et al. 2021).

## REFERENCES

Abouelata, Ahmed M. Awad, Adel Attia, and Gehan I. Youssef. 2022. "Electrochemical Polishing versus Mechanical Polishing of AISI 304: Surface and Electrochemical Study." *Journal of Solid State Electrochemistry* 26 (1). Springer Berlin Heidelberg: 121–129. doi:10.1007/s10008-021-05037-2

Bannard, J. 1977. "Electrochemical Machining." *Journal of Applied Electrochemistry* 7: 1–29.

Barr, A E De, and Donald Arthur Oliver. 1968. "Electrochemical Machining." *Macdonald Trends and Developments in Engineering Series* 7: 248.

Bhattacharyya, B., J. Munda, and M. Malapati. 2004. "Advancement in Electrochemical Micro-Machining." *International Journal of Machine Tools and Manufacture* 44 (15): 1577–1589. doi:10.1016/j.ijmachtools.2004.06.006

Bhattacharyya, Bijoy. 2015a. "Electrochemical Machining." *Electrochemical Micromachining for Nanofabrication, MEMS and Nanotechnology*, 25–52. ISBN: 9780323327374. Elsevier. April 10, 2015. doi:10.1016/b978-0-323-32737-4.00002-5

Bhattacharyya, Bijoy. 2015b. "Influencing Factors of EMM." *Electrochemical Micromachining for Nanofabrication, MEMS and Nanotechnology*, 123–143. ISBN: 9780323327374. Elsevier. April 10, 2015. doi:10.1016/b978-0-323-32737-4.00007-4

Bhattacharyya, Bijoy. 2015c. "Types of EMM." *Electrochemical Micromachining for Nanofabrication, MEMS and Nanotechnology*, 69–82. ISBN: 9780323327374. Elsevier. April 10, 2015. doi:10.1016/b978-0-323-32737-4.00004-9

Bi, Xiaolei, Yongbin Zeng, and Ningsong Qu. 2020. "Wire Electrochemical Micromachining of High-Quality Pure-Nickel Microstructures Focusing on Different Machining Indicators." *Precision Engineering* 61 (January). Elsevier: 14–22. doi:10.1016/J.PRECISIONENG. 2019.09.015

Bi, Xiaolei, Yongbin Zeng, and Ningsong Qu. 2021. "Micro-Shaping of Pure Aluminum by Intermittent Ultrasonic Oscillation Assisted Wire Electrochemical Micromachining with an Ultra-Low-Concentration Mixed Electrolyte." *Journal of The Electrochemical Society* 168 (11). IOP Publishing: 113503. doi:10.1149/1945-7111/AC377E

Chen, X. L., B. Y. Dong, C. Y. Zhang, M. Wu, and Z. N. Guo. 2018. "Jet Electrochemical Machining of Micro Dimples with Conductive Mask." *Journal of Materials Processing Technology* 257 (July). Elsevier: 101–111. doi:10.1016/j.jmatprotec.2018.02.035

Cliffon, D., Y. Imai, and J. A. Mcgeough. 1993. "Some Ultrasonic Effects in Machining Materials Encountered in The Offshore Industries." In Kochhar, A.K. (eds) *Proceedings of the Thirtieth International MATADOR Conference*. Palgrave, London. https://doi.org/10.1007/978-1-349-13255-3_16

Datta, M. 1998. "Microfabrication by Electrochemical Metal Removal." *IBM Journal of Research and Development* 42 (5): 655–670, Sep. doi:10.1147/rd.425.0655

Datta, M., and D. Lanwlt. 1981. "Electrochemical Machining Under Pulsed Current Conditions." *Electrochimica Acta* 26 (7), 899–907. doi:10.1016/0013-4686(81)85053-0

Datta, M., L. T. Romankiw, D. R. Vigliotti, and R. J. von Gutfeld. 1989. "Jet and Laser-Jet Electrochemical Micromachining of Nickel and Steel." *Journal of The Electrochemical Society* 136 (8). The Electrochemical Society: 2251–2256. doi:10.1149/1.2097282/XML

Datta, Madhav, and Lubomyr T. Romankiw. 1994. "Electrochemical Micromachining Tool and Process for Through-Mask Patterning of Thin Metallic Films Supported by Non-Conducting or Poorly Conducting Surfaces." Google Patents.

Fang, Xiaolong, Ningsong Qu, Yudong Zhang, Zhengyang Xu, and Di Zhu. 2014. "Effects of Pulsating Electrolyte Flow in Electrochemical Machining." *Journal of Materials Processing Technology* 214 (1). Elsevier: 36–43. doi:10.1016/j.jmatprotec.2013.07.012

Gao, Chuanping, and Ningsong Qu. 2019. "A Distinct Perception on Wire Electrochemical Micromachining of Pure Tungsten with Neutral Aqueous Solution." *Journal of The Electrochemical Society* 166 (13). The Electrochemical Society: E465–E472. doi: 10.1149/2.0051914JES/XML

Hang, Yusen, Yongbin Zeng, Tao Yang, and Lingchao Meng. 2020. "The Dissolution Characteristics and Wire Electrochemical Micromachining of Metallic Glass Ni82Cr7Si5Fe3B3." *Journal of Manufacturing Processes* 58 (October). Elsevier: 884–893. doi:10.1016/J.JMAPRO.2020.08.069

Harding, Frances J., Salvatore Surdo, Bahman Delalat, Chiara Cozzi, Roey Elnathan, Stan Gronthos, Nicolas H. Voelcker, and Giuseppe Barillaro. 2016. "Ordered Silicon Pillar Arrays Prepared by Electrochemical Micromachining: Substrates for High-Efficiency Cell Transfection." *ACS Applied Materials and Interfaces* 8 (43). American Chemical Society: 29197–29202. doi:10.1021/ACSAMI.6B07850/ASSET/IMAGES/LARGE/AM-2016-07850Z_0005.JPEG

Hopenfeld, J., and R. R. Cole. 1969. "Prediction of the One-Dimensional Equilibrium Cutting Gap in Electrochemical Machining." *Journal of Engineering for Industry* 91 (3). American Society of Mechanical Engineers Digital Collection: 755–763. doi:10.1115/1.3591683

Jain, and Ankit Kumar Chouksey. 2017. "A Comprehensive Analysis of Three-Phase Electrolyte Conductivity during Electrochemical Macromachining / Micromachining." *Proc IMechE Part B: J Engineering Manufacture.* doi:10.1177/0954405417690558

Jain, V. K. 2002. *Advanced Machining Processes*, New Delhi: Allied Publisher.

Jain, V. K. 2011. *Advanced Machining Processes.* International Journal of Manufacturing Technology and Management. Inderscience Enterprises.

Jiang, Kai, Xiaoyu Wu, Jianguo Lei, Zhaozhi Wu, Wen Wu, Wen Li, and Dongfeng Diao. 2018. "Vibration-Assisted Wire Electrochemical Micromachining with a Suspension of B4C Particles in the Electrolyte." *International Journal of Advanced Manufacturing Technology* 97 (9–12). Springer London: 3565–3574. doi:10.1007/s00170-018-2190-8

Kamaraj, Abishek B., and M. M. Sundaram. 2013. "Mathematical Modeling and Verification of Pulse Electrochemical Micromachining of Microtools." *International Journal of Advanced Manufacturing Technology* 68 (5–8): 1055–1061. doi:10.1007/s00170-013-4896-y

Kumar, Abhinav, Manjesh Kumar, Anupam Alok, and Manas Das. 2020. "Surface Texturing by Electrochemical Micromachining: A Review." *IOP Conference Series: Materials Science and Engineering* 804 (1). doi:10.1088/1757-899X/804/1/012011

Kunieda, Masanori, Ritsu Katoh, and Yasushi Moril. 1998. "Rapid Prototyping by Selective Electrodeposition Using Electrolyte Jet." *CIRP Annals* 47 (1): 161–164. doi:10.1016/S0007-8506(07)62808-X

Labib, A. W., V. J. Keasberry, J. Atkinson, and H. W. Frost. 2011. "Towards next Generation Electrochemical Machining Controllers: A Fuzzy Logic Control Approach to ECM."

*Expert Systems with Applications* 38 (6). Pergamon: 7486–7493. doi:10.1016/J.ESWA. 2010.12.074

Leese, Rebecca J., and Atanas Ivanov. 2016. "Electrochemical Micromachining: An Introduction." *Advances in Mechanical Engineering* 8 (1): 1–13. doi:10.1177/1687814 015626860

Lu, Yanfei, Manik Rajora, Pan Zou, Steven Y. Liang, George W. Woodruff, Xiaoliang Jin, and Dan Zhang. 2017. "Physics-Embedded Machine Learning: Case Study with Electrochemical Micro-Machining." *Machines 2017* 5 (1), Page 4. Multidisciplinary Digital Publishing Institute: 4. doi:10.3390/MACHINES5010004

Madore, C., and D. Landolt. 1997. "Electrochemical Micromachining of Controlled Topographies on Titanium for Biological Applications." *J. Micromech. Microeng.* Vol. 7: 270–275.

Mativenga, Paul. 2018. "Micromachining." *CIRP Encyclopedia of Production Engineering.* The International Academy for Production Engineering, Sami Chatti, Tullio Tolio. Berlin, Heidelberg: Springer, 1–5. doi:10.1007/978-3-642-35950-7_17-4

McGeough, J. A. 1974. *Principles of Electrochemical Machining.* London: Chapman and Hall.

Natsu, W., S. Ooshiro, and M. Kunieda. 2008. "Research on Generation of Three-Dimensional Surface with Micro-Electrolyte Jet Machining." *CIRP Journal of Manufacturing Science and Technology* 1 (1): 27–34. doi:10.1016/j.cirpj.2008.06.006

Özdemir, C. Hakan, and James G. Smith. 1992. "New Phenomena Observed in Electrochemical Micromachining of Silicon." *Sensors and Actuators A: Physical* 34 (1). Elsevier: 87–93. doi:10.1016/0924-4247(92)80143-Q

Park, Min Soo, and Chong Nam Chu. 2007. "Micro-Electrochemical Machining Using Multiple Tool Electrodes." *Journal of Micromechanics and Microengineering* 17 (8): 1451–1457. doi:10.1088/0960-1317/17/8/006

Patel, Divyansh Singh. 2021. "Sustainable Electrochemical Micromachining Using Atomized Electrolyte Flushing." *Journal of The Electrochemical Society* 168 (4). IOP Publishing: 43504. doi:10.1149/1945-7111/abf4e9

Patel, Divyansh Singh, Vyom Sharma, V. K. Jain, and J. Ramkumar. 2021. "Sustainable Electrochemical Micromachining Using Atomized Electrolyte Flushing." *Journal of The Electrochemical Society* 168 (4). IOP Publishing: 043504. doi:10.1149/1945-7111/abf4e9

Patel, Divyansh Singh, Vyom Sharma, Vijay Kumar Jain, and Janakarajan Ramkumar. 2020. "Reducing Overcut in Electrochemical Micromachining Process by Altering the Energy of Voltage Pulse Using Sinusoidal and Triangular Waveform." *International Journal of Machine Tools and Manufacture* 151 (February). Elsevier Ltd: 103526. doi:10.1016/j.ijmachtools.2020.103526

Pradeep, N., K. Shanmuga Sundaram, and M. Pradeep Kumar. 2019. "Performance Investigation of Variant Polymer Graphite Electrodes Used in Electrochemical Micromachining of ASTM A240 Grade 304." 35 (1). Taylor & Francis: 72–85. doi:10. 1080/10426914.2019.1697445. Https://Doi.Org/10.1080/10426914.2019.1697445

Praveena Gopinath, T., J. Prasanna, C. Chandrasekhara Sastry, and Sandeep Patil. 2021. "Experimental Investigation of the Electrochemical Micromachining Process of Ti-6Al-4V Titanium Alloy under the Influence of Magnetic Field." *Materials Science-Poland* 39 (1). Sciendo: 124–138. doi:10.2478/MSP-2021-0013

Qi, Xinxin, Xiaolong Fang, and Di Zhu. 2018. "Investigation of Electrochemical Micromachining of Tungsten Microtools." *International Journal of Refractory Metals and Hard Materials* 71 (February). Elsevier Ltd: 307–314. doi:10.1016/j.ijrmhm.2017.11.045

Qu, N. S., X. L. Fang, Y. D. Zhang, and D. Zhu. 2013. "Enhancement of Surface Roughness in Electrochemical Machining of Ti6Al4V by Pulsating Electrolyte." *International Journal of Advanced Manufacturing Technology* 69 (9–12): 2703–2709. doi:10.1007/ s00170-013-5238-9

Rosset, E., and D. Landolt. 1989. "Experimental Investigation of Shape Changes in Electrochemical Micromachining through Photoresist Masks." *Precision Engineering* 11 (2). Elsevier: 79–82. doi:10.1016/0141-6359(89)90056-1

Samanta, Suman, and Shankar Chakraborty. 2011. "Parametric Optimization of Some Non-Traditional Machining Processes Using Artificial Bee Colony Algorithm." *Engineering Applications of Artificial Intelligence* 24 (6). Pergamon: 946–957. doi:10.1016/J.ENGAPPAI.2011.03.009

Saxena, Krishna Kumar, Jun Qian, and Dominiek Reynaerts. 2020. "A Tool-Based Hybrid Laser-Electrochemical Micromachining Process: Experimental Investigations and Synergistic Effects." *International Journal of Machine Tools and Manufacture* 155 (August). Pergamon: 103569. doi:10.1016/J.IJMACHTOOLS.2020.103569

Schuster, Rolf, Viola Kirchner, Philippe Allongue, and Gerhard Ertl. 2000. "Electrochemical Micromachining." *Science*. 289: 98–101. doi:10.1126/science.289.5476.98

Sharma, Vyom, Divyansh Singh Patel, V. K. Jain, and J. Ramkumar. 2020. "Wire Electrochemical Micromachining: An Overview." *International Journal of Machine Tools and Manufacture* 155 (January). Elsevier Ltd: 103579. doi:10.1016/j.ijmachtools.2020.103579

Sharma, Vyom, Divyansh Singh Patel, V. K. Jain, J. Ramkumar, and Akash Tyagi. 2018. "Wire Electrochemical Threading: A Technique for Fabricating Macro/Micro Thread Profiles." *Journal of The Electrochemical Society* 165 (9): E397–E405. doi:10.1149/2.1181809jes

Shenoy, R. V., M. Datta, and L. T. Romankiw. 1996. "Investigation of Island Formation during Through-Mask Electrochemical Micromachining." *Journal of The Electrochemical Society* 143 (7). The Electrochemical Society: 2305–2309. doi:10.1149/1.1836997/XML

Singh Patel, Divyansh, Vishal Agrawal, J. Ramkumar, V. K. Jain, and Gaganpreet Singh. 2020. "Micro-Texturing on Free-Form Surfaces Using Flexible-Electrode through-Mask Electrochemical Micromachining." *Journal of Materials Processing Technology* 282 (July 2019). Elsevier: 116644. doi:10.1016/j.jmatprotec.2020.116644

Soundarrajan, M., and R. Thanigaivelan. 2021. "Electrochemical Micromachining of Copper Alloy through Hot Air Assisted Electrolyte Approach." *Russian Journal of Electrochemistry* 57 (2). Pleiades journals: 172–182. doi:10.1134/S1023193521020117

Tsai, Te Hui, Ming Yuan Lin, Zhi Wen Fan, and Hung Lin Lin. 2020. "Three-Dimensional Simulation of Performance in through-Mask Electrochemical Micromachining:" 234 (6). SAGE Publications. Sage UK: London: 523–32. doi:10.1177/0954408920934550. Https://Doi.Org/10.1177/0954408920934550

Wang, K. 2011. "Electrochemical Micromachining Using Vibratile Tungsten Wire for High-Aspect-Ratio Microstructures." *Surface Engineering and Applied Electrochemistry 2010 46:5* 46 (5). Springer: 395–399. doi:10.3103/S1068375510050017.

Wang, M. H., and D. Zhu. 2008. "Fabrication of Multiple Electrodes and Their Application for Micro-Holes Array in ECM." *The International Journal of Advanced Manufacturing Technology 2008 41:1* 41 (1). Springer: 42–47. doi:10.1007/S00170-008-1456-Y.

West, Alan C., Charles Madore, Michael Matlosz, and D. Landolt. 1992. "Shape Changes during Through-Mask Electrochemical Micromachining of Thin Metal Films." *Journal of The Electrochemical Society* 139 (2). The Electrochemical Society: 499–506. doi:10.1149/1.2069245/XML

Wu, Ming, Jiangwen Liu, Junfeng He, Xiaolei Chen, and Zhongning Guo. 2020. "Fabrication of Surface Microstructures by Mask Electrolyte Jet Machining." *International Journal of Machine Tools and Manufacture* 148 (October 2019). Elsevier Ltd: 103471. doi:10.1016/j.ijmachtools.2019.103471

Xu, Kun, Yongbin Zeng, Peng Li, Xiaolong Fang, and Di Zhu. 2016. "Effect of Wire Cathode Surface Hydrophilia When Using a Travelling Wire in Wire Electrochemical Micro Machining." *Journal of Materials Processing Technology* 235. Elsevier B.V.: 68–74. doi:10.1016/j.jmatprotec.2016.04.008

Yang, Tao, Yanliang Li, Zhengyang Xu, and Yongbin Zeng. 2021. "Electrochemical Cutting with Inner-Jet Electrolyte Flushing for Titanium Alloy (Ti-6Al-4V)." *International Journal of Advanced Manufacturing Technology* 112 (9–10). The International Journal of Advanced Manufacturing Technology: 2583–2592. doi:10.1007/s00170-020-06494-1

Zhang, Xifang, Ningsong Qu, and Xiaolei Chen. 2016. "Sandwich-like Electrochemical Micromachining of Micro-Dimples." *Surface and Coatings Technology* 302 (September). Elsevier B.V.: 438–447. doi:10.1016/j.surfcoat.2016.05.088

Zhang, Xifang, Ningsong Qu, and Xiaolong Fang. 2017. "Sandwich-like Electrochemical Micromachining of Micro-Dimples Using a Porous Metal Cathode." *Surface and Coatings Technology* 311 (February). Elsevier B.V.: 357–364. doi:10.1016/j.surfcoat.2017.01.035

## ABBREVIATIONS

| | |
|---|---|
| ANN | Artificial Neural Network |
| CELT | Confined Etchant Layer Technique |
| CNC | Computer Numeric Control |
| DC | Direct Current |
| EDL | Electric Double Layer |
| ECM | Electrochemical Machining |
| ECMM | Electrochemical Micromachining |
| EF | Etch factor |
| IEG | Inter-electrode Gap |
| MEMS | Micro-electromechanical systems |
| MRR | Material removal rate |
| TMECMM | Through-mask electrochemical micromachining |
| MJEM | Mask electrolyte jet machining |
| PECM | Pulse electrochemical machining |
| PECMM | Pulse electrochemical micromachining |
| PES | Process energy source |
| VSM | Voltage signal modulation |

## LIST OF SYMBOLS

| | |
|---|---|
| $A_e$ | surface area of anode immersed in a electrolyte |
| $D_f$ | Final diameter |
| $D_i$ | Initial diameter |
| $f$ | feed rate |
| $k$ | electrolyte conductivity |
| $k_0$ | initial electrolyte conductivity |
| $k_d$ | conductivity of the dispersed medium |
| $k_{eff}$ | effective conductivity |
| $m$ | amount of mass removed |
| $MRR_g$ | mass material removal rate in g/s |
| $MRR_v$ | volumetric removal rate in $mm^3/s$ |
| $r_e$ | radius of rotating electrode |
| $r_f$ | final radius |

| $r_i$ | initial radius |
| Symbols | Nomenclature |
| $t_{on}$ | pulse on time |
| $\rho_a$ | density of anode material |
| $\alpha$ | temperature coefficient of specific conductance |
| $\alpha_v$ | void fraction |
| $\Delta V$ | overpotential |
| n | current efficiency |

## CONSTANTS

| h | distance between the electrodes |
| $h_e$ | distance between the electrodes at equilibrium |
| I | current |
| J | current density |
| F | Faraday's constant |
| A | gram equivalent weight of anode material, |
| l | length of electrode immersed |
| t | time |
| T | temperature |
| z | valency |

# 2 Fabrication of Electrochromic Devices

## Mechanistic Insights, Components, and Applications

Chanchal Chakraborty and Susmita Roy

## CONTENTS

DOI: 10.1201/b23359-2

## 2.1   INTRODUCTION

The term "chromism" refers to a process of reversible color change of a compound that may be caused by a variety of different stimuli or reactions. In an area of chemistry, some compounds respond to a wide and varied chromism due to oxidation/ redox reaction, termed "electrochromic material". In the recent era, electrochromic materials (ECMs) have attained a huge interest in academics and industry because of their controllable transmittance, which potentially helps to save the primary energy (indirectly helps to reduce the usage of fossil fuels). The continuous increment of the world's energy demand and consumption strongly hinges on the fossil fuels, and generates various environmental issues like global warming, etc. Though there is enormous research going on in renewable energy sources, the research on power-efficient devices to use low or no power has become ever more necessary and urgent in parallel with renewable energy development. In this regard, a major revolution in energy technology is essential.

Besides industry and transport, building sectors use almost 30–40% of the primary energy worldwide [1]. Among these, nearly 50% of the entire building's energy is spent only for cooling, heating, ventilation, and lighting of the interior places [2–4]. This percentage is even more in the industrialized area. However, the building energy demand dominates the peak, especially the air conditioning has grown by ~17% per year in the European Union [5,6]. This significant fraction of building energy consumption counts only for our inability to control over passage and absorption of solar irradiation (UV, visible, NIR) through the transparent window materials and building blocks, respectively. The National Renewable Energy Laboratory (NERL) sources reported that the invisible NIR radiation (700 nm to 2,500 nm) contributes ~50% of the total solar energy reaching the earth's surface. This NIR radiation is mainly responsible for the surface and environmental heating of the earth (Figure 2.1). The NIR radiation doesn't contribute toward the daylight but to solar heating, whereas the visible radiation passage is accountable for daylight. The continuous passage and absorption of this solar irradiation through the window and building blocks lead to overheating the building indoors and cause unpleasant weather. Generally, the people from the industrial or corporate world spend 80–90% of daytime in indoor conditions

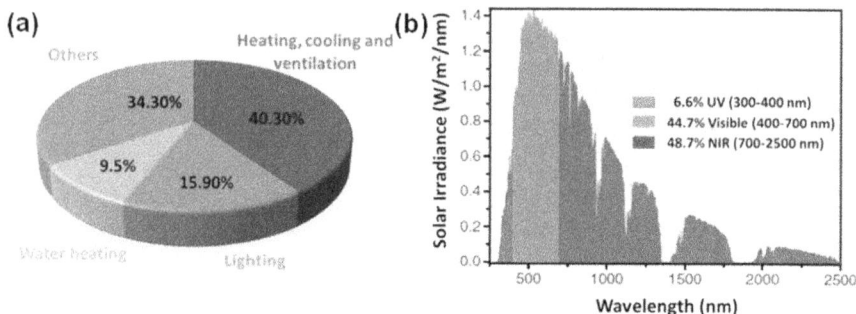

**FIGURE 2.1**   (a) U.S. buildings' energy end-use in 2008. (b) Solar spectrum on earth. Data are taken from National Renewable Energy Laboratory (NREL) database.

and as a result they need a comfortable weather condition with air conditioning either in a smart building or in vehicles. In that case, the controlling of solar radiation passing through the window is necessary and utilizing a smart window can reduce effectively the solar heat gain as well as can save the energy consumption and greenhouse effect and air pollution etc.

## 2.1.1   ECMs in Smart Window Technology

The focus on advanced civil architectures combines energy competence with decent interior comfort drives the majority of the current research towards "smart electrochromic technology". In building architecture, a window's primary function is to maintain the daylight and to make a visual relation between the inside and outside of the building, and not proposed to control energy. But the direct passage of solar radiation through the window causes interior overheating. Thus, an independent control on solar visible and NIR radiation by a window material without negotiating the visible transparency is important for advanced energy-efficient buildings. Several window materials have been developed in this regard. Most of them focus on employing external mechanical shutters, phase change materials (PCMs), aerogels, organic liquid crystals, [7–10] etc. Definitely, these processes are shortcoming in changing climate conditions. Chromogenic materials, on the other hand, are referred to as implemented on the window materials as they show adjustable visual responses under changed exterior stimuli. Some major chromic materials include 'photochromic' (light stimuli), 'thermochromic' (heat stimuli), 'solvatochromic' (solvent stimuli), 'vapochromic' (vapor stimuli), 'electrochromic' (EC) (electric bias as stimuli) [11,12] etc.

An ideal smart glass is a highly versatile product, which can be used almost anywhere that regular glass is used. ECMs are referred to use in ideal smart technology among all these chromogenic materials because they seem the most promising as they can regulate the solar irradiance effectively in various weather conditions. The EC display-based smart windows can efficiently adjust energy and daylighting indoors by regulating the light without losing optical contact with the outside [13].

## 2.1.2   ECMs and Their Applications

The ECMs are capable of reversible color modulation by varying their transmittance in an external applied electrical potential. The applied bias effectively changes the electrochemical redox state of the material, causing the change in the energy bandgap of the entire molecule. As a result, there would be several nanometers shift in the absorption spectrum, and a new color state would appear. So, at most, the core EC layer in ECD can change the optical properties between a transparent/bleached state and a colored state or between two colored states under potential tuning. However, suppose there is the existence of more than two redox states during the electrochemical changes. In that case, the ECM may exhibit several color shades and, therefore, are responsible for displaying multi-colored electrochromism.

Owing to this special color-changing ability of ECMs, they are applicable in various cutting-edge electronic applications, including EC smart windows, information displays, antiglare rear-view mirrors for automobiles, e-papers, e-skins and biosensors, military camouflage, billboards, flat-panel displays, etc. [14–20]. Unlike the light-emitting diode (LED) or liquid crystal displays (LCDs), EC displays do not use a bright backlight and are less harmful to human eyes. For this reason, EC display technology is considered to be the next-generation display technology for tomorrow's world.

Besides all these, electrochromic devices (ECDs) are slowly gaining a potential interest in lightweight and miniature electronic devices. Based on the EC function in ECDs, they can be integrated in opto-electronic devices to produce hybrid electrochromic-energy storage devices, and can be used in small electronics as well as in smart windows too [21].

## 2.2  FABRICATION OF ECDS

An ECD is nothing but an electrochemical cell that needs three necessary components: a working electrode (WE), electrolyte, and counter electrode (CE). Based on the application approach and working principle of ECMs, a typical ECD consists of the superimposed layers of all these components materials backed by transparent substrates. In brief, for the fabrication of an ECD, first, an ECM is coated on a transparent conductive surface (TCS); on top of this coated substrate, an electrolyte layer is placed. Then, the whole system is sandwiched by another TCS. The ECDs are mainly two types of fabrication styles: single-layer and complementary-layer ECDs, as shown in Figure 2.2. All the constituent components of an ECD are discussed below.

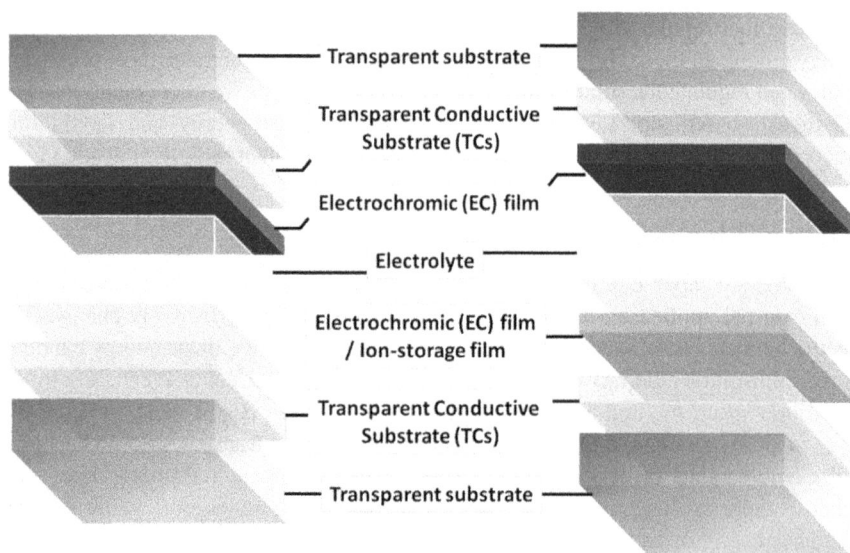

**FIGURE 2.2**  Schematic representation of single-layer (left) and complementary-layer (right) ECDs. (Figure is not in scale).

## 2.2.1 Transparent Conductors

For the fabrication of ECDs, a highly conductive transparent material with high optical transparency (>85% in the visible range) and electronic conductivity (resistivity ~$10^{-4}$ $\Omega \cdot$cm) is necessary. High visual transparency provides the high optical contrast variation of the ECDs. On the other hand, high electronic conductivity reduces the overpotential of the device. However, only a few transparent conducting oxide (TCO) materials are commonly used TCS in the industry, among which $In_2O_3$:Sn (ITO) is the best due to the high transparency, conductivity, and stability. The other TCOs are $In_2O_3$:M (M = Zn, Nb, Al), ZnO:M (M = Al, Ga, In, Si, B, F, Nb, Sc), $SnO_2$:M (M = F, Sb, Ta, W), indium gallium zinc oxide (IGZO), etc. [22].

These transparent conductive ITO materials serve as the electrodes and charge-balancing layers in ECDs. Deposition of this transparent conductive material like ITO or fluorine-doped tin oxide (FTO) on the non-conductive glass/PET transparent substrate provides mechanical support to the whole device. In single-layer ECDs, ITO on a glass substrate (ITO/glass) serves as the WE and helps to deliver the charges to the ECMs, whereas a bare ITO/glass serves as the CE acts as a charge balancing layer (Figure 2.2). On the contrary, in complementary-layer ECDs, instead of a bare ITO/glass, an ECM-coated ITO/glass serves as CE to balance the charges and reduce the potential gap between electrodes.

## 2.2.2 Electrolytes

The general function of an electrolyte material in an electrochemical cell is to act as the source ion for the charge balancing of two electrodes. It's usually an electron insulator to avert the short circuits in the device. However, an electrolyte must have high ionic conductivity.

In particular, for EC application, the electrolyte must possess high transparency in the Vis-to-NIR region, and high thermal and photostability under UV irradiation. The commonly used electrolytes in ECDs are gel and ionic liquids. However, the gel electrolytes are most widely applied as they possess good adhesive properties to firmly paste the two electrodes together for ECD fabrication. Two different polymer-based gel electrolytes are used in fabrication: polymer electrolyte and polyelectrolyte. In polymer-based electrolyte, the polymers are dissolved in solvents to form high viscous gel. Polymers like poly(methyl methacrylate) (PMMA), poly(vinyl alcohol) (PVA), and poly(ethylene oxide) (PEO) are used in this regard [23]. Ion conductivity in this gel-polymer was provided by adding some particular concentration of salt or acid solution ($LiClO_4$). In contrast, polyelectrolytes are polymers with ionic groups, and they can provide the gel form as well as the ion-conducting properties by their own. In this regard, poly(2-acrylamido-2-methylpropanesulfonic acid) (PAMPS) is anionic polyelectrolyte with gel forming property. However, polyelectrolytes are not recommended for ECD fabrication due to insufficient ionic conductivity and thermal instability. Like gel electrolytes, liquid electrolytes are also recommended for device fabrication. In liquid electrolytes (e.g. $LiClO_4$ in propylene carbonate), the salts or acid/bases are dissolved in a solvent or ionic liquid to provide ions for electrochemical processes. Ionic liquids like 1-butyl-3-methylimidazolium hexafluorophosphate

([BMIM][PF$_6$]) and 1-butyl-3-methylimidazolium tetrafluoroborate ([BMIM][BF$_4$] are the salts in a liquid state and can be directly used as electrolyte in this process [24]. The solution-based electrolytes hold the high ionic mobility and improve the response time of ECD compared to the gel-based electrolytes. At the same time, the low cost and good thermal stability of the ionic liquids make them a popular option as electrolytes. However, the leakage or solvent evaporation of liquid electrolytes hinders them in large-area application. In that case, the ECDs would always be completely sealed.

### 2.2.3 EC LAYERS

EC layers are the functional part of ECDs, which display optical modulations by switching their redox state to the applied external bias [25–30]. When a specific voltage is applied across an ECD, ion separation in the electrolyte happens, where cations move towards the cathode and anions move towards the anode. Consequently, the EC layer changes its redox state to maintain electroneutrality in the whole system. Based on the ion movement direction in the electrolyte layer towards the EC lattice, there are two kinds of EC layer materials available: cathodically coloring and anodically coloring materials. When a material shows the color change on cation insertion known as cathodically ECMs such as WO$_3$, TiO$_2$, Prussian blue etc. On the contrary, anodically ECM changes its hue on cation desertion such as NiO, thiophene moiety, pyrrole, polyamine, etc. There are plenty of ECMs available, which are classified into different generations. However, to fabricate the ECDs, ECM must be deposited or coated on conductive ITO/glass substrate by different methods and technology [31]. Some of the special methods include chemical bath deposition, solvothermal deposition, electrochemical deposition, chemical vapor deposition, sol-gel synthesis, layer-by-layer assembly, thermal evaporation, drop-casting, spin/spray coating, spray pyrolysis, etc. [32–37] Besides all these, a very recent method to direct a thin film of polymer material is an interfacial synthesis of ECM, which can directly be deposited on ITO by a solvent removal technique [38].

## 2.3 FUNCTIONAL PRINCIPLE BEHIND ELECTROCHROMISM

The basic principle of electrochromism in ECDs involves the change in optical properties of an ECM when a small *dc* potential is applied across the device. Behind the mechanism, electrolytes play a crucial role here. The electrolytes have good ionic conductivity with negligible or no electronic conductivity. However, in contact with an EC film surface, the ion conductor helps to diffuse both ions and electrons. The application of an electric potential to the electrodes of an EC cell provides the electromotive force to the cell (Figure 2.3a). Since the electrolyte itself is an electron-insulator, an electric charge layer will appear on the surface of the electrodes. At this point, positively charged ions in electrolytes are repelled and moved towards the cathode, while negatively charged ions move towards the cathode, known as electrophoresis. Nevertheless, as small cations have higher mobility, the transfer of cations becomes faster. So, to maintain the electroneutrality in the system, a charge-balancing redox reaction takes place where electrons are

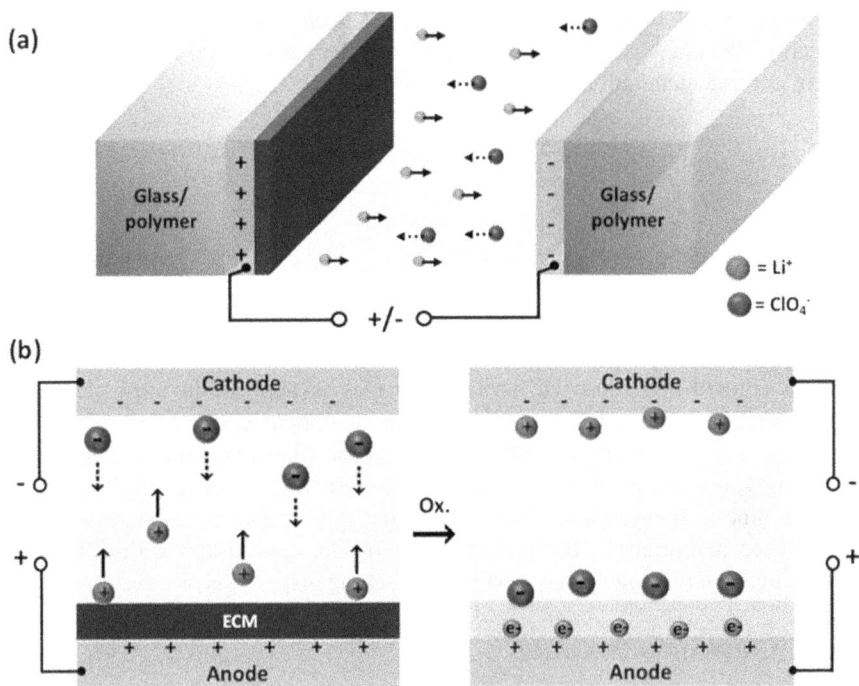

**FIGURE 2.3** (a) Schematic of an EC cell in presence of applied voltage. (b) Schematic illustration of mechanistic path of ion-transfer and redox reaction in an anodically colored ECD under a certain *dc* voltage.

exchanged on the surface of the electrode and anions slowly diffuse to the system from the electrolyte layer. This redox reaction of EC films results in a change in the Fermi level of the materials, and thus its optical properties (Figure 2.3b). The below figure shows a schematic mechanistic representation of an anodically coloring ECM. Reversal of the voltage leads to a change the electrons and ion's flow direction and thus brings back the original property of the ECM.

## 2.4 EC PARAMETERS

Despite evincing EC properties by many chemical species, only those with favorable EC parameters are acceptable for possible commercial utilizations. Typically, ECMs with high optical contrast, high coloration efficiency, distinct color modulations, and high cycle life are demandable. The essential performance parameters are discussed next.

### 2.4.1 OPTICAL CONTRAST

The optical contrast is quantified as the maximum change in transmittance of an ECM achieved at a wavelength of interest ($\lambda_{max}$) for a certain amount of inserted charge, Q. The optical contrast is measured as a percentage value (shown in

Equation 2.1). Alternatively, this performance sometimes may be given as a contrast ratio (CR) (shown in Equation 2.2), which refers to the ratio of the transmitted light of colored state and that of the bleached state at $\lambda_{max}$.

$$\Delta T = T_{bleached} - T_{colored} \tag{2.1}$$

$$CR = T_{colored}/T_{bleached} \tag{2.2}$$

### 2.4.2 Response Time

The response time of an ECM is stated as the time taken for the sample to switch between colored and bleached states when bias is applied. In most cases, the switching time is recorded for 95% change in optical contrast instead of full switching. Various factors can affect the response time like film thickness, film morphology, applied potential, nature of electrolyte, and so on. Typically, a faster response time is always desirable; however, the significance of response time as a performance measurement highly depends upon the desired application. In some applications like reading display, a slower switching response is also recommended.

### 2.4.3 Coloration Efficiency

Another key performance parameter is coloration efficiency (CE), which determines the power efficiency of an ECM for color change. Basically, it refers to the change in absorption ($\Delta A$) at a particular wavelength for inserted charges (Q) per unit area, which can be calculated by Equation 2.3. The unit of measurement for CE is $cm^2\ C^{-1}$; as derived from area (in $cm^2$) per unit of inserted charge Q (measured in coulombs, C). $\Delta A$ is determined by the transmittance between bleached ($\%T_{bleached}$) and colored ($\%T_{colored}$) states, as shown in Equation 2.4. This parameter quantitatively defines that the charges insertion of the materials stimulates the color change. The higher the CE value for an ECD, the more energy efficiency.

$$CE\ (\eta) = \Delta A/Q \tag{2.3}$$

$$\Delta A = log_{10}(\%T_{bleached}/\%T_{colored}) \tag{2.4}$$

### 2.4.4 Durability

For an ECD to be commercially and economically viable, it must have an adequate cycle life. That means the device can switch colored and bleached states for many hundreds or thousands of times without significant decrement of the device's performance. In repeating redox cycles, the EC performance may decline because of over-potential/over-charging, sudden loss of ionic conductivity of electrolyte, side reactions, irreversible reactions due to aerial oxidation, iR drops of the electrodes or the electrolyte, and so on. Thus, higher durability of an ECD is always necessary to be commercially applicable.

### 2.4.5 OPTICAL MEMORY

Like response time, optical memory is another parameter that highly depends on the desired application. Optical memory signifies the sustainability of EC material in a new colored/bleached state at open circuit potential (OCP). Most of the organic materials show less optical memory, which means they quickly return to the original state, even in OCP. However, some inorganic metal oxides can sustain for more than a day. This difference is based on the bistable redox states of the ECMs. The pristine state can be regenerated in OCP, owing to the diffusion of ions and trapped electrons through the system which cause the self-reduction/oxidation of the ECD even after the termination of the electric field [29].

### 2.4.6 OPTICAL MODULATION

Besides longevity and response time, optical modulation is also a crucial performance parameter, especially for EC windows. Regarding optical modulation, ECM requires not only transmittance change in a wide UV-Vis region, but also needs to have the optical modulation in NIR region to prevent solar heat transfer.

## 2.5 TYPES OF EC MATERIALS

The term "electrochromism" was first used by Platt in 1961 to designate a possible color change of organic dye molecule by an electric field [39]. In fact, the color change phenomenon by a redox reaction was observed in early 1815 by a Berzelius chemist. The study was witnessed the reduction induced color change from pale yellow to blue in $WO_3$ [13]. In 1840, Prussian Blue (PB) was applied for photography, initiating the redox-coloration process with electron transfer [13]. This redox-coloration phenomenon was observed till the 20th century in the materials like iodine, silver, $MoO_3$ etc. The first widely accepted and properly documented suggestion on ECD was attributed by Prof. S. K. Deb in 1969, where a $WO_3$ thin film displayed EC coloration upon $10^{-4}$ $V.cm^{-1}$ bias application [29]. In 1973, the invention of a viologen-based EC display device [29] garnered widespread attention towards finding other new ECMs, and thus, numerous ECMs have been investigated by many researchers. Based on the EC properties in a specified class of materials, ECMs are divided into several generations. A brief discussion and examples of these classes of materials are given below.

### 2.5.1 INORGANIC METAL OXIDE MATERIALS (FIRST-GENERATION ECMs)

Typical transition metal oxides are the potential candidates for this class of ECMs. Besides tungsten oxide ($WO_3$), the oxides of other transition metals like Ce, Cr, Co, Cu, Ir, Fe, Mn, Mo, Ni, Nb, Pd, Rh, Ru, Ta, Ti, and V show electrochromism. As we discussed earlier, depending on ion-movement and electron uptake/intake in the presence of electrical bias, two typical metal oxides existed: cathodic and anodic metal oxides. The EC oxides of W, Ti, Ta, Nb, Mo etc. show cathodically coloration, whereas the oxide layers of V, Cr, Mn, Fe, Co, Ni, Rh, Ir exhibit anodically coloration.

In transition metal oxide category, $WO_3$ is the most widely studied as ECM. It is an indirect bandgap semiconductor that exhibits the intercalation and deintercalation of small cations like $H^+$, $Li^+$, $K^+$, etc. under the applied potential/bias [40]. The lattice of $WO_3$ film with $W^{6+}$ sites is colorless or pale yellow. Cathodic polarization causes ion insertion and injection of electrons to the $WO_3$ lattice, reducing some $W^{6+}$ sites to $W^{5+}$ and appearing intense blue coloration. The blue hue arises due to the intervalence charge transfer (IVCT) transition between the adjacent $W^{5+}$ and $W^{6+}$ sites.

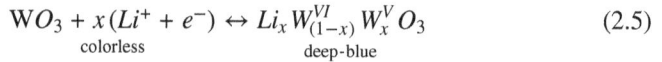

$$WO_3 + x(Li^+ + e^-) \leftrightarrow \underset{\text{deep-blue}}{Li_x W^{VI}_{(1-x)} W^V_x O_3} \qquad (2.5)$$
$$\underset{\text{colorless}}{}$$

At a low value of $x$, the film imparts deep blue color; however, at a higher value of $x$, the ion insertion irreversibly forms a metallic bronze that is golden or red. In addition, the $WO_3$ and its reduced state experience a non-metal-to-metal transition during redox reactions. At a fully oxidized $W^{6+}$ state ($x = 0$, insertion coefficient), the $WO_3$ becomes the insulator with low electronic conductivity. With increasing x, the electronic conductivity of $Li_x W^{VI}_{(1-x)} W^V_x O_3$ increases from non-metallic to metallic owing to the increasing number of delocalized transferable electrons. Thus, the insertion coefficient plays a vital role for metallic oxides, which varies from different cations and affects the EC process.

Besides the size of cations, the insertion coefficient highly depends on the structure and morphology of the metal oxides. The morphology of $WO_3$ depends on the synthetic methods. The electro- or thermal deposition yields the amorphous $WO_3$ (a-$WO_3$) thin film. The crystalline perovskite structure with monoclinic phase, $WO_3$ (c-$WO_3$) can be produced through sputtering, hydrothermal synthesis, and by pyrolysis process. The a-$WO_3$ contains large portion of defects than the c-$WO_3$ and the defects strongly effect the electronic structure. Since a-$WO_3$ is more electronically conductive than c-$WO_3$, the EC switching fastness of a-$WO_3$ can be better than that of c-$WO_3$. Again, the CE of the a-$WO_3$ can be better than that of c-$WO_3$ due to good electrical conductivity and good ion diffusion in the amorphous film.

As noted above, in a similar way, many other metal-oxides thin film such as $MoO_3$, $V_2O_5$ are also EC in nature due to the IVCT caused by the reduction [13].

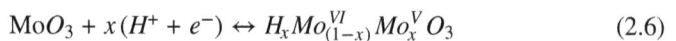

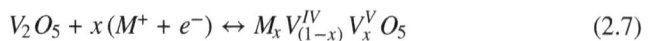

$$MoO_3 + x(H^+ + e^-) \leftrightarrow H_x Mo^{VI}_{(1-x)} Mo^V_x O_3 \qquad (2.6)$$

$$V_2O_5 + x(M^+ + e^-) \leftrightarrow M_x V^{IV}_{(1-x)} V^V_x O_5 \qquad (2.7)$$

Contrary to the cathodic metal oxides, anodic metal oxides revealed EC on electron ejection and cation deintercalation. For example, group VIII transition metal oxide or hydroxide like $Ir(OH)_3$ displayed blue-grey to colorless EC on oxidation. Though the mechanism is still unclear, both anion insertion and cation deintercalation were proposed as the following equations.

$$Ir(OH)_3 = IrO_2. H_2O + H^+ + e^- \tag{2.8}$$

$$Ir(OH)_3 + (OH)^- = IrO_2. H_2O + H_2O + e^- \tag{2.9}$$

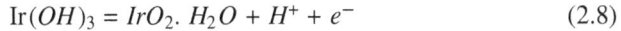

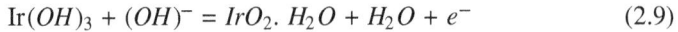

## 2.5.2 ORGANIC SMALL MOLECULES AND POLYMERS (SECOND-GENERATION ECMs)

Among the organic molecules, the molecules which can form a stable radical species impart EC in a small potential range. Both small organic molecules and conjugated polymers are capable of that. Among small organic molecules, viologens are the widely studied moiety [41]. Viologens are salts of quaternized 4,4'-bipyridine and widely used as redox indicator, electron transfer mediator, redox flow battery, etc. along with the EC applications. The prototype viologen is the methyl viologen (MV), which is a di-cation system, colorless in its original state. Subsequent reduction of MV forms the radical cation, and the stability of this monovalent cation radical is attributed to the delocalization of the radical electron throughout the $\pi$ framework of the bipyridyl nucleus. Once the radical state is generated, it imparts an intense blue color owing to the optical charge transfer between +1 valent to zero-valent nitrogen. Thus, suitable change in the nitrogen substituents affects the molecular orbital energy level, and thus, optical modulation in different spectral regions can be achieved.

Besides small organic molecules, conjugated polymers, also known as conducting polymers, contribute a major part to this class of materials. Due to the presence of $\pi$-delocalized framework, they have very interesting optical and electronic properties. The easy availability of monomers, ease of processing, and high optical contrast at different redox state of the conductive polymers have drawn significant interest to the researchers. Like metal oxides, the ion insertion/extraction and electron injection/ extraction in the polymer backbone results in modified electronic band structure and consequent for the change in optical properties. In other words, the change in the electronic bandgap ($E_g$) between the highest-occupied $\pi$-electron band (valance band) and the lowest-unoccupied band (conduction band) imparts the color change in a certain applied voltage. The ion-insertion to the polymer framework in a certain applied bias is termed as "doping". During oxidation, the EC layer creates a positive charge on the framework, which is balanced by counter anion insertion, termed p-doping, creates a delocalized $\pi$-electron band structure. The electrochemical reduction of the oxidized EC layer leads to the removal of the counter anion from the EC layer, and changes the cationic state to yield an undoped electrically neutral polymer structure.

The intrinsic color change or the optical contrast change between undoped and doped polymer is directly related to the bandgap ($E_g$) magnitude. The conjugated polymer thin films with $E_g > 3$ eV (~400 nm) are colorless and transparent in the undoped form. Once they undergo p-doing, due to the formation of a temporary doped band, the $E_g$ reduces to below 1.5 eV and imparts intense color in the visible region. However, those with $E_g \leq 1.7$–1.9 eV (~650–900 nm) are highly absorbing in the undoped form. However, after doping, the bandgap reduces so much that the absorption falls in the NIR spectral region. Polymers with intermediate bandgaps have distinct optical changes throughout the visible region and exhibit several colors.

Among the conjugated polymers, the most studied ECM is poly(thiophene)-based polymers. Thin polymeric films of the parent poly(thiophene) are blue ($\lambda_{max}$ = 730 nm) in the doped (oxidized) state and red ($\lambda_{max}$ = 470 nm) in the undoped form. Color tuning of this polymer can be achieved by changing the substituents at '3' or '4' or in both positions in the framework and a huge number of substituted thiophenes have been synthesized. The steric hindrance in the substitutes position can significantly affect the $E_g$ as well as the color modulation, as shown in Figure 2.4a [42]. An alkylenedioxy-substituted thiophene polymer,

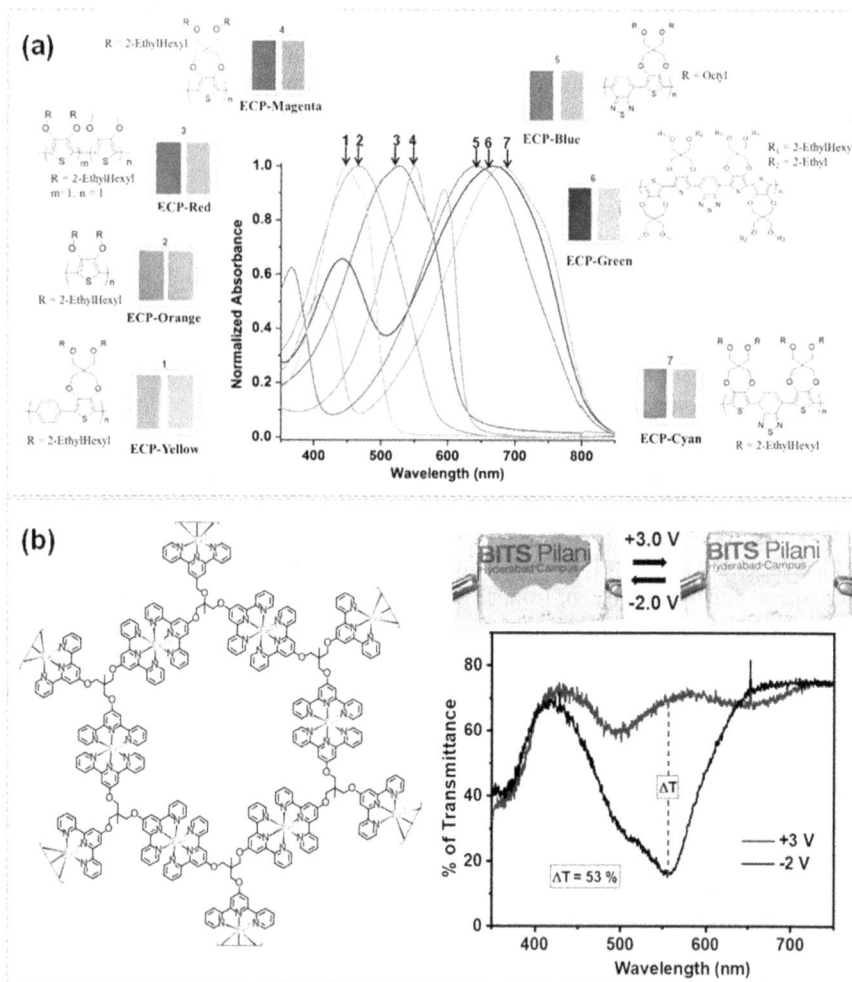

**FIGURE 2.4** (a) EC color and corresponding absorption spectra of different substituted poly(thiophene) polymer, adopted with permission from [42]. Copyright (2019) Royal Chemical Society. (b) EC color change of Fe(II)-based metallopolymer from pink to colorless under applied potential, adopted with permission from [48]. Copyright 2020 American Chemical Society.

poly[3,4-(ethylenedioxy)thiophene] (PEDOT) exhibits a deep-blue color in its neutral state and a light-blue transmissive state upon oxidation. It is a widely investigated polymer and derivatives of PEDOT that exhibit a huge optical modulation in the wide spectral region.

Besides polythiophene and its derivative, other conductive polymers like poly-aniline, polypyrrole, polycarbazole, polyazulene, polyindole, etc. showed electro-chromism covering broad spectral regions from visible to the NIR region [43].

## 2.5.3 Transition Metal Complexes (Third-Generation ECMs)

Metal coordination complexes are the potential candidates for this category because of their intense coloration, high stability, and good redox activity. The chromogenic color arises from the different electronic band transitions, especially for metal-to-ligand charge transfer (MLCT) and IVCT combined with intra-ligand charge transfer and related d-d transitions. As MLCT/IVCT transitions involve valence electrons, the chromism can disappear or alter upon oxidation or reduction of the complexes. Small metal-complexes including metallophthalocyanine, transition-metal polypyridyl complexes, Prussian blue (PB) analogous are the promising materials in this class.

PB is the earliest modern synthetic pigment with iron (III) hexacyanoferrate chemical composition. The intense blue color arises for the IVCT electronic transition between mixed-valence oxidation states. PB is a prototype member of polynuclear transition-metal hexa-cyanometallates form an important class of insoluble mixed-valence compounds. They have the general formula of $M'_k [M_m"(CN)_6]_l.nH_2O$, where M' and M" are the transition metals with different oxidation states. Among these PB-analogous, only ruthenium purple [Fe(III) hexacyanoruthenate(II)] and osmium purple [Fe(III) hexacyanoosmate(II)] displayed promising EC behaviour along with the PB.

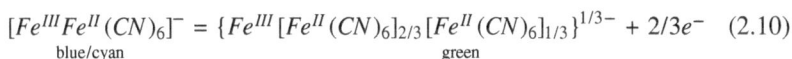

$$[Fe^{III}Fe^{II}(CN)_6]^- = \{Fe^{III}[Fe^{II}(CN)_6]_{2/3}[Fe^{II}(CN)_6]_{1/3}\}^{1/3-} + 2/3e^- \quad (2.10)$$
$$\text{blue/cyan} \qquad\qquad\qquad\qquad\qquad\qquad \text{green}$$

$$[Fe^{III}Fe^{II}(CN)_6]^- = [Fe^{III}Fe^{III}(CN)_6] + e^- \quad (2.11)$$
$$\text{blue/cyan} \qquad\qquad \text{yellow}$$

$$[Fe^{III}Fe^{II}(CN)_6]^- + e^- = [Fe^{II}Fe^{II}(CN)_6]^{2-} \quad (2.12)$$
$$\text{blue/cyan} \qquad\qquad \text{colorless}$$

The thin-film preparation of PB for ECD fabrication involves galvanostatic, po-tentiodynamic, and potentiostatic electrochemical reduction of solutions containing Fe(III) and hexacyanoferrate(III) ions. Due to the presence of two potentially switchable redox-active metal centers, the film can exhibit both oxidation or reduction and display different colors. PB chromophore itself is blue in its original state due to IVCT between Fe(II) and Fe(III) centers. Partial electrochemical oxi-dation of PB yields a green color state, whereas the fully oxidized state is golden yellow due to the electronic transition in $[Fe^{III}(CN)_6]^-$ fragment. However, PB yields colorless redox state under reduction, and results in the disappearance of IVCT transition [44].

### 2.5.4  METALLO-SUPRAMOLECULAR POLYMERS (FOURTH-GENERATION ECMs)

The metallo-supramolecular polymers (MSPs) are the newest class of ECMs, and have drawn immense attention towards smart EC technology because of their high optical contrast, high coloration ability, and low switching potential. The hybrid polymers are specifically the coordination complex polymers, comprising conjugated multidentate chelating ligand and metal ion monomers [45]. As chelating ligands have the property of trapping metal ions from solution by complex formation, the conjugated multidentate chelating ligands such as bis-terpyridine, tris-terpyridine, bis-phenanthrolene, etc. are potential candidates as ligand monomers [46]. These hybrid polymers impart intense colors especially due to the characteristic MLCT from metal d-HOMO to ligand $\pi$*-LUMO. This typical charge transfer band is easily altered by inserting different metal ions (change in HOMO level) or changing the spacer in the ligand framework (change in LUMO level). This causes the color-tuning in a particular polymer framework [47].

Most of the MSPs show prominent EC under positive bias. For an example, a Fe(II)-metallopolymer with tris-terpyridine ligand is intense pink in the original state (Figure 2.4b) [48]. The characteristic MLCT arises owing to the MLCT at $\lambda_{max} \sim 565$ nm. On application of +3 V external bias through solid-state ECD, oxidation of the metal center results in the formation of Fe(III) centers. In the new oxidized state, the bandgap of MLCT is very high, and consequently no MLCT happens and the film becomes colorless. The original state is again regained once the voltage is reverted back, leading to the formation of Fe(II) from Fe(III) by reduction at −2V.

The self-assembly between transition metal ions to ditopic bidentate or tridentate coordinated ligands leads to 1D nanorod/nanofibers-based MSPs. On the other hand, complexation of a metal ion with multitopic branched ligand e.g., tris-terpyridine in a bilayer surface (with slow, controlled diffusion of metal ions) assures the construction of 2D MSPs, whereas bulk self-assembly builds up 3D dendritic assembly-based MSPs. They can be grown from ITO substrate; nonetheless, a deposition of thin film by spin/spray coating of these bulk polymers is difficult for low solubility.

Another way to get the molecular assembly of the MSP structure is the electro-polymerization technique [49]. The EC film can be prepared directly on the transparent conductive surfaces by simple electrodeposition of metal complexes with polymerizable monomer moieties, initiated either by oxidation or reduction of the monomer. The films prepared in this technique are insoluble but stable, adherent, and electrochemically active. The polymerization proceeds through the formation of a radical intermediate anion followed by radical-radical coupling and chain propagation. The widely used electropolymerization technique is reductive-electropolymerization, where the monomer must have at least two vinyl groups, so that the cross-linking between the formed radicals is highly effective. Unlike the vinyl-containing complex monomers, which undergo reductive electropolymerization, metal-coordinated monomer complexes containing pyrrole, carbazole, or triphenylamine undergo oxidative electropolymerization and provide 1D or hyperbranched 3D cross-linked polymeric film.

## 2.6   CONCLUSION AND OUTLOOKS

Since the electrochromism was discovered, the smart window technology has dramatically progressed toward advanced and energy-efficient buildings. Vivid color range, long-term durability, light weight, and easy portability are the key factors for a successful product. Though the EC performance parameters of different generation ECMs have been tried to modify in various ways, the high-cost manufacturing issue has been kept aside.

The costly transparent conducting substrates are very necessary for the device fabrication. On the other hand, using glass as a substrate material makes the device heavy and brittle. Though polymer-based substrate materials are used nowadays for flexible, lightweight devices, finding an alternative transparent substrate is more necessary.

For a power-efficient building, the EC technology is combined with a solar panel to drive the ECDs through solar power. However, the generation of self-powered ECDs is required to overcome today's energy demand. On the other hand, a concentration on the dual functionality of ECD is required to get a hybrid electrochromic-energy storage device, which again helps to save primary energy sources. Moreover, in the smart buildings, a remote-controlled EC switching is recommended for a good commercial business, which needs a potential interest and development.

## REFERENCES

[1] P. Huovila, U. N. E. Programme, U. N. E. P. S. Consumption, R. Branch, Buildings and climate change: Status, challenges, and opportunities, united nations environment programme, sustainable consumption and production branch, 2007.

[2] U.S. Department of Energy, http://Buildingsdatabook.Eren.Doe.Gov/. 2010, 1, accessed: October, 2016.

[3] A. M. Omer, Energy, environment and sustainable development, *Renew. Sustain. Energy Rev.* 12 (2008) 2265–2300.

[4] J. Huang, K. R. Gurney, Impact of climate change on U.S. building energy demand: Sensitivity to spatiotemporal scales, balance point temperature, and population distribution, *Clim. Change.* 137 (2016) 171–185.

[5] A. Jäger-Waldau, PV Status Report 2004.

[6] M. Darwish, Energy efficient air conditioning: Case study for Kuwait, *Kuwait J. Sci. Eng.* 32 (2005) 209–222.

[7] H. Khandelwal, A. P. H. J. Schenning, M. G. Debije, Infrared regulating smart window based on organic materials, *Adv. Energy Mater.* 7 (2017) 1602209.

[8] E. L. Runnerstrom, A. Llordés, S. D. Lounis, D. J. Milliron, Nanostructured EC smart windows: Traditional materials and NIR-selective plasmonic nanocrystals, *Chem. Commun.* 50 (2014) 10555–10572.

[9] A. Sharma, V. V. Tyagi, C. R. Chen, D. Buddhi, Review on thermal energy storage with phase change materials and applications, *Renew. Sustain. Energy Rev.* 13 (2009) 318.

[10] S. Kim, J. Cha, S. Kim, K. W. Park, D. R. Lee, J. H. Jo, Improvement of window thermal performance using aerogel insulation film for building energy saving, *J. Therm. Anal. Calorim.* 116 (2014) 219.

[11] G. B. Smith, C. G. Granqvist, Green nanotechnology: Solutions for sustainability and energy in the built environment, *J. Nanophotonics.* 5 (2011) 1–2.

[12] R. Baetens, B. P. Jelle, A. Gustavsen, Properties, requirements and possibilities of smart windows for dynamic daylight and solar energy control in buildings: A state-of-the-art review, *Sol. Energy Mater. Sol. Cells.* 94 (2010) 87–105.

[13] C. G. Granqvist, *Handbook of Inorganic EC Materials*, Elsevier Science, 1995.

[14] A. M. Österholm, D. E. Shen, J. A. Kerszulis, R. H. Bulloch, M. Kuepfert, A. L. Dyer, J. R. Reynolds, Four shades of brown: Tuning of EC polymer blends toward high-contrast eyewear, *ACS Appl. Mater. Interfaces.* 7 (2015) 1413–1421.

[15] W.-H. Wang, J.-C. Chang, T.-Y. Wu, 4-(Furan-2-yl)phenyl-containing poly-dithienylpyrroles as promising electrodes for high contrast and CE EC devices, *Org. Electron.* 74 (2019) 23–32.

[16] T. G. Yun, M. Park, D.-H. Kim, D. Kim, J. Y. Cheong, J. G. Bae, S. M. Han, I.-D. Kim, All-transparent stretchable EC supercapacitor wearable patch device, *ACS Nano.* 13 (2019) 3141–3150.

[17] X. Li, T. Y. Yun, K.-W. Kim, S. H. Kim, H. C. Moon, Voltage-tunable dual image of electrostatic force-assisted dispensing printed, tungsten trioxide-based EC devices with a symmetric configuration, *ACS Appl. Mater. Interfaces.* 12 (2020) 4022–4030.

[18] S. Soylemez, H. Z. Kaya, Y. A. Udum, L. Toppare, A multipurpose conjugated polymer: EC device and biosensor construction for glucose detection, *Org. Electron.* 65 (2019) 327–333.

[19] P. R. Somani, S. Radhakrishnan, EC materials and devices: Present and future, *Mater. Chem. Phys.* 77 (2003) 117–133.

[20] J.-Y. Shao, C.-J. Yao, B.-B. Cui, Z.-L. Gong, Y.-W. Zhong, Electropolymerized films of redox-active ruthenium complexes for multistate near-infrared electrochromism, ion sensing, and information storage, *Chinese Chem. Lett.* 27 (2016) 1105–1114.

[21] C. G. Granqvist, M. A. Arvizu, İ. Bayrak Pehlivan, H.-Y. Qu, R.-T. Wen, G. A. Niklasson, EC materials and devices for energy efficiency and human comfort in buildings: A critical review, *Electrochim. Acta.* 259 (2018) 1170–1182.

[22] C. G. Granqvist, ECs for smart windows: Oxide-based thin films and devices, *Thin Solid Films.* 564 (2014) 1–38.

[23] V. K. Thakur, G. Ding, J. Ma, P. S. Lee, X. Lu, Hybrid materials and polymer electrolytes for EC device applications, *Adv. Mater.* 24 (2012) 4071–4096.

[24] R. Marcilla, F. Alcaide, H. Sardon, J. A. Pomposo, C. Pozo-Gonzalo, D. Mecerreyes, Tailor-made polymer electrolytes based upon ionic liquids and their application in all-plastic EC devices, *Electrochem. Commun.* 8 (2006) 482–488.

[25] R. J. Mortimer, EC materials, *Chem. Soc. Rev.* 26 (1997) 147–156.

[26] H. Yu, S. Shao, L. Yan, H. Meng, Y. He, C. Yao, P. Xu, X. Zhang, W. Hu, W. Huang, Side-chain engineering of green color EC polymer materials: Toward adaptive camouflage application, *J. Mater. Chem. C.* 4 (2016) 2269–2273.

[27] R. J. Mortimer, Organic EC materials, *Electrochim. Acta.* 44 (1999) 2971–2981.

[28] R. J. Mortimer, EC Materials, *Annu. Rev. Mater. Res.* 41 (2011) 241–268.

[29] C. J. Schoot, J. J. Ponjee, H. T. van Dam, R. A. van Doorn, P. T. Bolwijn, New EC memory display, *Appl. Phys. Lett.* 23 (1973) 64–65.

[30] B. A. Korgel, Composite for smarter windows, *Nature.* 500 (2013) 278–279.

[31] S.-I. Park, Y.-J. Quan, S.-H. Kim, H. Kim, S. Kim, D.-M. Chun, C. S. Lee, M. Taya, W.-S. Chu, S.-H. Ahn, A review on fabrication processes for EC devices, *Int. J. Precis. Eng. Manuf. Technol.* 3 (2016) 397–421.

[32] C. M. White, D. T. Gillaspie, E. Whitney, S.-H. Lee, A. C. Dillon, Flexible EC devices based on crystalline $WO_3$ nanostructures produced with hot-wire chemical vapor deposition, *Thin Solid Films*. 517 (2009) 3596–3599.

[33] Y.-S. Lin, H.-T. Chen, J.-Y. Lai, EC performance of PECVD-synthesized $WO_xC_y$ thin films on flexible PET/ITO substrates for flexible EC devices, *Thin Solid Films*. 518 (2009) 1377–1381.

[34] P. Egger, G. D. Sorarù, S. Diré, Sol-gel synthesis of polymer-YSZ hybrid materials for SOFC technology, *J. Eur. Ceram. Soc.* 24 (2004) 1371–1374.

[35] S. Pereira, A. Gonçalves, N. Correia, J. Pinto, L. Pereira, R. Martins, E. Fortunato, EC behavior of NiO thin films deposited by e-beam evaporation at room temperature, *Sol. Energy Mater. Sol. Cells*. 120 (2014) 109–115.

[36] S. H. Mujawar, A. I. Inamdar, C. A. Betty, R. Cerc Korošec, P. S. Patil, Electrochromism in composite $WO_3$–$Nb_2O_5$ thin films synthesized by spray pyrolysis technique, *J. Appl. Electrochem*. 41 (2011) 397–403.

[37] C. E. Patil, N. L. Tarwal, P. S. Shinde, H. P. Deshmukh, P. S. Patil, Synthesis of EC vanadium oxide by pulsed spray pyrolysis technique and its properties, *J. Phys. D. Appl. Phys*. 42 (2008) 25404.

[38] K. Takada, R. Sakamoto, S.-T. Yi, S. Katagiri, T. Kambe, H. Nishihara, EC bis (terpyridine)metal complex nanosheets, *J. Am. Chem. Soc*. 137 (2015) 4681–4689.

[39] J. R. Platt, Electrochromism, a possible change of color producible in dyes by an electric field, *J. Chem. Phys*. 34 (1961) 862–863.

[40] S. Cong, F. Geng, Z. Zhao, Tungsten oxide materials for optoelectronic applications, *Adv. Mater*. 28 (2016) 10518–10528.

[41] W. W. Porter, T. P. Vaid, A. L. Rheingold, Synthesis and characterization of a highly reducing neutral "extended viologen" and the isostructural hydrocarbon 4,4″-Di-n-octyl-p-quaterphenyl, *J. Am. Chem. Soc*. 127 (2005) 16559–16566.

[42] A. L. Dyer, E. J. Thompson, J. R. Reynolds, Completing the color palette with spray-processable polymer electrochromics, *ACS Appl. Mater. Interfaces* 3 (2011) 1787–1795.

[43] P. M. Beaujuge, J. R. Reynolds, Color control in π-conjugated organic polymers for use in EC devices, *Chem. Rev*. 110 (2010) 268–320.

[44] K. Itaya, K. Shibayama, H. Akahoshi, S. Toshima, Prussian-blue-modified electrodes: An application for a stable EC display device, *J. Appl. Phys*. 53 (1982) 804–805.

[45] P. R. Andres, U. S. Schubert, New functional polymers and materials based on 2,2′:6′,2″-Terpyridine metal complexes, *Adv. Mater*. 16 (2004) 1043–1068.

[46] M. Higuchi, EC organic–metallic hybrid polymers: Fundamentals and device applications, *Polym. J*. 41 (2009) 511–520.

[47] M. Higuchi, Stimuli-responsive metallo-supramolecular polymer films: Design, synthesis and device fabrication, *J. Mater. Chem. C*. 2 (2014) 9331–9341.

[48] S. Roy, C. Chakraborty, Interfacial coordination nanosheet based on nonconjugated three-arm terpyridine: A highly color-efficient EC material to converge fast switching with long optical memory, *ACS Appl. Mater. Interfaces*. 12 (2020) 35181–35192.

[49] L. Motiei, M. Altman, T. Gupta, F. Lupo, A. Gulino, G. Evmenenko, P. Dutta, M. E. van der Boom, Self-propagating assembly of a molecular-based multilayer, *J. Am. Chem. Soc*. 130 (2008) 8913–8915.

# 3 Mathematical Modelling of a Piezoelectric Wave Energy Converter Device Integrated with a Vertical Breakwater over a Stepped Seabed

*Vipin Valappil and Santanu Koley*

## CONTENTS

## 3.1 INTRODUCTION

The movement to phase out fossil fuels has been gaining historic momentum in the last decade. The pollution they produce includes anything from climate-damaging greenhouse gases to health-endangering particulates. To solve these social and environmental problems, long-term potential actions for sustainable development are necessary. Renewable energy resources have a significate role to solve these problems effectively. In renewable energy resources, wind energy and solar energy developed significantly in recent years, whereas wave energy is still in its immature stage. Wave energy is immensely available irrespective of the seasonality and weather conditions; still, wave energy is not explored because of the huge offshore equipment, expensive manufacturing, and high cost of maintenance. The significance of the piezoelectric wave energy converter device is emerging in these scenarios. PWEC devices are low cost in manufacturing and maintenance, ease of deployment, and the piezoelectric material can be bounded with various materials irrespective of the size and shape. The PWEC devices are commonly used to operate low-power electronic devices such as LED, various sensors, and etc. The PWEC device is composed of a flexible substrate with piezoelectric materials bonded on both faces of the substrate. The working mechanism behind the PWEC device is the piezoelectric effect. As the PWEC device is placed on the ocean surface, wave tides hit the PWEC plate, and consequently, the piezoelectric materials in both phases of the flexible plate are excited and a net electric charge will be generated.

A substantial number of works on the modelling of the PWEC device were done in the last few years in which one of the significant works of [1] investigated the bending problem of the piezoelectric plate with circular holes. The study shows that the existence of the piezoelectric effect increases the field concentration and piezoelectric plates with holes increase the field concentration even more. [2] showed that the nano-piezoelectric harvester has a better ability to absorb wave power and also shows that the natural frequency of the piezoelectric plates is depending on the size of the plate. [3] developed a wave energy harvester that harvests wave energy from the transverse wave motion using piezoelectric patches and studied the influence of the design parameters of the device and the wave parameters to optimize the performance of the wave energy harvester. [4] studied a PWEC device made up of several piezoelectric cantilevers integrated with a floating buoy. The study depicts that an increase in the ratio of the wavelength to the length of the cantilever decreases the power generated by the device. Further, the investigation shows that the power generation increases by increasing the length and reducing the thickness of the piezoelectric cantilevers. [5] investigated the power generated by the PWEC device and depicts that the wave period is the most influencing parameter in the wave power generation. [6] studied a breakwater-integrated PWEC device using eigenfunction expansion and showed the influence of the boundary edge conditions of the PWEC device.

In this chapter, the performance of a submerged PWEC device integrated with a vertical rigid and impermeable breakwater with the stepped bottom is investigated. The boundary element method (BEM) is used to study power generation. The variation of the power generated by the PWEC device ($P_{ext}(\mathrm{Wm}^{-1})$) is studied for various parameters related to the water waves, bottom profile, and the PWEC device. The write-up of the chapter begins with the mathematical formulation, and then the solution methodology using the BEM methodology, the results and discussion in a data analytic perspective, and subsequently, the conclusions are provided.

## 3.2  MATHEMATICAL FORMULATION

In this chapter, the power generated by a PWEC device is studied subjected to regular waves of amplitude $A$ and angular frequency $\omega$. The side view of the problem is provided in Figure 3.1. The PWEC device consists of a flexible plate with both the faces of the plate bounded with the piezoelectric material. The thickness of the flexible plate and piezoelectric materials are considered to be very less as compared to the incident wavelength. A 2-D Cartesian coordinate system is taken to model the boundary value problem (BVP), where the direction of incident

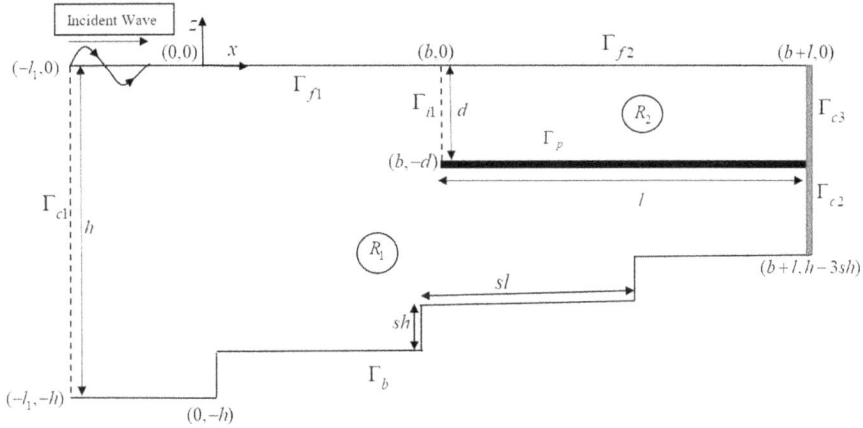

**FIGURE 3.1**  Illustration of the physical problem.

wave is represented on the x-axis and the vertically downward direction from the average free surface is denoted by the z-axis. Due to water waves, external pressure will act on the piezoelectric material. Consequently, the piezoelectric plate gets excited and power will be generated by the PWEC device. The PWEC device of length $l$ is placed at a submergence depth $d$ from $z = 0$, which cover the region $b \leq x \leq b + l$, $z = -d$. The front edge of the PWEC device satisfies either the fixed edge condition or the free edge condition. Furthermore, the lee side of the PWEC device is attached to the rigid and impermeable vertical breakwater by means of fixed edge condition or the moored edge condition. The bottom of the seabed is considered to be stepped type and occupies the region $0 \leq x \leq b + l$. Maximum number of steps are considered as three. For modelling, the water region of the problem is divided into two; $R_2 = \{b \leq x \leq b + l\} \cup \{-d \leq z \leq 0\}$, and the rest of the domain is denoted by $R_1$. The velocity potentials in both the regions are denoted by $\Phi_j(x, z, t)$, $j = 1, 2$ and they are of the form $\mathrm{Re}\{\phi_j(x, z)e^{-i\omega t}\}$, and satisfies

$$\left(\frac{\partial^2}{\partial x^2} + \frac{\partial^2}{\partial z^2}\right)\phi_j = 0, \quad j = 1, 2. \tag{3.1}$$

Here, the subscript $j$ denotes the fluid domain $R_j$.

The boundary condition (bcs) on $z = 0$ is given by

$$\frac{\partial \phi_j}{\partial z} = K\phi_j, \quad \text{on} \quad \Gamma_{fj}, \quad \text{for} \quad j = 1, 2, \tag{3.2}$$

where $K = \omega^2/g$. The bottom bcs on $\Gamma_b$ is given by

$$\frac{\partial \phi_1}{\partial n} = 0. \tag{3.3}$$

The boundary conditions on the rigid and impermeable breakwater are given by

$$\frac{\partial \phi_j}{\partial n} = 0, \quad \text{on } \Gamma_{cj+1}, \quad j = 1, 2. \tag{3.4}$$

The BCs on $\Gamma_{i1}$ is given by

$$\phi_1 = \phi_2, \quad \frac{\partial \phi_1}{\partial n} = -\frac{\partial \phi_2}{\partial n}. \tag{3.5}$$

The kinematic BC on $\Gamma_p$ is given by

$$\frac{\partial \phi_1}{\partial n} = -\frac{\partial \phi_2}{\partial n} = -i\omega\xi, \quad \text{on } \quad b < x < b + l, \quad z = -d, \tag{3.6}$$

where $\xi(x, t) = \mathrm{R}\{\xi(x)e^{-i\omega t}\}$ is the plate deflection. The dynamic boundary condition on $\Gamma_p$ is given by

$$g\chi\left[1 + \frac{\beta^2 \zeta\omega}{i + \zeta\omega}\right]\frac{\partial^4\xi}{\partial x^4} - \omega^2\gamma\xi = i\omega(\phi_1(x, -d^-) - \phi_2(x, -d^+)), \quad \text{on } \Gamma_p. \tag{3.7}$$

The parameters used in Equation (3.7) are given by

$$\chi = \frac{B}{\rho g}, \quad \beta = \frac{\theta}{\sqrt{BC}}, \quad \zeta = \frac{C}{V}, \quad \gamma = \frac{I_b}{\rho},$$

where $B$ denotes the flexural rigidity of the PWEC device, $\theta$ denotes the piezoelectric coupling factor, and $C$ and $V$ are the electrical surface capacitance and the surface conductance, respectively. Further, $I_b$ and $\rho$ represent the water density and surface density, respectively (see [5] and [6]). The edge conditions used to model the plate edges are either fixed, free, or moored edge conditions, and are given as

$$\frac{\partial\xi}{\partial x} = 0, \quad \xi = 0 \quad \text{at} \quad x = b, \; x = b + l, \tag{3.8}$$

$$\frac{\partial^2\xi}{\partial x^2} = 0, \quad \frac{\partial^3\xi}{\partial x^3} = 0, \quad \text{at} \quad x = b, \, b + l, \tag{3.9}$$

$$\frac{\partial^2\xi}{\partial x^2} = 0, \quad \rho g\chi\left[1 + \frac{\beta^2 \zeta\omega}{i + \zeta\omega}\right]\frac{\partial^3\xi}{\partial x^3} = q\xi, \quad \text{at} \quad x = b, \, b + l. \tag{3.10}$$

The BC on $\Gamma_{C_1}$ is given by

$$\frac{\partial(\phi_1 - \phi')}{\partial x} + ik_0(\phi_1 - \phi') = 0, \tag{3.11}$$

where the incident wave potential $\phi'(x, z) = e^{ik_0x}f_0(k_0, z)$.

## 3.3  SOLUTION METHODOLOGY

The boundary value problem (BVP) described is examined. Firstly, the problem associated with $\phi$ transform into Fredholm integral equations of second kind and convert them into a system of algebraic equations utilizing BEM. Employing Green's second identity to the $\phi$ and the fundamental solution $G(x, z; \xi, \eta)$, we obtain

$$\frac{1}{2}\phi(\xi, \eta) = \int_{\Gamma} \left( \phi(x, z) \frac{\partial G(x, z; \xi, \eta)}{\partial n} - G(x, z; \xi, \eta) \frac{\partial \phi(x, z)}{\partial n} \right) d\Gamma(x, z). \tag{3.12}$$

Here, $(\xi, \eta)$ denotes the source point and $(x, z)$ denotes the field point. The source point given in Equation (3.12) lies on the boundary of the physical problem. Here, $G(x, z; \xi, \eta)$ satisfies Equation (3.1) (see [7,8] and [9] for details) is given by

$$G(x, z; \xi, \eta) = \frac{1}{2\pi} \ln(r), \quad \text{where } r = \sqrt{(x - \xi)^2 + (z - \eta)^2}. \tag{3.13}$$

Also, $G(x, z; \xi, \eta)$ satisfies

$$\Delta^2 G(x, z; \xi, \eta) = \delta(x - \xi)\delta(z - \eta), \quad \Delta^2 \equiv \left( \frac{\partial^2}{\partial x^2} + \frac{\partial^2}{\partial z^2} \right). \tag{3.14}$$

The normal derivatives of the fundamental function are given by (see [10] and [11] for details)

$$\frac{\partial G}{\partial n} = \frac{1}{2\pi r} \frac{\partial r}{\partial n} = \frac{1}{2\pi r} \left( n_x \frac{\partial r}{\partial x} + n_z \frac{\partial r}{\partial z} \right), \tag{3.15}$$

where $n_x$ and $n_z$ are the components of the unit normal vector along the $x$ and $z$ directions. Applying the boundary conditions in Equations (3.2)–(3.6) and Equation (3.11) into Equation (3.12), the following integral equations are formed corresponding to both of the regions $R_1$ and $R_2$ as (see [12] and [13])

$$
-\frac{1}{2}\phi_1 + \int_{\Gamma_{c1}} \phi_1\left(\frac{\partial G}{\partial n} - ik_0 G\right)d\Gamma + \int_{\Gamma_b} \phi_1\frac{\partial G}{\partial n}d\Gamma + \int_{\Gamma_{c2}} \phi_1\frac{\partial G}{\partial n}d\Gamma
$$

$$
+ \int_{\Gamma_p}\left(\phi_1\frac{\partial G}{\partial n} + i\omega\xi G\right)d\Gamma + \int_{\Gamma_{i1}}\left(\phi_2\frac{\partial G}{\partial n} + G\frac{\partial\phi_2}{\partial n}\right)d\Gamma \tag{3.16}
$$

$$
+ \int_{\Gamma_{f1}} \phi_1\left(\frac{\partial G}{\partial n} - KG\right)d\Gamma = \int_{\Gamma_{c1}} G\left(\frac{\partial\phi'}{\partial n} - ik_0\phi'\right)d\Gamma,
$$

$$
-\frac{1}{2}\phi_2 + \int_{\Gamma_{i1}}\left(\phi_2\frac{\partial G}{\partial n} - G\frac{\partial\phi_2}{\partial n}\right)d\Gamma + \int_{\Gamma_p}\left(\phi_2\frac{\partial G}{\partial n} - i\omega\xi G\right)d\Gamma + \int_{\Gamma_{c3}} \phi_2\frac{\partial G}{\partial n}d\Gamma
$$

$$
+ \int_{\Gamma_{f2}} \phi_2\left(\frac{\partial G}{\partial n} - KG\right)d\Gamma = 0 . \tag{3.17}
$$

Now, following the procedure as discussed in [14], Equations (3.16) and (3.17) are modified into a system of algebraic equations as follows

$$
\sum_{j=1}^{N_{c1}} \phi_{1j}(H^{ij} - ik_0 G^{ij})|_{\Gamma_{c1}} + \sum_{j=1}^{N_b} \phi_{1j}H^{ij}|_{\Gamma_b} + \sum_{j=1}^{N_{c2}} \phi_{1j}H^{ij}|_{\Gamma_{c2}}
$$

$$
+ \sum_{j=1}^{N_p}\left(\phi_{1j}H^{ij} + i\omega\xi_j G^{ij}\right)|_{\Gamma_p} + \sum_{j=1}^{N_{i1}}\left(\phi_{2j}H^{ij} + G^{ij}\frac{\partial\phi_{2j}}{\partial n}\right)|_{\Gamma_{i1}} + \sum_{j=1}^{N_{f1}} \phi_{1j}(H^{ij} - KG^{ij})|_{\Gamma_{f1}}
$$

$$
= \sum_{j=1}^{N_{c1}} G^{ij}\left(\frac{\partial\phi'_j}{\partial n} - ik_0\phi'_j\right)|_{\Gamma_{c1}},
$$

$$
\tag{3.18}
$$

$$
\sum_{j=1}^{N_{i1}}\left(\phi_{2j}H^{ij} - G^{ij}\frac{\partial\phi_{2j}}{\partial n}\right)|_{\Gamma_{i1}} + \sum_{j=1}^{N_p}\left(\phi_{2j}H^{ij} - i\omega\xi_j G^{ij}\right)|_{\Gamma_p} + \sum_{j=1}^{N_{c3}} \phi_{2j}H^{ij}|_{\Gamma_{c3}}
$$

$$
+ \sum_{j=1}^{N_{f2}} \phi_{2j}(H^{ij} - KG^{ij})|_{\Gamma_{f2}} = 0, \tag{3.19}
$$

where $H^{ij} = \frac{1}{2}\delta_{ij} + \int_{\Gamma_j}\frac{\partial G}{\partial n}d\Gamma$, and $G^{ij} = \int_{\Gamma_j} Gd\Gamma$ and the procedure to evaluate the coefficients is given in [15,16] and [14]. The method of collocation is used to equate the number of equations and the number of unknowns in the system (see [17]). To derive Equations (3.18) and (3.19), all the bcs are used except the dynamic bcs of the plate as given in Equation (3.7). Equations (3.18) and (3.19) contain the plate deflection $\xi_j$, which is unknown. So, to solve Equations (3.18) and (3.19), proper coupling with Equation (3.7) needs to be done. Utilizing the central difference formula, the discretized form of the Equation (3.7) over the $j^{th}$ element in $\Gamma_p$ is given by (see [18])

$$
\left(\frac{\xi_{j+2} - 4\xi_{j+1} + 6\xi_j - 4\xi_{j-1} + \xi_{j-2}}{\Delta^4}\right) + A\xi_j = B\left(\phi_1^j - \phi_2^j\right), \tag{3.20}
$$

where $A = -\gamma\omega^2/\Pi$ and $B = i\omega/\Pi$ with $\Pi = g\chi\left[1 + \frac{\beta^2\zeta\omega}{i+\zeta\omega}\right]$. Now the unknowns, $\phi$, $\frac{\partial\xi}{\partial x}$ and $\xi$, can be solved using Equations (3.18)–(3.20).

## 3.4 RESULTS AND DISCUSSIONS

Results corresponding to the power generated by the PWEC device and related discussions in data analytic perspective are provided. The analysis of parameters of the PWEC device as functions of the structural parameters associated with the bottom topography is also given. See [18] for the parameters related with the PWEC device, incident wave, and step-bottom. The $P_{ext}(Wm^{-1})$ of the PWEC device can be evaluated (see [6]) as

$$P_{ext} = \frac{i\rho\omega}{4} \int_{-h}^{0} \left(\phi\frac{\partial\phi^*}{\partial x} - \frac{\partial\phi}{\partial x}\right)\Bigg|_{x=-l_1} dz. \qquad (3.21)$$

Now, by using Equation (3.11), the $P_{ext}(Wm^{-1})$ of the PWEC device is obtained as

$$P_{ext} = \frac{\rho g^2 A^2 [\sinh(k_0 h)\cosh(k_0 h) + k_0 h]}{4\omega \cosh^2(k_0 h)}(1 - |R_c|^2), \qquad (3.22)$$

where $|R_c|$ is the reflection coefficient.

### 3.4.1 EFFECT OF THE PWEC PLATE LENGTH AND THE PLATE EDGE TYPE

In this subsection, the performance of the PWEC device is analyzed for the change in length of the PWEC plate $l/h$ for various plate edge conditions. The power generation is evaluated for a wave period $T_0$ that varies from 4.0 sec to 9.0 sec. In Figure 3.2(a), the average power generated by the PWEC device $P_{ext}(Wm^{-1})$ is plotted as a function of the PWEC plate length $l/h$ when the lee side of the PWEC plate is fixed on the rigid and impermeable vertical breakwater and the front edge of the PWEC plate is also fixed.

It is clearly seen that the power generated by the PWEC plate first increases and after reaching a threshold value, the power generation deceases gradually. Under the above-mentioned edge condition, a PWEC device with a plate length $l/h$ varies between 2.2 to 2.8 is the ideal PWEC plate length to obtain maximum power. The maximum power generated by the PWEC device is obtained to be 2600 $Wm^{-1}$ and it is obtained for a PWEC plate length $l/h = 2.3$. Figure 3.2(b) shows the average power generated by the PWEC device as a function of the plate length were the lee edge of the PWEC plate is fixed with the vertical breakwater and front edge of the PWEC device is considered to be free. From the comparative study with Figure 3.2(a) it is seen that the average power generated by the PWEC device reduced irrespective of the plate length. The maximum power generated by the PWEC device under the mentioned edge condition is 2,100 $Wm^{-1}$ and it is obtained for a pWEC plate of length 2.2 $l/h$. The ideal length of the PWEC plate for the

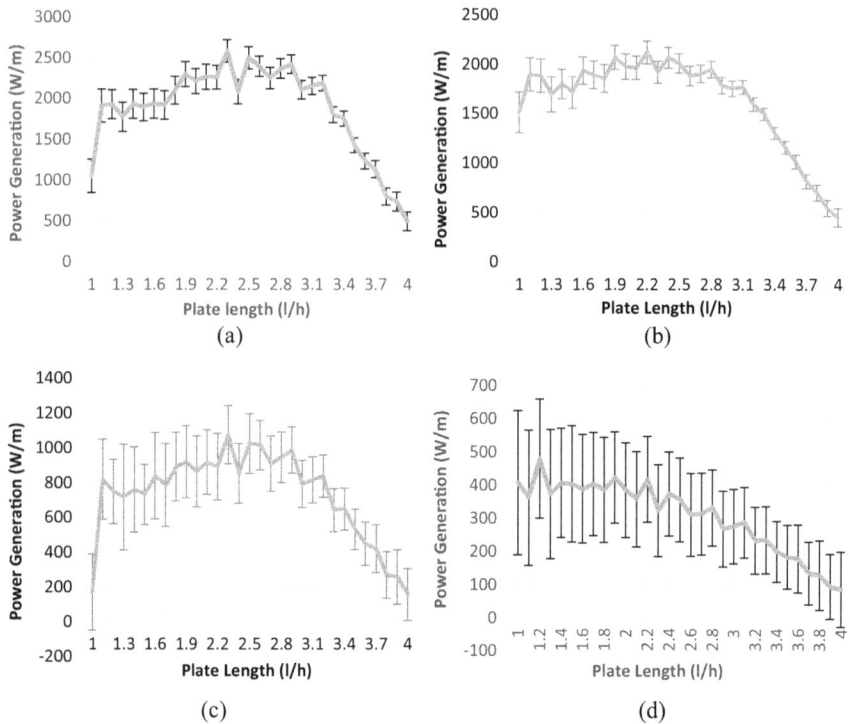

**FIGURE 3.2**   Variation of power generated $P_{ext}(Wm^{-1})$ with respect to the change in the PWEC plate length $l/h$ for various boundary edge conditions.

mentioned boundary edge condition varies from 2 $l/h$ to 2.4 $l/h$. Figure 3.2(c) and Figure 3.2(d) depict the variation of the average of $P_{ext}(Wm^{-1})$ of the PWEC device corresponding to the PWEC plate length for the lead edge is moored with the vertical breakwater and front edge boundary condition is fixed or free, respectively. It clearly seen that the average power generated by the PWEC device decreased significantly by changing the lee edge boundary condition from a fixed edge condition to moored edge condition. A measure of dispersion called coefficient of variation (CV) is also provided in Figure 3.2.

$$\text{Coefficient of Variation(CV)} = \frac{\text{Standard Deviation}(\sigma)}{\text{Mean}(\mu)} * 100 \qquad (3.23)$$

This study depicts that the power generated by the PWEC device is more consistent if the lee edge of the PWEC plate is fixed in nature.

From this discussion it is clearly seen that the boundary edge condition dramatically influences the power generated by the PWEC device. The fixed edge condition on both edges of the PWEC plate is the most suitable plate edge condition to obtain the maximum power generation by the PWEC device, in which the PWEC

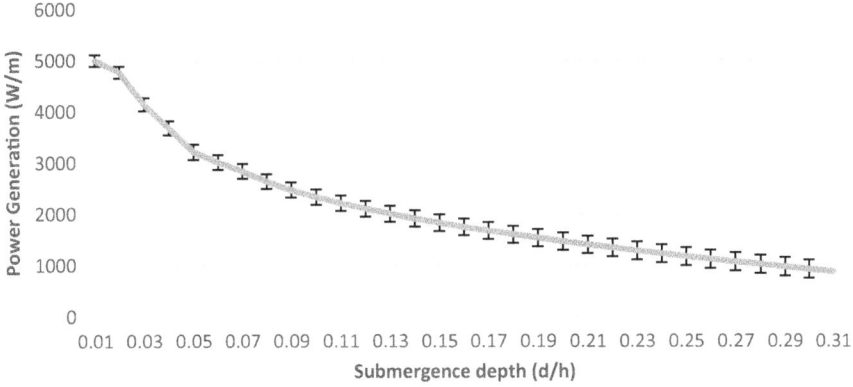

**FIGURE 3.3** Variation of the average power generated $P_{ext}(Wm^{-1})$ vs $d/h$.

plate with lengths 2.2 to 2.8 gives the optimum power generation. Hence, in the rest of the figures we considered both edges of the PWEC plate as fixed in nature.

### 3.4.2 Effect of the Submergence Depth of the PWEC Device

Figure 3.3 depicts the variation of the average of $P_{ext}(Wm^{-1})$ of the PWEC device as a function of d/h of the PWEC device. It is illustrated that as the submergence depth increases, the power generated by the PWEC device decreases gradually. The maximum power generated by the PWEC device is 5000 $Wm^{-1}$ is obtained at the lowest submergence depth of 0.01 d/h. Hence, the submergence depth of the PWEC device is one of the vital parameters influencing the power generated by the PWEC device.

### 3.4.3 Effect of the Stepped-Bed-Profile

Figure 3.4 depicts the effect of the stepped bottom profile in the $P_{ext}(Wm^{-1})$ of the PWEC device. Figure 3.4(a) shows the variation of the $P_{ext}(Wm^{-1})$ of the PWEC

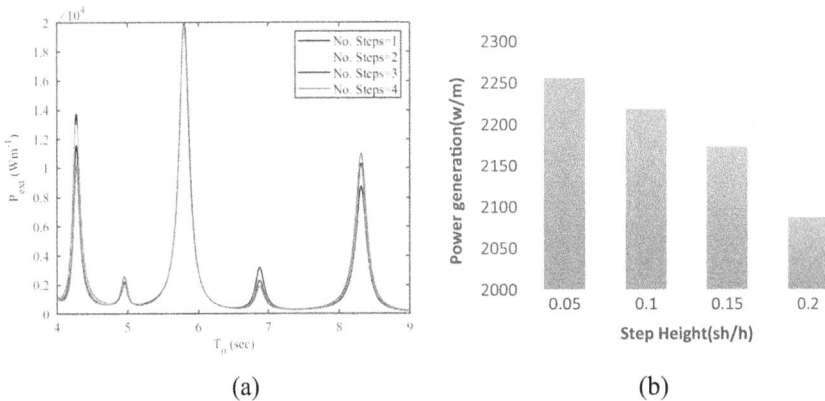

(a)                                                      (b)

**FIGURE 3.4**  (a) $P_{ext}$ vs $T_0$ for different number of steps. (b) Average $P_{ext}$ vs step height $sh/h$.

device with respect to $T_0(sec)$ for the number of steps. It is illustrated that there are three significant resonating peaks in the short-wave regime, long wave regime, and one in the intermediate wave regime. In the intermediate wave regime, the variation of the power generated by the PWEC device is insignificant to the number of steps, whereas an opposite trend appeared for long wave and short-wave regimes. Hence, the variation in the average wave power absorbed by the PWEC device is insignificant of the number of steps. The step height $sh/h$ has a significant influence on the power generated by the PWEC device. From Figure 3.4(b), it is noticed that as the height of the step-bottom increases, the power generated by the PWEC device decreases gradually. This is because most of the wave power is dissipating if the step-height of the bottom topography is high.

## 3.5 CONCLUSIONS

The work investigates the performance of the PWEC device in terms of the power generation of the PWEC device. Analysis of various parameters associated with the physical problem is provided. It is observed that the PWEC device with both edges of the PWEC plate is of fixed type generates higher wave power as compared to other plate edge conditions mentioned. While considering the structural parameters of the PWEC device, it is seen that both parameters, PWEC plate length and sub-mergence depth, are high influence parameters. The analysis on the PWEC plate length shows that the moderate length of the PWEC plate varies from 2.0 to 2.8 generated higher wave power for a wider range of wave periods. Also, the average of the maximum power generated by the PWEC device is obtained to be $2,600\ Wm^{-1}$ and is obtained for a plate length l/h = 2.2 with fixed plate edge condition on both the edges of the PWEC plate. Further, the CV is less if the lee edge boundary of the PWEC plate is fixed as compared to a moored boundary edge. So, the power generated by the PWEC plate is consistent if the lee edge boundary of the PWEC plate is fixed. It is observed that as the submergence depth of PWEC device decreases, the wave power generated by the PWEC device increases significantly. This is due to the high wave power concentration in the free surface. The analysis regarding the stepped seabed reveals that the number of steps in the bottom topography does not have much influence in the power generated by the PWEC device, whereas it is seen that as an increase in the step height of the bottom topography decreases, the power generated by the PWEC device.

## ACKNOWLEDGMENT

"CSIR-SRF File no: 09/1026(0030)/2019-EMR-I", and "DST project: DST/INSPIRE/04/2017/002460".

## REFERENCES

[1] Xu, S. P. and Wang, W. 2009. Bending of piezoelectric plates with a circular hole. *Acta Mechanica*, 203(3): 127–135.

[2] Zhang, C., Chen, W. and Zhang, C. 2013. Two-dimensional theory of piezoelectric plates considering surface effect. *European Journal of Mechanics-A/Solids*, 41: 50–57.

[3] Xie, X. D., Wang, Q. and Wu, N. 2014. Energy harvesting from transverse ocean waves by a piezoelectric plate. *International Journal of Engineering Science*, 81: 41–48.

[4] Wu, N. Wang, Q. and Xie, X. 2015. Ocean wave energy harvesting with a piezoelectric coupled buoy structure. *Applied Ocean Research*, 50: 110–118.

[5] Renzi, E. 2016. Hydroelectromechanical modelling of a piezoelectric wave energy converter. *Proceedings of the Royal Society A: Mathematical, Physical and Engineering Sciences*. 472(2195): 20160715.

[6] Zheng, S., Meylan, M., Zhang, X., Iglesias, G. and Greaves, D. 2021. Performance of a plate-wave energy converter integrated in a floating breakwater. *IET Renewable Power Generation*, 15: 3206–3219.

[7] Koley, S., Behera, H. and Sahoo, T. 2015. Oblique wave trapping by porous structures near a wall. *Journal of Engineering Mechanics*, 141(3): 04014122.

[8] Koley, S., Kaligatla, R. B. and Sahoo, T. 2015. Oblique wave scattering by a vertical flexible porous plate. *Studies in Applied Mathematics*, 135(1): 1–34.

[9] Koley, S., Sarkar, A. and Sahoo, T. 2015. Interaction of gravity waves with bottom-standing submerged structures having perforated outer-layer placed on a sloping bed. *Applied Ocean Research*, 52: 245–260.

[10] Kar, P., Koley, S. and Sahoo, T. 2018. Scattering of surface gravity waves over a pair of trenches. *Applied Mathematical Modelling*, 62: 303–320.

[11] Koley, S. and Sahoo, T. 2017. Scattering of oblique waves by permeable vertical flexible membrane wave barriers. *Applied Ocean Research*, 62: 156–168.

[12] Trivedi, K. and Koley, S. 2021. Mathematical modeling of breakwater-integrated oscillating water column wave energy converter devices under irregular incident waves. *Renewable Energy*, 178: 403–419.

[13] Koley, S. and Trivedi, K. 2020. Mathematical modeling of oscillating water column wave energy converter devices over the undulated sea bed. *Engineering Analysis with Boundary Elements*, 117: 26–40.

[14] Koley, S. and Sahoo, T. 2017. Wave interaction with a submerged semicircular porous breakwater placed on a porous seabed, *Engineering Analysis with Boundary Elements*, 80: 18–37.

[15] Koley, S., Panduranga, K., Almashan, N., Neelamani, S. and Al-Ragum, A. 2020. Numerical and experimental modeling of water wave interaction with rubble mound offshore porous breakwaters. *Ocean Engineering*, 218: 108218.

[16] Koley, S. 2019. Wave transmission through multilayered porous breakwater under regular and irregular incident waves. *Engineering Analysis with Boundary Elements*, 108: 393–401.

[17] Koley, Santanu. 2016. Integral equation and allied methods for wave interaction with porous and flexible structures. PhD dissertation, IIT Kharagpur, India.

[18] Koley, S. 2020. Water wave scattering by floating flexible porous plate over variable bathymetry regions. *Ocean Engineering*, 214: 107686.

# 4 (p)ppGpp Mediated Biofilm Formation and Estimation on Chip

*Yogeshwar Devarakonda, Reddi Lakshmi, Neethu R. S., Mitul Bhalerao, and Kirtimaan Syal*
Department of Biological Sciences, Center for Genetics and Molecular Microbiology, Institute of Eminence, Birla Institute of Technology and Sciences-Pilani, Hyderabad, Telangana, India

## CONTENTS

## 4.1  INTRODUCTION

Microorganisms tend to adhere to living and non-living surfaces to form communities known as biofilms (Costerton, Stewart, and Greenberg 1999). These microbial communities are found in diverse environments like industrial wastes, sewage, natural aquatic systems, and medical devices. Biofilms have beneficial applications as well as hazardous health effects. On one hand, it could assist in degradation of toxic substances in soil, wastewater treatment, and the commercial production of chemicals. On the other hand, these biofilms could also be harmful to humankind because of their role in infections, contamination, and biofouling (Karunakaran et al. 2011). Biofilm formation is a complex multi-step transformation of a planktonic cell or group of planktonic cells to multicellular community. In the beginning, planktonic cells attach and form microcolonies on the surface and are irreversibly attached. At a later stage, the bacterial cells are embedded in the extracellular polymeric substrate (EPS) that includes the biofilm matrix (Flemming et al. 2016) and displays an altered phenotype in terms of growth rate and gene transcription. EPS plays a crucial role in physical, social interaction and antibiotic tolerance. It consists of proteins, cellulose, extracellular teichoic acid, nucleic acid, lipids, polysaccharides, phospholipids, extracellular DNA, and other organic compounds. About 85% of the biomass of EPS constitutes the niche for biofilm cells, which resembles the "city of microbes" (Nikolaev Iu and Plakunov 2007).

The physical and physiological property of biofilm makes it resilient to antimicrobial agents. Though the defensive strategy of the biofilm is not clearly understood, some of the mechanisms responsible for resistance involves delayed penetration of the antimicrobial agents, altered growth rate, and physiological changes due to biofilm growth (Donlan and Costerton 2002, Maiti et al. 2017, Syal et al. 2016, Syal et al. 2015). Bacteria also communicate with their neighbouring organisms to perform various activities such as bioluminescence production, biofilm development, activation of virulence factors, and exoenzyme secretion. This cell-to-cell communication occurs through a mechanism known as quorum sensing (QS) (Miller and Bassler 2001), which is controlled at the molecular level by chemical signalling molecules called autoinducers (AIs). The accumulation of the signalling molecules produced by the community members helps the bacteria recognize the population density. These signalling molecules accumulate in the extracellular environment and activate the response only at higher cell density (Syal and Chatterji 2018, Syal et al. 2015, Naresh et al. 2012).

Bacteria use purine derivatives like guanosine pentaphosphate (pppGpp), guanosine tetraphosphate (ppGpp), and cyclic diguanylate monophosphate (c-di-GMP), etc., as intracellular signalling molecules (Syal et al. 2015). These small signalling molecules help monitoring of the intracellular and extracellular environmental conditions to modulate the growth and multiplication of the cells in response to the availability of nutrient sources and ecological stress. The c-di-GMP is a secondary messenger that plays a crucial role in regulating the biological processes such as quorum sensing, dormancy, and virulence in bacteria. Its synthesis and hydrolysis enzymes are extensively distributed in bacteria; diguanylate cyclases (DGC) involved in synthesis and phosphodiesterases (PDE) in hydrolysis of ci-di-GMP (Hong et al. 2013). Numerous studies have indicated that c-di-GMP negatively regulates the properties of planktonic bacteria such as flagellum medicated swimming while promoting adherent phenotypes such as biofilm (Purcell and Tamayo 2016).

The other secondary messenger molecule that Cashel identified over 50 years ago, known as (p)ppGpp, plays an essential key role in the stringent response under starvation of nutrients and other stressful conditions in bacteria (Syal, Rs, and Reddy 2021). He discovered that the autoradiograms derived from the extracts of the *Escherichia coli* culture stressed with amino acid starvation revealed a "magic spot" (Cashel and Gallant 1969). Later, several other studies showed that stringent response (p)ppGpp molecules modulate the bacterial multiplication rate, survival during starvation conditions, exposure to antibacterial compounds, and osmotic stress (Hauryliuk et al. 2015, Irving and Corrigan 2018). It has been shown that intracellular levels of (p)ppGpp modulate the bacterial physiological processes by regulating the activities of RNA polymerase (RNAP), DNA primase (DnaG), and several metabolic enzymes in *E. coli* (Potrykus et al. 2011). The (p)ppGpp is a derivative of the guanosine nucleosides, where guanosine triphosphate (GTP) pyrophosphokinases (RelA/SpoT/RSH) transfer a pyrophosphate moiety from ATP to the 3'-OH position of the GTP or GDP. At standard conditions in a nutrient-rich growth medium, the cellular level of (p)ppGpp in *E. coli* was less than 0.2 mM (Mechold et al. 2013). Under stress conditions, the (p)ppGpp level may rise from 10- to 100-fold depending on the type of stress involved (Kalia et al. 2013, Syal, Rs, and Reddy 2021). The role of (p)ppGpp is not only known for a stringent response, but it also plays crucial roles in biofilm formation, antibiotic resistance, tolerance, modulating bacterial virulence, gene expression, and persistence (Syal and Chatterji 2018, Syal et al. 2017a, Syal, Bhardwaj, and Chatterji 2017, Syal and Chatterji 2015). It has been shown that stringent response positively modulates biofilm formation in *Escherichia coli*, *Vibrio cholera*, and *Mycobacterium smegmatis* (He et al. 2012, Teschler et al. 2015, Syal et al. 2017a). The efficacy of the antibiotics is reduced by (p)ppGpp mediated biofilm formation in both gram-negative and gram-positive bacteria. In this article, various novel methods have been discussed for quantifying and imaging the biofilm, such as a microfluidic system, which has been developed for analyzing biofilm growth. This microfluidic platform known as a biofilm chip is used to study new anti-biofouling strategies, including drug susceptibility testing for different species by using electrical impedance spectroscopy (Blanco-Cabra et al. 2021). These biofilm chips allow homogeneous biofilm development and co-culture of polymicrobial biofilm, resembling the natural biofilm infections found in complex matrices in diseases like tuberculosis and skin infection.

## 4.2   (P)PPGPP AND STRESS RESPONSE

The enzymes that are responsible for synthesizing and degradation of (p)ppGpp molecules are classified into three major groups: long Rel/SpoT homologue proteins (RSH) having both synthetase and hydrolase domains, small alarmone synthetases (SAS), and small alarmone hydrolases (SAH), which shows one type of activity based on the presence of synthetase and hydrolase domain, respectively (Atkinson, Tenson, and Hauryliuk 2011). These enzymes are widely spread in bacteria; for example, *E. coli* (gram-negative bacteria) has two RSHs: RelA and SpoT, *Bacillus subtilis* (gram-positive bacteria) has one RSH and two SASs: RelA, RelP and RelQ. Interestingly a gene Mesh1 encoding for (p)ppGpp hydrolases have been discovered in *Drosophila melanogaster* [Figure 4.1A], and it was observed to be crucial in starvation conditions (Sun et al. 2010). Remarkably, a regulating enzyme named

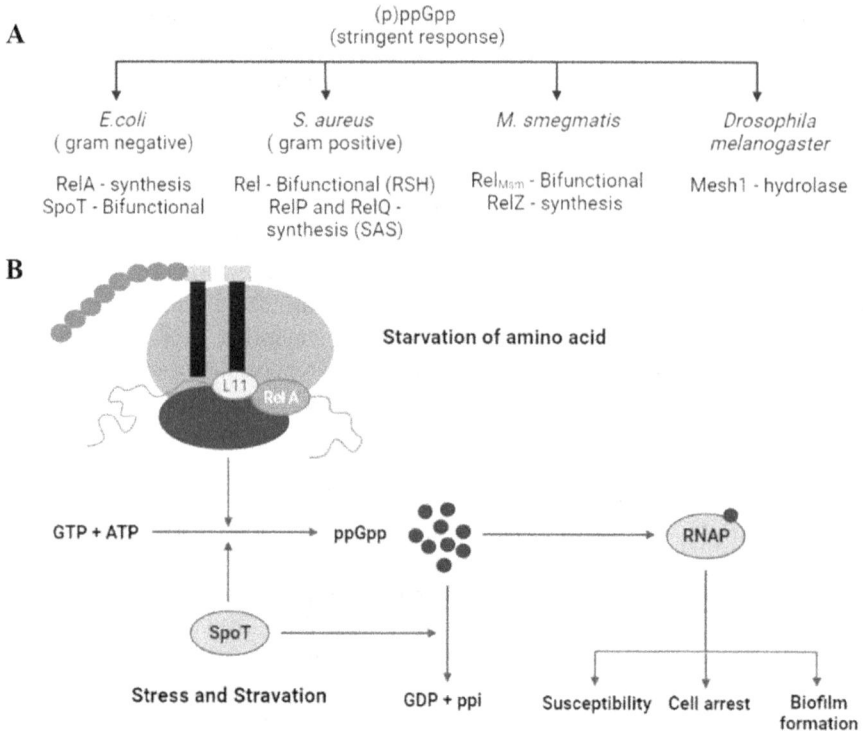

**FIGURE 4.1**   A: (p)ppGpp stringent response in different bacteria. The proteins involved in synthesis and hydrolysis of (p)ppGpp have been mentioned. B: Mechanism of synthesis and hydrolysis of ppGpp by RelA and SpoT. ppGpp binds to RNAP further involved in stress resistance and survival.

MS_RHII-RSD (renamed as RelZ) has been identified with dual role in (p)ppGpp synthesis and RNase HII activities identified in *Mycobacterium smegmatis*, suggesting the link for DNA replication, repair, and transcription in the stress survival pathway (Murdeshwar and Chatterji 2012). Most bacterial species have at least one protein from the RSH superfamily, but Planctomycete, Verrucomicrobia, and Chlamidiale superphylum (PVC) of bacteria do not encode for any RSH (Atkinson, Tenson, and Hauryliuk 2011).

The stringent response in *E. coli* during amino acid starvation mediated by small molecules (p)ppGpp is because of the downregulation of the stable RNA synthesis (rRNA and tRNA) and ribosome production (Stent and Brenner 1961, Syal, Rs, and Reddy 2021). In *E. coli*, the two regulatory proteins involved in the synthesis of (p) ppGpp are RelA and SpoT. In the amino acid starvation condition, the RelA protein synthesizes (p)ppGpp by recognizing and binding to the ribosomes with an uncharged tRNA wrapped in an A-site (Wendrich et al. 2002). SpoT is a bifunctional protein that synthesizes (p)ppGpp in stress conditions, the activation of this protein is not completely understood (Syal et al. 2015). The increased levels of (p)ppGpp inside the cell for a longer period is toxic. Thus, it is vital to hydrolyze it. Here, SpoT protein is responsible for hydrolysis in *E. coli* (Gentry and Cashel 1996). It has been shown that the mutant lacking both RelA and SpoT proteins does not synthesize (p)ppGpp molecules in any growth conditions and is unable to grow in minimal media as a result of the need for large amounts of the amino acid (auxotrophy) (Xiao et al. 1991). The accumulation of (p)ppGpp in the cytoplasm of *E. coli* leads to the expression of approximately 500 genes, further activating RpoS and RpoE (stress response sigma factor) (Costanzo and Ades 2006). The (p)ppGpp perturb rRNA synthesis and the activity of DNA primase (Gaca, Colomer-Winter, and Lemos 2015). In gram-negative bacteria (p)ppGpp binds to the RNA polymerase [Figure 4.1B], which modulates an allosteric signal to the catalytic Mg2+ site, which further decreases transcription and regulates the expression of genes; all these changes in the cell lead to slow growth or dormancy in most cells (Liu, Bittner, and Wang 2015).

Gram-positive bacteria such as *Streptococcus mutants, Bacillus subtilis,* and *Staphylococcus aureus* have one long bifunctional RSH protein named Rel, which contains both (p)ppGpp synthase and hydrolase activities (Lemos et al. 2007, Nanamiya et al. 2008, Geiger et al. 2014, Syal et al. 2015). It also includes two SAS proteins, RelP and RelQ, which have a synthesis domain for (p)ppGpp production in response to stress. After activation of the stringent response, the intracellular level of (p)ppGpp increases within the cell with concomitant decrease in GTP levels. This facilitates reduction in cellular transcription and degradation of macromolecules such as phospholipids, ribosomes, and amino acids. When the conditions become normal, the (p)ppGpp levels are downregulated and cell growth is restored. Together, (p)ppGpp as a stringent response molecule, plays an essential role in regulation of all the vital processes, which include DNA replication, transcription, and translation (Steinchen and Bange 2016).

In *Mycobacterium smegmatis*, the stringent response is similar to a gram-positive bacterium where (p)ppGpp is controlled by a single bifunctional protein $Rel_{Msm}$, which synthesizes as well as hydrolyses (p)ppGpp (Jain et al. 2006, Jain, Saleem-Batcha, and Chatterji 2007). A *rel*-deficient strain ($\Delta rel_{Msm}$) was expected to show (p)ppGpp null phenotype, and still showed the magic spot. This lead to the discovery of second (p)ppGpp synthetase also known as RelZ in mycobacteria. (Murdeshwar and Chatterji 2012).

## 4.3 BIOFILMS

Biofilms are the collection of microbial cells attached to the surface or to each other, embedded within a self-produced extracellular polymeric matrix and qualitatively having different properties compared to planktonic population of similar size (Syal 2017). This matrix consists of biopolymers including proteins, polysaccharides, and extracellular DNA, creating a definite environment. Biofilm formation protects the microbial community from environmental stresses such as UV radiation, limited nutrients, extreme pH, extreme temperature, high salt concentrations, and antimicrobial agents (Hall-Stoodley, Costerton, and Stoodley 2004). Based on the current understanding of the biofilm life cycle in unicellular organisms involves both sessile and motile stages. The biofilm formation is initiated by attachment of cells to the surface and followed by recruitment of cells from the surrounding

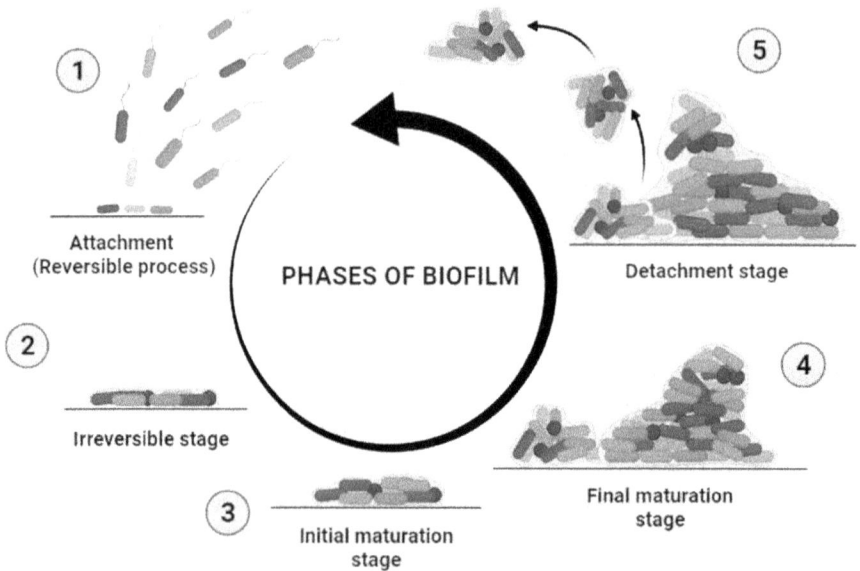

**FIGURE 4.2** Diagrammatic representation of developmental stages of biofilm.

environment (Syal et al. 2016). Cells in the biofilm form microcolonies with production of extracellular matrix and develop 3D biofilm structure at the maturation phase [Figure 4.2]. Finally, detachment or dispersal of the motile cells are released from the biofilm (Penesyan et al. 2021).

### 4.3.1 ATTACHMENT TO SURFACE

The initial step of biofilm formation is initiated by surface attachment of a few planktonic cells, which occurs in two stages: reversible and irreversible attachment. The initial attachment of the cells is achieved by the effects of non-specific physical forces such as electrostatic forces, hydrophobic interactions and Lifshitz–van der Waals interactions (Carniello et al. 2018). During this stage, the surface contact by rod-shaped flagellated cells is mediated by flagella. Additionally, pilli (type I, IV), curli fibres, and antigen 43 have been shown to play a key role in attachment of cells to the surface (Kostakioti, Hadjifrangiskou, and Hultgren 2013). In the case of non-motile Gram-positive bacteria, the surface contact is via pili and adhesins (examples of adhesins include SagA and Acm of *Enterococcus faecium*). Once cells are irreversibly attached to the surface, then they stop moving and initiate matrix production, likely to stick themselves to each other or to the surface. Hydrophobicity may play a crucial role in attachment of the microbes to the surface, because force of repulsion between the bacteria and the surface is reduced. Microorganisms attach more likely with hydrophobic and non-polar surfaces like plastics and teflon, compared to hydrophilic and polar surfaces like glass (Krasowska and Sigler 2014). QS is used by bacteria to synthesize and release chemical signals (autoinducers AIs). Both Gram-positive and Gram-negative bacteria regulate biofilm formation via a cell-to-cell signalling mechanism. Gram-positive bacteria use both acyl homoserine lactones (AHLs) and autoinducers, whereas Gram-negative bacteria employs universal autoinducers only. (Sun et al. 2004, Miller and Bassler 2001).

### 4.3.2 Maturation of Biofilm

At this stage, the required cell density is obtained by cells via cell signalling molecules and conduct to formation of microcolonies and maturation of biofilms. Following the stage of maturation, this induces certain gene expressions and the products of these genes are utilized for production of EPS, which acts as glue between embedded cells. The three-dimensional structure of biofilm is mainly composed of EPS; some interspatial gaps are produced in the matrix. These channels are filled with water, which act as circulatory systems by distributing nutrients to communities and removing waste products in biofilm (Parsek and Singh 2003). The motility is strictly restricted during the process of maturation within the microcolonies, and the gene expression pattern of sessile cells significantly differs from planktonic cells. More than 57 biofilm-associated proteins have been detected in Pseudomonas *aeruginosa* micro-colonies that were not present in planktonic cells (Heras et al. 2014). QS enables communication between the same or different species through secretion and detection of AIs. These signalling molecules are used by bacteria in order to sense the density and to regulate gene expression in response to changes in the population size. QS and AIs have a crucial role in maintaining existing biofilms (Yan and Bassler 2019).

### 4.3.3 Detachment of Biofilm

The last stage in biofilm development is dispersal or detachment. The biofilm distribution is closely related to the size of the building structure. Dispersion starts with spatial differentiation, described as the differential localization of motile and non-motile bacteria in biofilm structure when biofilm reaches a critical network. The motile bacteria are located in the mushroom cavity, whereas non-motile bacteria are situated at the stalk and walls of the structure. These dispersal mechanisms involve ECM degradation and autolysis of a biofilm sub-population, and these can also induce by environmental clues, pH, and various chemicals etc. These detachment mechanisms are essential for forming biofilms in new niches (Rumbaugh and Sauer 2020).

## 4.4 BIOFILM AND HUMAN PATHOGENS

Most pathogenic bacteria survive inside the host system by formation of biofilm-like structures with small communities attached to the surface and protect themselves from the host environment. Additionally a extracellular polymeric substrate obstructs the entry of antibiotics and biofilm ECM also facilitates the horizontal transmission of resistant genes within the biofilm. The chances for better survival of the bacteria is increased by inactivation of antibiotics by high metal ion concentration and presence of persistent cells (Kostakioti, Hadjifrangiskou, and Hultgren 2013).

Biofilms also exist as in multispecies community, for example in dental plaque. In spite of harsh environmental conditions like nutrient availability and exposure to cleansing agents, still bacteria persist in biofilm. The primary colonizers in the plaque formation are mostly *Streptococci*, which attaches to the surfaces (O'Toole, Kaplan, and Kolter 2000). These organisms are capable of interacting with their own genera and also with other genera. The growth rate increases once the threshold density of biofilm is reached, and further microbes secrete ECM and build up a niche.

## 4.5   SECOND MESSENGER AND FORMATION OF BIOFILM

The role of second messenger c-di-GMP has been studied in both Gram-positive and Gram-negative bacteria and is known for its function in biofilm formation, virulence, and cell cycle. Additionally, increased intracellular c-di-GMP concentrations negatively interfere with motility processes and positively regulate extracellular matrix production, which further promotes biofilm formation (Sharma, Petchiappan, and Chatterji 2014, Valentini and Filloux 2016). In contrast, low levels of c-di-GMP enable biofilm dispersal and standard mode (planktonic) of bacterial life. The second messenger alarmones, such as ppGpp, pppGpp, help to survive during stress conditions by interacting with RNAP. Studies have also shown that (p)ppGpp signalling cascades play a crucial role in mediating virulence, biofilm formation, colonization, and antibiotic tolerance [Figure 4.3]. The part of (p)ppGpp for the formation of biofilm and virulence has been shown in *Mycobacterium smegmatis, Enterococcus faecalis, Escherichia coli,* and *Pseudomonas aeruginosa* (Kim and Davey 2020, Nguyen et al. 2011, Chávez de Paz et al. 2012, Syal et al. 2017a, Syal, Bhardwaj, and Chatterji 2017). The molecular mechanisms that involve interlink between (p)ppGpp and persistence are not well understood (Syal, Rs, and Reddy 2021).

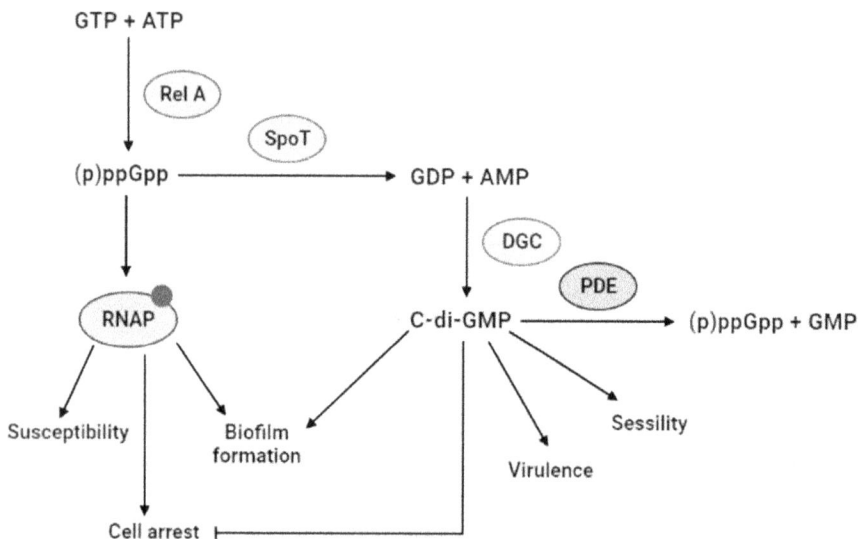

**FIGURE 4.3**   The interplay of (p)ppGpp and c-di-GMP involved in biofilm formation, motility, and their different functions.

Efflux pumps are transport proteins allows bacteria in removing toxic substances like clinically relevant antibiotics from within cells into the environment could constitute one of the linkers of (p)ppGpp and persistence. The efflux pumps are upregulated in cells embedded in biofilm (Syal et al. 2015). This upregulation in biofilm cells can be triggered by different signals, such as (p)ppGpp, ROS response, and QS (Syal, Rs, and Reddy 2021). Efflux pumps are critical for biofilm formation and multidrug resistance in bacteria; in *Helicobacter pylori*, it has been shown that Hp1174 (glucose/galactose transporter), an efflux pump encoded by gluP gene, is highly expressed in biofilm formation and is upregulated by SpoT. A mutant strain lacking gluP and its product Hp1174 has been observed with unstructured biofilm with a damaged matrix, revealing that the SpoT enzyme upregulates Hp1174 in persistent biofilm-forming cells (Ge et al. 2018).

ROS are produced as a natural response to regular metabolism of oxygen and perform functions in homeostasis and cell signalling. When cells are exposed to environmental stress, the increased ROS levels could damage DNA, lipids, proteins, and may lead to cell death. Molecules such as glutathione and vitamin C are capable of eliminating ROS (Syal, Bhardwaj, and Chatterji 2017). It has been shown that survival of multidrug tolerant persisters in the biofilm of *P. aeruginosa* was majorly dependent on catalase or SOD enzymes, which are controlled by (p)ppGpp signals (Nguyen et al. 2011).

Several compounds have been developed to inhibit the (p)ppGpp production and inhibit persistence. An immunomodulatory peptide IDR (innate defence regulator)-1018 has been shown to specifically target and kill biofilm cells. Low levels of peptide have been shown to disperse biofilm, whereas higher doses triggered biofilm cell death. The peptide inhibited stress response, mediating (p)ppGpp synthesis by targeting RelA enzyme (de la Fuente-Núñez et al. 2014). Relacin, a novel compound, has been shown to inhibit the RelA *in vitro* and *in vivo*. It impedesentry into stationary phase in gram positive bacteria and results in lower cell viability (Wexselblatt et al. 2012). Interestingly, one of us showed that a novel synthetic (p)ppGpp analogue can inhibit $Rel_{Msm}$ (a bifunctional enzyme) from *M. smegmatis*, which subsequently showed the effect of long-term persistence, biofilm formation, and biofilm disruption in *M. smegmatis* (Syal et al. 2017b). This compound is not toxic to human red blood cells and showed good permeability across the cell membrane of human lung epithelial cells. Another study has shown that when *M. smegmatis* cells were treated with vitamin C compound, the levels of (p)ppGpp molecules were decreased [Table 4.1] (Syal, Bhardwaj, and Chatterji 2017). In-vitro studies revealed that vitamin C at higher concentrations can inhibit the synthesis of (p)ppGpp and its effects in the long-term survival and biofilm formation has been confirmed in *M. smegmatis* (Syal, Bhardwaj, and Chatterji 2017).

## 4.6   BIOFILM QUANTIFICATION AND IMAGINING

Historically, many techniques have been used for characterizing and quantifying biofilms, ranging from traditional methods like the tube method to more recent approaches like ATP bioluminescence and fluorescence labelling of biofilms in conjunction with mathematical prediction modelling like COMSTAT. Some of

**TABLE 4.1**

**List of Compounds That Can Inhibit (p)ppGpp Synthesis (Selected)**

| Year | Author | Compound | Strategy for inhibition | Reference |
|------|--------|----------|-------------------------|-----------|
| 2012 | Wexselblatt and colleagues | Relacin | Relacinblocks sporulation by mediating stringent response and also affects biofilm communities in *Bacillus subtilis*. | (Wexselblatt et al. 2012) |
| 2014 | de la Fuente-Núñez and colleagues | Peptide-1018 | 1018 mediates (p)ppGpp synthesis by targeting RelA and SpoT. While showing biofilm dispersal and biofilm cell death. | (de la Fuente-Núñez et al. 2014) |
| 2017 | Syal and colleagues | Novel (p)ppGpp analogue | (p)ppGpp analogue inhibits Rel protein (a bifunctional synthetase and hydrolase) and effects biofilm formation in *M. smegmatis*. | (Syal et al. 2017b) |
| 2017 | Syal and colleagues | Vitamin C | *M. smegmatis* treated with vitamin C effects (p)ppGpp levels, further inhibits long time survival and biofilm formation. | (Syal, Bhardwaj, and Chatterji 2017) |

these methods are well known and extensively used for biofilm quantification and have been described here.

### 4.6.1 Crystal Violet Assay

Gram staining is a widely used and well-optimized method for identifying and visualizing bacteria in microbiology. Crystal violet is the main component and most commonly used dye in gram staining. The crystal violet assay works by staining cells on cell culture plates. Dead unattached cells are washed away during the experiment. Crystal violet is used to stain the remaining adherent living cells. The crystal violet dye is solubilized and quantified by absorbance at 570 nm after a wash step (O'Toole 2011). The amount of crystal violet staining in the assay is proportional to the amount of adherent cell biomass on the plate. Cell biomass can be used to estimate cell viability and cytotoxicity levels. In this assay, cells adhered to the plate are considered to be alive and quantified (O'Toole 2011, Syal 2017).

### 4.6.2 Solvent-Based Method

Biofilm quantification approaches that rely on direct observation to quantify the intended parameter are known as natural methods (number of cells, total biofilm volume, etc.). The Syal method provides a unique THF solvent-based approach for biofilm estimation that prevents biofilm layer loss especially in the case of aerobic biofilms. Quantification of aerobic biofilms like in the case of *M. smegmatis* is not possible through crystal violet method. Many bacteria build biofilms primarily at the liquid-air interface. The biofilms that form at the liquid-air interface are called pellicles. This novel Syal method was found to be useful for the estimation of early phase biofilm and aerobic biofilm layer formed at the liquid–air interphase, which was not possible by crystal violet method (Syal 2017).

### 4.6.3 TUBE-BASED METHOD

The tube method, is one of the qualitative assays, involves detection of biofilm layer by staining with safranine. Here, isolated colonies are grown in polystyrene test tube and biofilm layer formed on the surface of the test tube stained with safranin. Here, planktonic cells and excess stain are washed out by phosphate buffer saline (PBS). Number of washes can be standardized as per the amount, age and the type of biofilm. Subsequently, air drying is performedand biofilm can be directly observed at the bottom of the tube (Christensen et al. 1985).

### 4.6.4 ATP BIOLUMINESCENCE METHOD

In the food industry and biomedical communities, ATP bioluminescence testing is a well-established test for detecting microbial contamination on surfaces. Bioluminescence refers to the process by which organisms convert chemical energy to light in presence of ATP. ATP bioluminescence assay utilizes the enzyme luciferase which is responsible for light production in fireflies. At low ATP concentrations, luciferase is a reaction that produces light in proportion to the amount of ATP present in solution. Therefore, the amount of light can be used for interpretation of biofilm viability and biomass. The basic reaction is divided into two steps; the first is complexing of Luciferase, Luciferin, and ATP to create luciferyl adenylate complex. The second step is oxidation of luciferyl adenylate with oxygen into oxyluciferin, which results in the emission of a photon detected at approximately 550 to 570 nm (Sánchez et al. 2013).

### 4.6.5 MASS SPECTROMETRY

The expressed proteins that are located in the EPS matrix can be detected, characterized, and quantified by using mass spectrometry (MS). Here, biomolecules involved in biofilm process if quantified by MS can potentially act as a biomarker. Both, electrospray ionization and matrix assisted laser desorption ionization (MALDI)-Tof can be used for such quantification . In case of time of flight (TOF) mass spectrometer, the mass of the molecule is analyzed by ions desorbed in vacuum chamber. In MALDI-TOF, depending on the mass/charge ratio, TOF is determined. The surface proteins expressed on bacteria and exoenzymes that are secreted out of the cells can also be monitored, and quantified by the application of mass spectrometry (Kırmusaoğlu 2019). Here, amount of biofilm can be directly correlated to such proteins (as per the type of biofilm) present in the EPS.

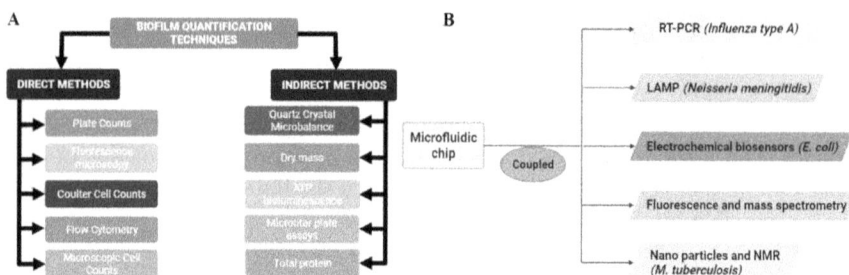

**FIGURE 4.4** A) Various techniques available for biofilm quantification classified into direct and indirect methodologies. B) Schematic of microfluidic chips coupled with other techniques and results in detection of microbes.

### 4.6.6 OTHER METHODS

Cells in biofilms have been quantified using a variety of diversed approaches. The plate counts, microscopic cell counts, Coulter cell counting, flow cytometry, and fluorescence microscopy are examples of different counting methods that can be used to count the number of cells in the biofilm. Indirect measurement methods involve quantification of mass, total organic carbon, microtiter plate tests, total protein, and quartz crystal microbalance; both direct and indirect procedures need homogenization of the biofilm to disperse cells in a liquid medium before analysis, which can be accomplished using a commercially available homogenizer and vortex mixing [Figure 4.4A] (Wilson et al. 2017).

Biofilm growth is frequently measured indirectly with a proxy marker that infers biofilm amount. Dry mass, total protein content, DNA, RNA, polysaccharides, and metabolites are examples of such markers. The essential assumption in all indirect quantification approaches is that the substance or characteristic to be quantified correlates with the number of cells or that the amount of protein/DNA/mass is constant from cell to cell (Choi and Morgenroth 2003). This assumption has been confirmed for biofilms, making these approaches particularly beneficial because indirect methods are merely proxy quantification based on metabolic function and biomolecule synthesis, which may vary depending on the organism, culture circumstances, and age, and therefore requires re-evaluation using direct methods (Trulear and Characklis 1982). Bio-volume, surface area coverage, biofilm thickness distribution, mean biofilm thickness, micro-colony volume, fractal dimension, roughness coefficient, average and maximum distance, and surface-to-volume ratio have been calculated using computer programmes (COMSTAT). The basic knowledge of biological systems with complicated physical properties, such as biofilms, is improved by mechanistic mathematical models connected with computational systems biology (Heydorn et al. 2000).

### 4.7  BIOFILM CHIP

A biofilm chip is a device that allows a novel insight into the bacterial culture of clinical origin. It could enable monitoring of clinical biofilm samples using confocal microscopy or electrical impedance spectroscopy (EIS), wherein the interdigitated sensor attaches in an irreversible and homogenous manner. Conventional screening techniques like MBEC Assay® (Harrison et al. 2010, Ceri et al. 1999), biofilm ring test (Olivares et al. 2016, Chavant et al. 2007), microtiter plate method (Au - O'Toole 2011), and the Lubbock system (Sun et al. 2008) rely on colourimetric measurements, which results in endpoint destruction of the sample. Therefore, the live monitoring of the same sample is not possible. Conventionally, antibiotic efficacy and susceptibility are determined by bacteria in the planktonic state rather than the ones in biofilm because of the complex and intricate equipment involved. Also, the clinical biofilm samples have colonies of different bacterial species, while the standard susceptibility tests are done on single isolates; consequently, meaningful information about other species and their interaction is missed out (Cendra et al. 2019, Hall-Stoodley and Stoodley 2009). The setup for the biofilm chip involves one inlet for bacteria and the other one for growth media, which merges into a pre-chamber, three interdigitated electrodes (chambers) on which the biofilm grows, are attached to a microfluidic chamber. EIS is integrated onto this biofilm chip which helps biomass evaluation during biofilm formation and tests antimicrobial drugs. It reduces the cost of using sophisticated technologies like confocal microscopy; with the low manufacturing cost, the individual price of the components makes it ideal for microbiological laboratories and industry. One of the several advantages includes microfluidics to mimic growth and naturally occurring biofilm much more efficiently. It also helps in reducing the consumption of reagent and the growth media in performing these biofilm experiments. The prechamber in the setup design prevents any disturbance caused to the biofilm via manual inoculation resulting in a suddenly increased flow rate into the chamber. The efficient prechamber design facilitates the more homogenous media in the three chambers, resulting in uniformity of the biofilm growth. To mimic biofilms similar to the in-vivo ones, continuous biofilm cultures are used to study antibiotic susceptibility (Reichhardt and Parsek 2019), whose analysis can be very tedious and expensive and involves handling by professionals. The EIS integration in the biofilm chip makes the analysis process simpler and quicker without any prerequisite training (Blanco-Cabra et al. 2021). Biofilm chip is being used for sputum samples from cystic fibrosis, bronchitis patients, samples from urinary tract infections, medical devices or the food industry, etc. This thus makes monitoring the growth of polymicrobial biofilms from the clinical samples possible without the possibility of any species getting excluded from the culture.

## 4.8  OTHER CHIP-BASED METHODS

Microfluidic chips are coupled with some analytical methods for detecting and identifying microorganisms [Figure 4.4B]. A single-use microfluidic chip has been developed, integrating solid phase transition and molecular amplification via reverse transcription-polymerase chain reaction (RT-PCR) to amplify influenza virus type A RNA (Cao et al. 2012). This microfluidic chip has more sensitivity compared to rapid immunoassays. A device with versatile and cost-effective polydimethylsiloxane is integrated with loop-mediated isothermal amplification (LAMP) for fast, sensitive, and instrumental free detection for mainly meningitis-causing bacteria (Dou et al. 2014). An automated microfluidic device was developed with electrochemical biosensors to detect and quantify *E. coli* in water samples. The quantification of *E. coli* was investigated using this device, and the results showed a limit of detection as low as 50 CFU/mL (Altintas et al. 2018). Some of the other microdevices combined with fluorescence spectrometry and mass spectrometry have also been developed (Măriuţa et al. 2020, Oedit et al. 2015). A new and straightforward nanoparticle-based platform has been constructed to detect pathogens in the biological samples. In this method, the microfluidic chamber with bacteria is incubated with magnetic nanoparticles and detected by nuclear magnetic resonance (NMR) (Lee, Yoon, and Weissleder 2009). This diagnostic platform has been evaluated by detecting tuberculosis with speed and sensitivity; 20 CFU in 1 mL of sputum in less than 30 min. These microfluidic chips are fast, easy to operate, and portable instruments. The newly developed platforms can serve as quick diagnostic tools for the detection of microorganisms and their biofilms with rapid and susceptible results.

## 4.9   DISCUSSION

Biofilm infections are highly tolerant to the antibiotics and other physical treatments. The strategies that are involved in showing tolerance to antibiotics include adaptive response, persistent cells, and low penetration of drugs (Ciofu et al. 2022). These infectious biofilms cause human diseases, leading to tissue damage and permanent pathology. The intracellular and extracellular signalling molecules play an essential role in cell communication, stringent response, biofilm formation, virulence, and host-pathogen interaction (Syal, Rs, and Reddy 2021). Quorum sensing plays a major role in cell-cell communication and gene expression to promote pathogenesis. The intracellular signalling molecules also regulate cell processes during stress conditions and other external factors. c-di-GMP and (p)ppGpp are the intracellular secondary signalling molecules that control the cell activities such as motility, biofilm cycle, and stringent response in many bacteria (Syal et al. 2015, Bhardwaj, Syal, and Chatterji 2018, Syal and Chatterji 2015). The enzymes involved in activating or repressing (p)ppGpp signals and metabolism have been studied in different bacteria, suggesting an essential role in starvation, antibiotic treatment, and other stress conditions that help microbes survive. (p)ppGpp also regulates DNA replication, transcription, and protein synthesis (Hobbs and Boraston 2019). The role of signalling cascades involved in mediating the biofilm maturation and communication between the communities play a significant role in existing biofilms (Syal et al. 2015, Naresh et al. 2012). Directly or indirectly, the second messengers trigger the biofilm formation and the available methods for quantification and analysis of biofilm do not measure the actual physiological conditions and changes in them. To study the exact physiological changes and life cycle of biofilm, a novel biofilm chip has been developed. This biofilm chip may serve as the ideal tool for detecting microbes, studying biofilm, imaging, and testing antibiotic susceptibility (Blanco-Cabra et al. 2021). Some of the other microfluidic platforms are also available, including low cost, a small number of reagents, rapid detection, and easy operation. Recent developments in microfluidic platforms can help human health in underdeveloped regions with inadequate health infrastructures. However, cost and availability of such methodologies remain a major challenge.

## REFERENCES

Altintas, Z., M. Akgun, G. Kokturk, and Y. Uludag. 2018. "A fully automated microfluidic-based electrochemical sensor for real-time bacteria detection." *Biosens Bioelectron* 100:541–548. doi: 10.1016/j.bios.2017.09.046

Atkinson, G. C., T. Tenson, and V. Hauryliuk. 2011. "The RelA/SpoT homolog (RSH) super-family: Distribution and functional evolution of ppGpp synthetases and hydrolases across the tree of life." *PLoS One* 6 (8):e23479. doi: 10.1371/journal.pone.0023479

Au - O'Toole, G. A. 2011. "Microtiter dish biofilm formation assay." *JoVE* (47):e2437. doi: doi:10.3791/2437.

Bhardwaj, N., K. Syal, and D. Chatterji. 2018. "The role of omega-subunit of Escherichia coli RNA polymerase in stress response." *Genes Cells* 23 (5):357–369. doi: 10.1111/gtc.12577

Blanco-Cabra, N., M. J. López-Martínez, B. V. Arévalo-Jaimes, M. T. Martin-Gómez, J. Samitier, and E. Torrents. 2021. "A new BiofilmChip device for testing biofilm formation and antibiotic susceptibility." *NPJ Biofilms and Microbiomes* 7 (1):62. doi: 10.1038/s41522-021-00236-1

Cao, Q., M. Mahalanabis, J. Chang, B. Carey, C. Hsieh, A. Stanley, C. A. Odell, P. Mitchell, J. Feldman, N. R. Pollock, and C. M. Klapperich. 2012. "Microfluidic chip for molecular amplification of influenza A RNA in human respiratory specimens." *PLOS ONE* 7 (3):e33176. doi: 10.1371/journal.pone.0033176

Carniello, V., B. W. Peterson, H. C. van der Mei, and H. J. Busscher. 2018. "Physico-chemistry from initial bacterial adhesion to surface-programmed biofilm growth." *Advances in Colloid and Interface Science* 261:1–14. doi: 10.1016/j.cis.2018.10.005

Cashel, M., and J. Gallant. 1969. "Two compounds implicated in the function of the RC gene of Escherichia coli." *Nature* 221 (5183):838–841. doi: 10.1038/221838a0

Cendra, M. D. M., N. Blanco-Cabra, L. Pedraz, and E. Torrents. 2019. "Optimal environmental and culture conditions allow the in vitro coexistence of Pseudomonas aeruginosa and Staphylococcus aureus in stable biofilms." *Sci Rep* 9 (1):16284. doi: 10.1038/s41598-019-52726-0

Ceri, H., M. E. Olson, C. Stremick, R. R. Read, D. Morck, and A. Buret. 1999. "The calgary biofilm device: New technology for rapid determination of antibiotic susceptibilities of bacterial biofilms." *Journal of Clinical Microbiology* 37 (6):1771–1776. doi: 10.1128/JCM.37.6.1771-1776.1999

Chavant, P., B. Gaillard-Martinie, R. Talon, M. Hébraud, and T. Bernardi. 2007. "A new device for rapid evaluation of biofilm formation potential by bacteria." *J Microbiol Methods* 68 (3):605–612. doi: 10.1016/j.mimet.2006.11.010

Chávez de Paz, L. E., J. A. Lemos, C. Wickström, and C. M. Sedgley. 2012. "Role of (p)ppGpp in biofilm formation by Enterococcus faecalis." *Appl Environ Microbiol* 78 (5):1627–1630. doi: 10.1128/aem.07036-11

Choi, Y. C., and E. Morgenroth. 2003. "Monitoring biofilm detachment under dynamic changes in shear stress using laser-based particle size analysis and mass fractionation." *Water Sci Technol* 47 (5):69–76.

Christensen, G. D., W. A. Simpson, J. J. Younger, L. M. Baddour, F. F. Barrett, D. M. Melton, and E. H. Beachey. 1985. "Adherence of coagulase-negative staphylococci to plastic tissue culture plates: A quantitative model for the adherence of staphylococci to medical devices." *Journal of Clinical Microbiology* 22 (6):996–1006. doi: 10.1128/jcm.22.6.996-1006.1985

Ciofu, O., C. Moser, P. Ø. Jensen, and N. Høiby. 2022. "Tolerance and resistance of microbial biofilms." *Nature Reviews Microbiology*. doi: 10.1038/s41579-022-00682-4

Costanzo, A., and S. E. Ades. 2006. "Growth phase-dependent regulation of the extracytoplasmic stress factor, sigmaE, by guanosine 3',5'-bispyrophosphate (ppGpp)." *J Bacteriol* 188 (13):4627–4634. doi: 10.1128/jb.01981-05

Costerton, J. W., P. S. Stewart, and E. P. Greenberg. 1999. "Bacterial biofilms: A common cause of persistent infections."*Science* 284 (5418):1318–1322. doi: 10.1126/science. 284.5418.1318.

de la Fuente-Núñez, C., F. Reffuveille, E. F. Haney, S. K. Straus, and R. E. Hancock. 2014. "Broad-spectrum anti-biofilm peptide that targets a cellular stress response." *PLoS Pathog* 10 (5):e1004152. doi: 10.1371/journal.ppat.1004152

Donlan, R. M., and J. W. Costerton. 2002. "Biofilms: Survival mechanisms of clinically relevant microorganisms." *Clin Microbiol Rev* 15 (2):167–193. doi: 10.1128/cmr.15.2. 167-193.2002

Dou, M., D. C. Dominguez, X. J. Li, J. Sanchez, and G. Scott. 2014. "A versatile PDMS/paper hybrid microfluidic platform for sensitive infectious disease diagnosis." *Analytical Chemistry* 86 (15):7978–7986. doi: 10.1021/ac5021694

Flemming, H. C., J. Wingender, U. Szewzyk, P. Steinberg, S. A. Rice, and S. Kjelleberg. 2016. "Biofilms: An emergent form of bacterial life." *Nat Rev Microbiol* 14 (9):563–575. doi: 10.1038/nrmicro.2016.94

Gaca, A. O., C. Colomer-Winter, and J. A. Lemos. 2015. "Many means to a common end: The intricacies of (p)ppGpp metabolism and its control of bacterial homeostasis." *J Bacteriol* 197 (7):1146–1156. doi: 10.1128/jb.02577-14

Ge, X., Y. Cai, Z. Chen, S. Gao, X. Geng, Y. Li, Y. Li, J. Jia, and Y. Sun. 2018. "Bifunctional enzyme spoT is involved in biofilm formation of helicobacter pylori with multidrug resistance by upregulating efflux pump Hp1174 (gluP)." *Antimicrobial Agents and Chemotherapy* 62 (11):e00957- 18. doi: 10.1128/AAC.00957-18

Geiger, T., B. Kästle, F. L. Gratani, C. Goerke, and C. Wolz. 2014. "Two small (p)ppGpp synthases in Staphylococcus aureus mediate tolerance against cell envelope stress conditions." *J Bacteriol* 196 (4):894–902. doi: 10.1128/jb.01201-13

Gentry, D. R., and M. Cashel. 1996. "Mutational analysis of the Escherichia coli spoT gene identifies distinct but overlapping regions involved in ppGpp synthesis and degradation." *Mol Microbiol* 19 (6):1373–1384. doi: 10.1111/j.1365-2958.1996.tb02480.x

Hall-Stoodley, L., J. W. Costerton, and P. Stoodley. 2004. "Bacterial biofilms: From the natural environment to infectious diseases." *Nat Rev Microbiol* 2 (2):95–108. doi: 10. 1038/nrmicro821

Hall-Stoodley, L., and P. Stoodley. 2009. "Evolving concepts in biofilm infections." *Cell Microbiol* 11 (7):1034–1043. doi: 10.1111/j.1462-5822.2009.01323.x

Harrison, J. J., C. A. Stremick, R. J. Turner, N. D. Allan, M. E. Olson, and H. Ceri. 2010. "Microtiter susceptibility testing of microbes growing on peg lids: A miniaturized biofilm model for high-throughput screening." *Nat Protoc* 5 (7):1236–1254. doi: 10. 1038/nprot.2010.71

Hauryliuk, V., G. C. Atkinson, K. S. Murakami, T. Tenson, and K. Gerdes. 2015. "Recent functional insights into the role of (p)ppGpp in bacterial physiology." *Nat Rev Microbiol* 13 (5):298–309. doi: 10.1038/nrmicro3448

He, H., J. N. Cooper, A. Mishra, and D. M. Raskin. 2012. "Stringent response regulation of biofilm formation in Vibrio cholerae." *J Bacteriol* 194 (11):2962–2972. doi: 10.1128/ jb.00014-12

Heras, B., M. Totsika, K. M. Peters, J. J. Paxman, C. L. Gee, R. J. Jarrott, M. A. Perugini, A. E. Whitten, and M. A. Schembri. 2014. "The antigen 43 structure reveals a molecular velcro-like mechanism of autotransporter-mediated bacterial clumping."*PNAS* 111 (1):457–462. doi: 10.1073/pnas.1311592111

Heydorn, A., A. T. Nielsen, M. Hentzer, C. Sternberg, M. Givskov, B. K. Ersbøll, and S. Molin. 2000. "Quantification of biofilm structures by the novel computer program COMSTAT." *Microbiology (Reading)* 146 (Pt 10):2395–2407. doi: 10.1099/00221287-146-10-2395

Hobbs, J. K., and A. B. Boraston. 2019. "(p)ppGpp and the stringent response: An emerging threat to antibiotic therapy." *ACS Infectious Diseases* 5 (9):1505–1517. doi: 10.1021/acsinfecdis.9b00204

Hong, Y., X. Zhou, H. Fang, D. Yu, C. Li, and B. Sun. 2013. "Cyclic di-GMP mediates Mycobacterium tuberculosis dormancy and pathogenecity." *Tuberculosis* 93 (6):625–634. doi: 10.1016/j.tube.2013.09.002

Irving, S. E., and R. M. Corrigan. 2018. "Triggering the stringent response: signals responsible for activating (p)ppGpp synthesis in bacteria." *Microbiology (Reading)* 164 (3):268–276. doi: 10.1099/mic.0.000621

Jain, V., R. Saleem-Batcha, and D. Chatterji. 2007. "Synthesis and hydrolysis of pppGpp in mycobacteria: a ligand mediated conformational switch in Rel." *Biophys Chem* 127 (1-2):41–50. doi: 10.1016/j.bpc.2006.12.003

Jain, V., R. Saleem-Batcha, A. China, and D. Chatterji. 2006. "Molecular dissection of the mycobacterial stringent response protein Rel." *Protein Sci* 15 (6):1449–1464. doi: 10.1110/ps.062117006

Kalia, D., G. Merey, S. Nakayama, Y. Zheng, J. Zhou, Y. Luo, M. Guo, B. T. Roembke, and H. O. Sintim. 2013. "Nucleotide, c-di-GMP, c-di-AMP, cGMP, cAMP, (p)ppGpp signalling in bacteria and implications in pathogenesis." *Chem Soc Rev* 42 (1):305–341. doi: 10.1039/c2cs35206k

Karunakaran, E., J. Mukherjee, B. Ramalingam, and C. A. Biggs. 2011. ""Biofilmology": a multidisciplinary review of the study of microbial biofilms." *Appl Microbiol Biotechnol* 90 (6):1869–1881. doi: 10.1007/s00253-011-3293-4

Kim, H. M., and M. E. Davey. 2020. "Synthesis of ppGpp impacts type IX secretion and biofilm matrix formation in Porphyromonas gingivalis." *NPJ Biofilms and Microbiomes* 6 (1):5. doi: 10.1038/s41522-020-0115-4

Kırmusaoğlu, S. 2019. "The methods for detection of biofilm and screening antibiofilm activity of agents." In *Antimicrobials, Antibiotic Resistance, Antibiofilm Strategies and Activity Methods* 1-17. doi: 10.5772/intechopen.84411

Kostakioti, M., M. Hadjifrangiskou, and S. J. Hultgren. 2013. "Bacterial biofilms: Development, dispersal, and therapeutic strategies in the dawn of the postantibiotic era." *Cold Spring Harb Perspect Med* 3 (4):a010306. doi: 10.1101/cshperspect.a010306

Krasowska, A., and K. Sigler. 2014. "How microorganisms use hydrophobicity and what does this mean for human needs?" *Frontiers in Cellular and Infection Microbiology* 4:112 1–7. doi: 10.3389/fcimb.2014.00112

Lee, H., T. J. Yoon, and R. Weissleder. 2009. "Ultrasensitive detection of bacteria using core–shell nanoparticles and an NMR-filter system." *Angewandte Chemie International Edition* 48 (31):5657–5660. doi: 10.1002/anie.200901791

Lemos, J. A., V. K. Lin, M. M. Nascimento, J. Abranches, and R. A. Burne. 2007. "Three gene products govern (p)ppGpp production by Streptococcus mutans." *Mol Microbiol* 65 (6):1568–1581. doi: 10.1111/j.1365-2958.2007.05897.x

Liu, K., A. N. Bittner, and J. D. Wang. 2015. "Diversity in (p)ppGpp metabolism and effectors." *Curr Opin Microbiol* 24:72–79. doi: 10.1016/j.mib.2015.01.012

Maiti, K., K. Syal, D. Chatterji, and N. Jayaraman. 2017. "Synthetic arabinomannan heptasaccharide glycolipids inhibit biofilm growth and augment isoniazid effects in Mycobacterium smegmatis." *Chembiochem* 18 (19):1959–1970. doi: 10.1002/cbic.201700247

Mӑriuṭa, D., S. Colin, C. Barrot-Lattes, S. Le Calvé, J. G. Korvink, L. Baldas, and J. J. Brandner. 2020. "Miniaturization of fluorescence sensing in optofluidic devices." *Microfluidics and Nanofluidics* 24 (9):65. doi: 10.1007/s10404-020-02371-1

Mechold, U., K. Potrykus, H. Murphy, K. S. Murakami, and M. Cashel. 2013. "Differential regulation by ppGpp versus pppGpp in Escherichia coli." *Nucleic Acids Res* 41 (12):6175–6189. doi: 10.1093/nar/gkt302

Miller, M. B., and B. L. Bassler. 2001. "Quorum sensing in bacteria." *Annu Rev Microbiol* 55:165–199. doi: 10.1146/annurev.micro.55.1.165

Murdeshwar, M. S., and D. Chatterji. 2012. "MS_RHII-RSD, a dual-function RNase HII-(p) ppGpp synthetase from Mycobacterium smegmatis." *J Bacteriol* 194 (15):4003–4014. doi: 10.1128/jb.00258-12

Nanamiya, H., K. Kasai, A. Nozawa, C. S. Yun, T. Narisawa, K. Murakami, Y. Natori, F. Kawamura, and Y. Tozawa. 2008. "Identification and functional analysis of novel (p) ppGpp synthetase genes in Bacillus subtilis." *Mol Microbiol* 67 (2):291–304. doi: 10.1111/j.1365-2958.2007.06018.x

Naresh, K., P. G. Avaji, K. Maiti, B. K. Bharati, K. Syal, D. Chatterji, and N. Jayaraman. 2012. "Synthesis of beta-arabinofuranoside glycolipids, studies of their binding to surfactant protein-A and effect on sliding motilities of M. smegmatis." *Glycoconj J* 29 (2-3):107–118. doi: 10.1007/s10719-012-9369-2

Nguyen, D., A. Joshi-Datar, F. Lepine, E. Bauerle, O. Olakanmi, K. Beer, G. McKay, R. Siehnel, J. Schafhauser, Y. Wang, B. E. Britigan, and P. K. Singh. 2011. "Active starvation responses mediate antibiotic tolerance in biofilms and nutrient-limited bacteria." *Science (New York, N.Y.)* 334 (6058):982–986. doi: 10.1126/science.1211037

Nikolaev Iu, A., and V. K. Plakunov. 2007. "Biofilm –"City of microbes" or an analogue of multicellular organisms?." *Mikrobiologiia* 76 (2):149–163.

O'Toole, G., H. B. Kaplan, and R. Kolter. 2000. "Biofilm formation as microbial development." *Annu Rev Microbiol* 54:49–79. doi: 10.1146/annurev.micro.54.1.49

O'Toole, G. A. 2011. "Microtiter dish biofilm formation assay." *Journal of visualized experiments: JoVE* (47):2437. doi: 10.3791/2437

Oedit, A., P. Vulto, R. Ramautar, P. W. Lindenburg, and T. Hankemeier. 2015. "Lab-on-a-Chip hyphenation with mass spectrometry: Strategies for bioanalytical applications." *Current Opinion in Biotechnology* 31:79–85. doi: 10.1016/j.copbio.2014.08.009

Olivares, E., S. Badel-Berchoux, C. Provot, B. Jaulhac, G. Prévost, T. Bernardi, and F. Jehl. 2016. "The biofilm ring test: A rapid method for routine analysis of pseudomonas aeruginosa biofilm formation kinetics." *J Clin Microbiol* 54 (3):657–661. doi: 10.1128/jcm.02938-15

Parsek, M. R., and P. K. Singh. 2003. "Bacterial biofilms: An emerging link to disease pathogenesis." 57 (1):677–701. doi: 10.1146/annurev.micro.57.030502.090720

Penesyan, A., I. T. Paulsen, S. Kjelleberg, and M. R. Gillings. 2021. "Three faces of biofilms: A microbial lifestyle, a nascent multicellular organism, and an incubator for diversity." *npj Biofilms and Microbiomes* 7 (1):80. doi: 10.1038/s41522-021-00251-2

Potrykus, K., H. Murphy, N. Philippe, and M. Cashel. 2011. "ppGpp is the major source of growth rate control in E. coli." *Environ Microbiol* 13 (3):563–575. doi: 10.1111/j.1462-2920.2010.02357.x

Purcell, E. B., and R. Tamayo. 2016. "Cyclic diguanylate signalling in Gram-positive bacteria." *FEMS Microbiol Rev* 40 (5):753–773. doi: 10.1093/femsre/fuw013

Reichhardt, C., and M. R. Parsek. 2019. "Confocal laser scanning microscopy for analysis of Pseudomonas aeruginosa biofilm architecture and matrix localization." *Front Microbiol* 10:677. doi: 10.3389/fmicb.2019.00677

Rumbaugh, K. P., and K. Sauer. 2020. "Biofilm dispersion." *Nat Rev Microbiol* 18 (10):571–586. doi: 10.1038/s41579-020-0385-0

Sánchez, M. C., A. Llama-Palacios, M. J. Marín, E. Figuero, R. León, V. Blanc, D. Herrera, and M. Sanz. 2013. "Validation of ATP bioluminescence as a tool to assess antimicrobial effects of mouthrinses in an in vitro subgingival-biofilm model." *Medicina oral, patologia oral y cirugia bucal* 18 (1):e86–e92. doi: 10.4317/medoral.18376

Sharma, I. M., A. Petchiappan, and D. Chatterji. 2014. "Quorum sensing and biofilm formation in mycobacteria: Role of c-di-GMP and methods to study this second messenger." *IUBMB Life* 66 (12):823–834. doi: 10.1002/iub.1339

Steinchen, W., and G. Bange. 2016. "The magic dance of the alarmones (p)ppGpp." *Mol Microbiol* 101 (4):531–544. doi: 10.1111/mmi.13412

Stent, G. S., and S. Brenner. 1961. "A genetic locus for the regulation of ribonucleic acid synthesis." *Proceedings of the National Academy of Sciences of the United States of America* 47 (12):2005–2014. doi: 10.1073/pnas.47.12.2005

Sun, D., G. Lee, J. H. Lee, H. Y. Kim, H. W. Rhee, S. Y. Park, K. J. Kim, Y. Kim, B. Y. Kim, J. I. Hong, C. Park, H. E. Choy, J. H. Kim, Y. H. Jeon, and J. Chung. 2010. "A metazoan ortholog of SpoT hydrolyzes ppGpp and functions in starvation responses." *Nat Struct Mol Biol* 17 (10):1188–1194. doi: 10.1038/nsmb.1906

Sun, J., R. Daniel, I. Wagner-Döbler, and A. P. Zeng. 2004. "Is autoinducer-2 a universal signal for interspecies communication: A comparative genomic and phylogenetic analysis of the synthesis and signal transduction pathways." *BMC Evolutionary Biology* 4 (1):36. doi: 10.1186/1471-2148-4-36

Sun, Y., S. E. Dowd, E. Smith, D. D. Rhoads, and R. D. Wolcott. 2008. "In vitro multispecies Lubbock chronic wound biofilm model." *Wound Repair Regen* 16 (6):805–813. doi: 10.1111/j.1524-475X.2008.00434.x

Syal, K. 2017. "Novel method for quantitative estimation of biofilms." *Curr Microbiol* 74 (10):1194–1199. doi: 10.1007/s00284-017-1304-0

Syal, K., N. Bhardwaj, and D. Chatterji. 2017. "Vitamin C targets (p)ppGpp synthesis leading to stalling of long-term survival and biofilm formation in Mycobacterium smegmatis." *FEMS Microbiol Lett* 364 (1). doi: 10.1093/femsle/fnw282

Syal, K., and D. Chatterji. 2015. "Differential binding of ppGpp and pppGpp to E. coli RNA polymerase: Photo-labeling and mass spectral studies." *Genes Cells* 20 (12):1006–1016. doi: 10.1111/gtc.12304

Syal, K., and D. Chatterji. 2018. "Vitamin C: A natural inhibitor of cell wall functions and stress response in mycobacteria." *Adv Exp Med Biol* 1112:321–332. doi: 10.1007/978-981-13-3065-0_22

Syal, K., K. Flentie, N. Bhardwaj, K. Maiti, N. Jayaraman, C. L. Stallings, and D. Chatterji. 2017a. "Synthetic (p)ppGpp analogue is an inhibitor of stringent response in mycobacteria." *Antimicrob Agents Chemother* 61 (6). doi: AAC.00443-17 [pii] 10.1128/AAC.00443-17

Syal, K., H. Joshi, D. Chatterji, and V. Jain. 2015. "Novel pppGpp binding site at the C-terminal region of the Rel enzyme from Mycobacterium smegmatis." *FEBS J* 282 (19):3773–3785. doi: 10.1111/febs.13373

Syal, K., K. Maiti, K. Naresh, P. G. Avaji, D. Chatterji, and N. Jayaraman. 2016. "Synthetic arabinomannan glycolipids impede mycobacterial growth, sliding motility and biofilm structure." *Glycoconj J* 33 (5):763–777. doi: 10.1007/s10719-016-9670-6. 10.1007/s10719-016-9670-6 [pii].

Syal, K., K. Maiti, K. Naresh, D. Chatterji, and N. Jayaraman. 2015. "Synthetic glycolipids and (p)ppGpp analogs: Development of inhibitors for mycobacterial growth, biofilm and stringent response." *Adv Exp Med Biol* 842:309–327. doi: 10.1007/978-3-319-11280-0_20

Syal, K., N. Rs, and M. Reddy. 2021. "The extended (p)ppGpp family: New dimensions in Stress response." *Curr Res Microb Sci* 2:100052. doi: 10.1016/j.crmicr.2021.100052. S2666-5174(21)00033-X [pii].

Syal, K., K. Flentie, N. Bhardwaj, K. Maiti, N. Jayaraman, C. L. Stallings, and D. Chatterji. 2017b. "Synthetic (p)ppGpp analogue is an inhibitor of stringent response in mycobacteria." *Antimicrobial agents and chemotherapy* 61 (6):e00443- 17. doi: 10.1128/AAC.00443-17

Teschler, J. K., D. Zamorano-Sánchez, A. S. Utada, C. J. Warner, G. C. Wong, R. G. Linington, and F. H. Yildiz. 2015. "Living in the matrix: Assembly and control of Vibrio cholerae biofilms." *Nat Rev Microbiol* 13 (5):255–268. doi: 10.1038/nrmicro3433

Trulear, M. G., and W. G. Characklis. 1982. "Dynamics of biofilm processes." *Journal (Water Pollution Control Federation)* 54 (9):1288–1301.

Valentini, M., and A. Filloux. 2016. "Biofilms and cyclic di-GMP (c-di-GMP) signalling: Lessons from Pseudomonas aeruginosa and other bacteria." *The Journal of Biological Chemistry* 291 (24):12547–12555. doi: 10.1074/jbc.R115.711507

Wendrich, T. M., G. Blaha, D. N. Wilson, M. A. Marahiel, and K. H. Nierhaus. 2002. "Dissection of the mechanism for the stringent factor RelA." *Mol Cell* 10 (4):779–788. doi: 10.1016/s1097-2765(02)00656-1

Wexselblatt, E., Y. Oppenheimer-Shaanan, I. Kaspy, N. London, O. Schueler-Furman, E. Yavin, G. Glaser, J. Katzhendler, and S. Ben-Yehuda. 2012. "Relacin, a novel antibacterial agent targeting the stringent response." *PLOS Pathogens* 8 (9):e1002925. doi: 10.1371/journal.ppat.1002925

Wilson, C., R. Lukowicz, S. Merchant, H. Valquier-Flynn, J. Caballero, J. Sandoval, M. Okuom, C. Huber, T. D. Brooks, E. Wilson, B. Clement, C. D. Wentworth, and A. E. Holmes. 2017. "Quantitative and qualitative assessment methods for biofilm." Growth: A Mini-review." *Res Rev J Eng Technol* 6 (4).

Xiao, H., M. Kalman, K. Ikehara, S. Zemel, G. Glaser, and M. Cashel. 1991. "Residual guanosine 3',5'-bispyrophosphate synthetic activity of relA null mutants can be eliminated by spoT null mutations." *J Biol Chem* 266 (9):5980–5990.

Yan, J., and B. L. Bassler. 2019. "Surviving as a community: Antibiotic tolerance and persistence in bacterial biofilms." *Cell Host & Microbe* 26 (1):15–21. doi: 10.1016/j.chom.2019.06.002

# 5 A Review of State-of-the-Art Miniaturized Electrochemical Devices for Environmental Applications
## Monitoring, Detection and Remediation

*Prathiksha P Prabhu, Koustav Pan, and
Jegatha Nambi Krishnan*
Department of Chemical Engineering, BITS Pilani, K K Birla
Goa Campus, Goa, India

## CONTENTS

DOI: 10.1201/b23359-5

## 5.1   INTRODUCTION

Global environmental pollution associates with the crucial issues escalated by exploitation of earth and its resources mainly in developing countries. Rapid industrialization has brought about environmental concerns as toxic pollutants and wastes are being discharged into the surroundings. Better environmental control is increasingly necessitated by community concern and regulation. Additionally, the establishment of more suitable analytical processes is required due to the increase in the variety of analytes to detect. Sample specimens must be directed to a lab for typical "off-site" analytical testing. Conventional methods achieve the highest precision with the smallest detection limits but they are costly, require highly skilled workers and are time-consuming (Tothill 2001). The current trend of performing on-site observation has prompted the advancement of microfabricated electrochemical devices such as electrochemical sensors as novel analytical tools capable of providing quick, accurate and sensitive readings at a lower cost; designed towards on-site analysis. In this aspect, these devices wouldn't be in a position to challenge current analytical techniques, but they might be used by regulatory agencies as well as the business community to provide sufficient data for standard sample testing and screening (Hanrahan et al. 2004). In the domain of industrial, clinical, agricultural, and environmental analysis such devices own a wide range of essential applications. For decades, electrochemical devices have been used in the field to monitor a range of water quality indicators (for instance conductivity, dissolved oxygen and pH). A broader range of applications in the environment resulting from these are detection of trace metals and microplastics (e.g., polypropylene, polyethylene, polyethylene terephthalate or nylon) in natural waters, carcinogen monitoring (e.g., aromatic amines or N-nitroso compounds), environmental protection, clean energy conversion and organic pollutants (e.g., phenols, pesticides) in underground water.

When paper-based microfluidics is integrated with electrochemical methods a new cornerstone emerges in the field of electrochemical devices. Owing to their excellent recognition sensitivity and low limit of detection (LOD) at a low price, electrochemical detection technologies have become widely investigated (Mettakoonpitak et al. 2016; Yamada et al. 2017). Handheld potentiostats, for example HOME-Stat (Das et al. 2021) and eSTAT (Anshori et al. 2022), have been widely accessible for on-site investigation and the electrodes may be easily downsized and produced on paper (Noviana et al. 2020; Lillehoj et al. 2013).

The relevance of microfabricated electrochemical devices (MEDs) in contemporary environmental monitoring efforts is critically examined in this research. In terms of analytical advancements, distant communication capabilities and microfabrication, this review emphasizes at length significant developments in the electrochemical sensor field. The use of microfluidic integration and submersible devices for remote and continuous monitoring are recent applications and emerging developments in electrochemical devices will be highlighted.

## 5.2   WORKING PRINCIPLES

Electrical variables including current, potential, and charge are detected and also how they relate to chemical parameters using electroanalytical sensors are analyzed.

Depending on the analyte, sample matrix, and sensitivity and selectivity criteria, the maximum number of electrochemical devices being used in environmental monitoring fall into one of four types. As a result of redox processes involving the analyte, the potential difference between the electrodes or the current flowing through the system are utilized to estimate the analyte in the sample. The logarithm of the material's electrochemically active concentration and the electrical potential difference are proportional. These gadgets deliver a more reliable output, are more sensitive, react more quickly and are less prone to interference. The majority of sensing applications involve electrochemical measurements. Based on the electrical properties that they measure, electrochemical sensors are further divided into a number of categories. They include voltametric, conductimetric, amperometric, potentiometric and amperometric sensors (Brett 2001; Rodriguez-Mozaz et al. 2004; Tetyana et al. 2021).

### 5.2.1  VOLTAMETRIC SENSORS

Voltammetry is an electroanalytical approach that measures current as a function of potential to obtain information about one or more analytes. Voltammetry can be used to obtain data in a variety of ways, including stripping voltammetry, cyclic voltammetry and square wave voltammetry to mention a few typical methods. Guo et al. created an ionic liquid modified graphene composite by fusing the conductivity of graphene with ionic liquids' (IL) capacity for dispersing graphene and their surface-to-volume ratio. In comparison to both ionic liquid-graphite paste electrodes and liquid CNT, the IL-graphene paste electrode had a lower background current, 1.65 A $cm^2$ per ppb of higher sensitivity, 0.5 ppb lower detection limit and better reproducibility for 2,4,6-trinitrotoluene (TNT) detection. (Guo et al. 2010).

### 5.2.2  POTENTIOMETRIC SENSORS

For a long time, potentiometric sensors have been regarded as a zero-current method for determining the potential of an interface, most frequently a membrane. Potentiometric devices for in-situ pH, $pCO_2$, and $pO_2$ monitoring are typical examples. A novel potentiometric sensor on the basis of sensing of dissolved oxygen was created and verified by Zhuiykov et al., utilizing a $Cu_2O$-doped $RuO_2$ sensing electrode. Within the 9–30°C temperature range, from 0.5 to 8.0 ppm, $Cu_2O$-$RuO_2$ showed a linear response to DO (4.73 to 3.59, log[$O_2$]). Improved sensing properties and less fouling were achieved due to this doping method. These findings suggest that doping $RuO_2$ electrode with a low concentration of ruthenium increases sensitivity and selectivity while leaving the ruthenium chemical state unchanged (Zhuiykov et al. 2010).

### 5.2.3  AMPEROMETRIC SENSORS

The amperometric technique is of particular interest in applications like non-enzymatic glucose sensors, detection of several reactive $O_2$ and $N_2$ species and hydrogen peroxide ($H_2O_2$) sensors. An intact shallow-dipped graphite microelectrode array for sub-micromolar nitrite detection was described by Khairy et al. using a more conventional amperometric approach. Screen printed microelectrode arrays are particularly promising because of their disposable nature and low cost. This might become a powerful detection tool for a variety of analytes with the modification of these microelectrode arrays (Khairy et al. 2010).

## 5.3 DESIGN AND FABRICATION CONSIDERATIONS

When designing an electrochemical sensor for environmental monitoring, a number of key design conditions should be scrutinized such as analytical response, microfabrication, biofouling, portability, reversibility, power consumption and sensitivity selectivity. Ultimately factors including monitoring requirements, manufacturing processes, and cell configuration dictate the type of fabrication employed. Production of miniature sensor arrays and electrochemical sensors, particularly for remote network monitoring operations, has advanced significantly as a result of the ability of microfabrication to replace bulky electrodes and conventional electrochemical cells with simple sensing devices (Bourgeois et al. 2003). Electrochemical sensors for environmental monitoring have been designed using revolutionary approaches such as silicon-based techniques, and photolithography in recent studies. Micromachined technology has major benefits with respect to current environmental efforts (data temporal resolution and metal speciation measurements). Furthermore, the utilization of lower sample quantities (mL) as well as lower total power consumption is made possible by micromachined devices than typical electrochemical cells (Tercier-Waeber et al. 2005).

## 5.4 DEVICES FOR DETECTION, MONITORING AND REMEDIATION OF ENVIRONMENTAL POLLUTANTS

### 5.4.1 WATER POLLUTANTS

Water pollution is a cause of great concern today, jeopardizing human health. The number of potable water sources is limited and accounts for less than one percent of the world's fresh water (Melissa Denchak 2022), while approximately 71% of the world's population faces water scarcity to varying degrees, according to recent estimates (Nemiwal & Kumar 2021). Transport, heavy industry, petrochemical industry and agricultural operations contribute enormously to water pollution, releasing substantial amounts of hazardous chemicals, including heavy metals, pesticides and petrochemicals, into the atmosphere and aquatic environment (Abu-Ali et al. 2019).

#### 5.4.1.1 Inorganic Pollutants

Heavy-metal poisoning of groundwater poses a serious threat to many creatures due to their severe toxicity and growing environmental levels on their usage in industrial processes. The measurement of traces of heavy metals such as Cd, Cu, Pb, Zn and Hg in the environment is critical due to the metals' ability to bioaccumulate in living species, especially marine organisms (Rodriguez-Mozaz et al. 2004). Several studies and publications have been published regarding on-site and in-situ detection of heavy-metal pollutants utilizing techniques of electrochemical detection. Early approaches employed an electrochemical sensor in conjunction with a made-to-order potentiostat to provide a simple analysis system on-site. Such systems, however, required the use of signal processing, system management and a laptop computer for data acquisition, making routine environmental on-site monitoring in large field regions difficult. Yun et al. put forward and executed the miniaturized remote environmental wireless monitoring system that uses electrochemical detection to monitor heavy-metal ion pollution in water and wirelessly conducts to a base station the detected signal. This proposed monitoring system could be placed in vast open domains such as a river or a coast and consisted of an electrochemical sensor constructed on a substrate of silicon utilizing a custom radio frequency (RF) communication module and microelectromechanical system (MEMS) technology optimized for consumption of low power, with an antenna, and a potentiostat. In order for dispersed sensor systems to function, the base station creates and sends signal instructions. When the base station requests the analysis, the electrochemical sensor with potentiostat analyses heavy-metal ions in the water, and the RF module transmits the measured signals to the base station (Yun et al. 2004).

The distributed sensor system as shown in Figure 5.1 consists of three main modules: an RF communication module for wireless transmission, electrochemical detection using and the RF module transmits the measured signals to the base station, and customized potentiostats for sensor control, signal processing and readout.

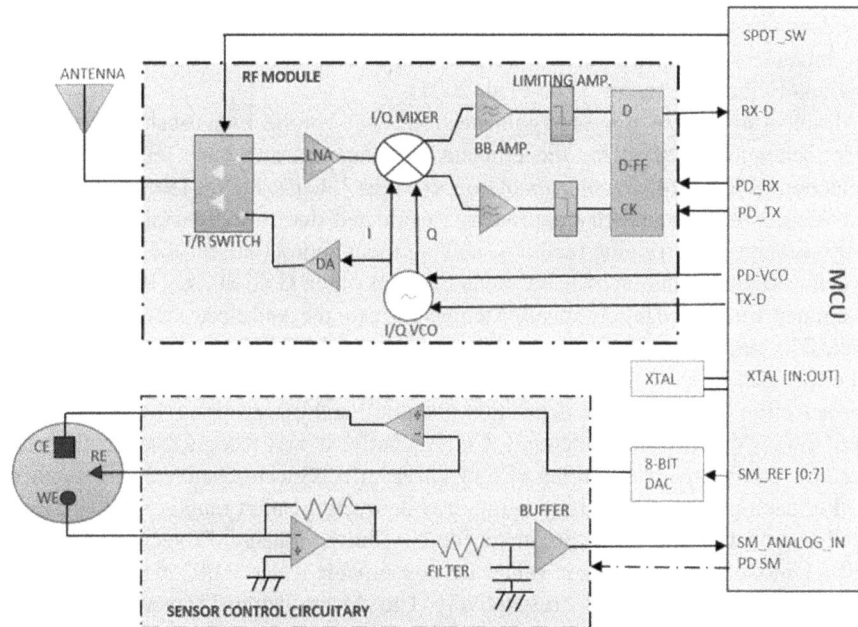

**FIGURE 5.1**   Schematic representation of the distributed sensor system comprising the sensor module, potentiostat, and RF module with an antenna. Modified from (Yun et al. 2004).

A customized potentiostat was designed and constructed by combining commercial low-noise op-amps in a hybrid configuration. At the digital-to-analog converters (DACs) eight-bit digital signals produced by the multipoint control unit (MCU) are transformed into square wave signals that are utilized to specify the power amid the reference and working electrode in order to start an electrochemical reaction on the working electrode's surface. The current-to-voltage amplifier step amplifies the reaction current that flows through the working electrode before sending it to the MCU via buffer stage and the high-pass filter. The analogue signal is subsequently transformed into a digital signal by the analog-to-digital converter (ADC). Lastly, the collected signal is analyzed in the MCU before being relayed back to the base station.

The proposed distributed sensor network features two low-power operating modes: wake-up and sleep mode. When the unit is in wake-up mode and the instruction signals are picked up, the MCU enables the system and then powers the other modules. The MCU progressively allows the operator as required and suitably deactivates the system's unnecessary components in this wake-up mode to reduce power usage. Instruction signals are wirelessly delivered from a nearby base station and received by the sensor system through an antenna. The monitoring system's MCU/DAC unit creates a signal that is applied to the working electrode for SWASV through DACs and amplifiers. First, the potential is maintained constant for limited amount of time. In the meantime, heavy metals ions build up and are then removed from the solution to produce a steady flow of current at the relevant reduction or oxidation potentials (stripping). After being modulated in the RF module, the MCU then relays the detection signal returning to the base station from the sensor module. Direct FSK modulation is used to modulate the signal output from the MCU before it is sent via antenna to the base station (Yun et al. 2004).

Using a single screen-printed gold electrode (SPE) probe, Elena et al. developed an electrochemical method for the efficient simultaneous and quick electrochemical detection of ions of heavy metals of concern ($Hg^{2+}$, $Cu^{2+}$, $Pb^{2+}$). The anodic square wave stripping voltammetry approach is evaluated due to the excellent analytical responsiveness of stripping methods, making them appropriate for the instantaneous synchronized recognition of heavy metal traces in water. Gold SPEs of two types were examined, that varied in solidification temperature for the gold electrode denoted low or high. The electrode treated at lower temperature (Au SPE-LT) has greater roughness. The Au electrode cured at a higher temperature (Au SPE-HT) was shown to be reliable for detection of $Cu^{2+}$ (0.0 V deposition potential) and inappropriate for determining $Pb^{2+}$ (−0.5 V deposition potential). Consequently, it was recognized that the gold surfaces cure temperature during SPE synthesis affects electrochemical performance.

For each pollutant, the study shows a wide quantification range (5–300 g/L), as well as detection limits that are under the U.S. Environmental Protection Agency's (EPA) maximal contamination values for consumable water: 0.002, 0.015, 1.3 mg/L and for $Hg^{2+}$, $Pb^{2+}$ and $Cu^{2+}$, respectively. The electrochemical sensors were put to the test in remote areas of the Amazon River that were extensively contaminated with heavy metals from mining under extremely high temperatures and humidity levels. The field data were compared to conventional analytical methods used in labs, and it was discovered that they were incredibly accurate (Bernalte et al. 2020).

Xuan et al. constructed a sensor with a commercial Pt counter electrode as working electrodes and Ag/AgCl reference electrode were immersed in a solution containing 0.1 mol/L potassium chloride as a supportive electrolyte and 2 mmol/L $K_3[Fe (CN)_6]$. The electrochemical analysis of Pb and Cd ions was carried out using the suggested flexible MEMS sensor by simultaneously accumulating 400 ppb bismuth (post improvement) and varying concentrations of Pb and Cd onto the working electrode. By in-situ electroplating Bi film, the synthesized sensor displayed distinct and distinct stripping peaks for lead ($Pb^{+2}$) and cadmium ($Cd^{+2}$) ions, correspondingly.

By increasing surface area of the electrode, the CNT was combined with the rGO to improve sensor performance. To achieve the highest stripping performance, a number of experimental factors, including electrodeposition circumstances and electrolyte environment, were thoroughly tuned. High sensitivity of 926 $nA/ppbcm^2$ ($Pb^{+2}$) and 262 $nA/ppbcm^2$ ($Cd^{+2}$) were attained under ideal circumstances, with detection limits of $Pb^{+2} = 0.2$ ppb and $Cd^{+2} = 0.6$ ppb. The produced Au/rGOCNT/Bi altered electrode greatly improved the determination efficiency toward target ions due to the increased electrode surface area. The developed sensor has good reliability, stability, and sensitivity in detecting the heavy metal ions (Xuan & Park 2018).

A novel surfactant with urea activity that has been produced has been immobilized, and based on this, 1-(2, 4-dinitrophenyl)-Dodecanoyl thiourea (DAN), with soil fertility-improving properties, Anum et al. developed a very sensitive electrochemical sensor for the identification of Hg (II). The peak current value of mercury rose considerably when this surfactant was attached to the glassy carbon electrode as linked to the unmodified bare electrode demonstrating that the DAN surfactant enhances the electrode's sensing capability for mercury ions. Various electrochemical methods, including cyclic voltammetry, square wave and differential pulse stripping voltammetry, were employed to optimize the experimental conditions in a systematic sequence of experiments.

Using square wave voltammetry, a calibration plot with high linearity up to 2 g/L, sensitivity of 0.164 AL/g and detection limit of 0.64 g/L in doubly deionized water at a buildup time of 360 sec was achieved under appropriate experimental circumstances. Additionally, samples of drinking and tap water were quantified using the suggested approach, with positive recovery and a relative variance of <3.5%. Computational investigations of the interaction of DAN and mercury were also carried out, the theoretical results correlating well with that of actual data (Zahid et al. 2016).

### 5.4.1.2   Organic Pollutants

Xu et al. designed an electrochemical system formulated on smartphone operation for mapping of nitrite contamination and rapid quantification comprising an improved screen-printed carbon electrode (SPCE), a smartphone and a handheld detector, as shown in Figure 5.2. An Android app based on Java was written to monitor the quantification, exhibit the outcomes instantaneously and transfer them to the virtual storage. As per convenience, the user could input the relevant experimental constraints in the app's graphical user interface and select the appropriate electrochemical procedures (cycle voltammetry, differential pulse voltammetry, or chronoamperometry). The modified SPCE would receive excitation signals from the hand-held detector, which will then transfer the response data to the smartphone over Bluetooth.

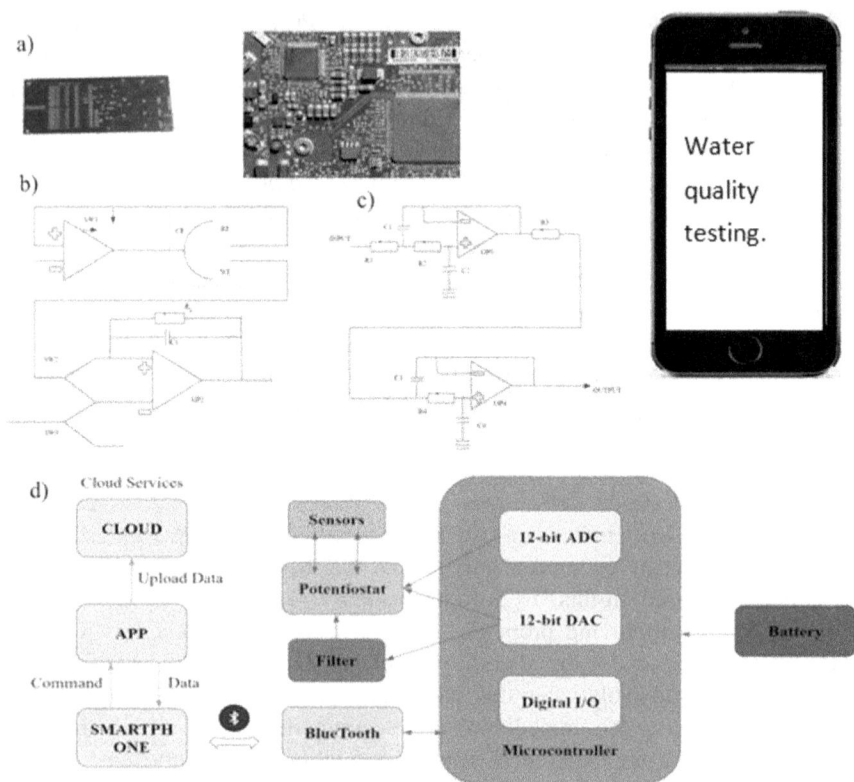

**FIGURE 5.2** a. The smartphone-based electrochemical system and the essential modules of the detector: i) microcontroller unit. ii) power management module. iii) electrode socket. iv) Bluetooth module. v) potentiostat module; b. the circuit design of the potentiostat; c. the circuit design of the hardware filter; d. an illustrative description of the smartphone-based electrochemical system. Modified from (Xu et al. 2020).

The detector included several critical elements, including a module for power management, one Bluetooth module, microcontroller unit, one potentiostat unit and an electrode socket. The detector's central processing unit interprets and executes the instructions from the smartphone, was an ARM STM32 microcontroller. An electrode port offered connection between the SPCE to the detector, whereas a Bluetooth module was used to link the smartphone and the detector. Two operational amplifiers that could sustain a constant voltage amid the reference and working electrodes were used in the design of the potentiostat unit.

Au nanoparticles (NP), NiO NP and rGO were electrochemically in-situ deposited on the SPCE in sequence utilizing cyclic voltammetry and chronoamperometry techniques by the system, which was used as a sensor for nitrite detection. The SPCE was first covered with 100 μL of a GO suspension containing 1 mg/mL. Through the use of the CV approach, electrochemical sequential reduction to rGO resulted in the production of rGO/SPCE. Second, using the CA technique, 100 μL of 5 mM Ni $(NO_3)^2$ was electrodeposited for 150 s at −1.1 V onto the rGO/SPCE. Third, until a steady-state current was produced, the Ni/rGO/SPCE obtained was transformed to NiO/rGO/SPCE using the CV approach. This was done in a solution of 0.05 M NaOH. Lastly, to make the Au/NiO/rGO/SPCE, the CA method was used and 100 L of 1 mM HAuCl4 was sprinkled onto the electrodes (deposition time: 300 s; deposition potential: −0.2 V).

By introducing 20 μM nitrite along with a variety of potential interfering chemicals, the selectivity of the smartphone-based system with Au/NiO/rGO/SPCE was investigated. A hundred-fold increase in $K^+$, $Na^+$, $NO_3^{3-}$, $Cl^-$, $Ca^{2+}$ and $SO_4^{2-}$ ions had just a low (5%) impact on nitrite detection. Investigations into the interference caused by electroactive substances revealed that a 100-fold concentration of ascorbic acid, $H_2O_2$, and glucose had a negligible (5%) impact on the nitrite response. Uric acid with 10-fold concentration only caused a small change in current, whereas a uric acid with 100-fold concentration caused a 7.40% change (4.64%). As a result, the system powered by smartphones showed great selectivity.

Using the electrochemical characterization, integrated techniques, quantitative detection, and electrode modification may all be carried out on the exact smartphone-based device. The practical usefulness of the smartphone-based system was demonstrated by evaluating the recovery of tap water after being diluted two times with 0.1 M PBS solution (pH 7.0). The measurements were completed using the standard addition process three times. The recoveries were found to be between 94.8–100.7% (Xu et al. 2020).

Zhu et al. synthesized a miniature electro-chemical detection platform with the use of Pt wire, an Ag/AgCl electrode built inside a pipette tip incorporating carboxylated multi-walled carbon nanotube/ionic liquid (MWCNTs-COOH/IL) altered graphite electrodes. Additional benefit was the pipette tip and MWCNTs-COOH/IL/PGE's low cost and disposability, which allowed for practical in-situ measurements while reducing time and preventing a loss in sensitivity brought on by the renewal process. Strong electrocatalytic activity for RhB was facilitated by the synergistic interaction of MWCNTs-COOH and IL. Based on an electrochemical device, differential pulse voltammetry (DPV) was used to recognize RhB quantitatively under the recommended ideal test circumstances. The oxidation peak current increased linearly with RhB concentration in the concentration ranges of 0.005–2.0 M and 2.0–60.0, and 1.0 nM was found to be the lowest detection limit. The MWCNTs-COOH/IL/PGE approach has a lower detection limit and a wider linear range than earlier electrochemical techniques, making it a more promising instrument for the sensitive measurement of RhB. The MWCNTs-COOH/IL/PGE approach offered a more promising instrument for the precise determination of RhB since it had a wider linear range and a lower detection limit than earlier electrochemical techniques.

RhB detection in fruit juice and water samples was used to examine the electrochemical system's usefulness. The suggested strategy's correctness was looked at utilizing the conventional addition technique. The recoveries ranged from 95.0% to 101.6%, proving the effectiveness of the suggested electrochemical method (Zhu et al. 2018).

Mohan et al. developed a three-electrode device built on droplet paper that can be used as a miniature platform for hydrazine electrochemical detection. The platform utilized a carbon ink counter electrode, an electrodeposited CuO cluster on a sheet of graphite as working electrode (Grp@CuO) with a 9 mm$^2$ (9 mm × 1 mm) and 3 mm$^2$ (3 mm × 1 mm) active surface area (reaction zone) and silver/silver chloride ink as a reference electrode. A CuO cluster being electro-deposited on the graphite sheet's surface was discovered by the physico-chemical characterization of the Grp@CuO.

In PBS at pH 7.0, Grp@CuO demonstrated selective electrocatalytic activity for hydrazine detection. At a potential range of 0.6–0.2 V against Ag/AgCl, a continuous rise in oxidation current that is linear is seen from 1 M to 7 M range, with good current sensitivity and a smaller detection range (0.3482 M). The built-in miniature droplet-based sensor was unaffected by additional chemical substances. Using the traditional addition process, actual water samples from lakes and taps were analyzed, and significant recovery values were obtained (Mohan et al. 2020). A compendium of detection limits and linear ranges of novel and commercial electrodes for common analytes are shown in Table 5.1. By selecting an internal electrolyte with low primary ion activity and avoiding its leak, detection limits can be reduced down to the picomolar level up to six orders of magnitude. Bromide cadmium, nitrate and fluoride as well as gases in solution such $CO_2$, $O_2$, $NO_2$ and $NH_4$ may all be assessed with ISE. Commercially accessible ISEs such as solid-state ISE, liquid membrane ISE and gas permeable ISE offer a simple and inexpensive solution to measure environmental parameters accurately. As a result, ISEs have lately been used to quantify trace metals and organic pollutants in relevant environmental matrices (Püntener et al. 2004).

## 5.4.2 AIR POLLUTANTS

Ozone ($O_3$) measurements are essential for the management of air quality and for most laboratory investigations of chemicals conducted at atmospheric conditions. Conventional ozone monitoring devices based on absorption of ultra violet (UV) are time-consuming, expensive and limited to fixed infrastructures by relatively high-power consumption. Pang et al. integrated a pair of portable $O_3$ measuring equipment with an "in-house" data acquisition system (LabJack and LabVIEW) using miniaturized electrochemical $O_3$ sensors (OXB421). The $O_3$ sensor was combined with a gas hood and supporting circuit board, and through its electrodes, the pre-configured circuit board outputs a low noise and high-resolution signal.

Separate studies were conducted to explore the effects of gas sample flow rate and relative humidity (RH) on sensor baseline, sensor sensitivity, and sensor gain. Simultaneous calibration curves show that sensor achievements were nearly equal even at diverse flow rates and RHs after resetting to counteract the sensor baseline shifts. Rapid RH fluctuations (20%/min) cause considerable and instantaneous variations in the sensor signal requiring up to 40 minutes to restore their previous values following such a swift RH shift. Sluggish RH variations (0.1%/min) on the other hand had minimal influence on sensor response. The flow rate of gas had a distinct inverse correlation with the sensor precision that decreased proportionally.

To assess the performance of the sensors, $O_3$ was quantitatively measured using the sensors in different laboratory tests and air quality measurement. By measuring ozone level at the entrance and exit of a reactor tank, in which $O_3$ interacted

**TABLE 5.1**
**Table Summarizing the Analytical Merit Figures and Electrodes for Common Analytes**

| Analyte | Electrode | LOD | Linear Range | Reference |
|---|---|---|---|---|
| **Inorganic Pollutants** | | | | |
| $Pb^{2+}$ and $Cd^{2+}$ | Solid-state Ag/AgCl reference electrodes and mercury working electrodes | $Pb^{2+}$: 1 μg/L, $Cd^{2+}$: 0.5 μg/L | – | (Yun et al. 2004) |
| | Solid-state commercial electrode | $Pb^{2+}$: 1,000 μg/L, $Cd^{2+}$: 200 μg/L | $Pb^{+2}$: $10^{-1} - 5*10^{-6}$ M, $Cd^{+2}$: $10^{-1} - 10^{-6}$ M | (Hanrahan et al. 2004) |
| $Pb^{2+}$, $Hg^{2+}$, $Cu^{2+}$ | Gold SPGE-LT | $Pb^{2+}$: 2.2 μg/L, $Hg^{2+}$: 1.3 μg/L, $Cu^{2+}$: 1.5 μg/L | 5–300 μg/L | (Bernalte et al. 2020) |
| $Cd^{2+}$ and $Pb^{2+}$ | Solid-state commercial electrode | $Pb^{2+}$: 1,000 μg/L | $Pb^{+2}$: $10^{-1} - 5*10^{-6}$ M | (Hanrahan et al. 2004) |
| | Carbon nanotube (CNT) composite electrode and reduced graphene oxide (rGO) | $Pb^{2+}$: 0.2 μg/L, $Cd^{2+}$: 0.6 μg/L | 20–200 μg/L | (Xuan & Park 2018) |
| | Solid-state commercial electrode | $Pb^{2+}$: 1,000 μg/L, $Cd^{2+}$: 200 μg/L | $Pb^{+2}$: $10^{-1} - 5*10^{-6}$ M, $Cd^{+2}$: $10^{-1} - 10^{-6}$ M | (Hanrahan et al. 2004) |
| $Hg^{2+}$ | Surfactant coated glassy carbon electrode | 0.64 μg/L | 2–22 μg/L | (Zahid et al. 2016) |
| **Organic pollutants** | | | | |
| Nitrite | Screen-printed carbon electrode | 0.2 μM | 1–500 μM | (Xu et al. 2020) |
| TNT | Gold electrodes | 0.37 μg/mL | 1–10 μg/mL | (Yu et al. 2017) |
| RhB | Ionic liquid modified pencil-graphite electrode and carboxylated multi-walled carbon nanotube | 1 nM | 0.005–60.0 μM | (Zhu et al. 2018) |
| Hydrazine | Grp@CuO electrode | 0.3482 μM | 1–7 μM | (Mohan et al. 2020) |

variedly with linoleic acid at the saltwater's microlayer surface, $O_3$ sensors were utilized in the lab to calculate the amount of ozone that was absorbed by seawater. To determine the $O_3$ absorption coefficient, the variation in $O_3$ loss by surface interaction with linoleic acid was believed to be the primary cause of the difference in $O_3$ composition varies in the closed reaction chamber across two places. On a sea microlayer surface, the volume of linoleic acid was discovered to determine how much ozone saltwater absorbed, and the projected absorption coefficients grounded on sensor data were found to be reasonably close to those from earlier studies.

An ozone sensor was used to measure surrounding ozone over the course of 18 days in order to assess the quality of the air. A medium-steel diaphragm metal bellows pump with a flow rate of 1 L/min was used to introduce ambient air from the primary sample inlet into the exhaust hood of the ozone sensor. The flow of air sample was concurrently carried through a UV photometric $O_3$ analyzer for reference. During the 18-day air standard measurement period, the modified values from the $O_3$ sensor agreed well with those acquired by the reference UV $O_3$ analyzer (Pang et al. 2017).

Zhao et al. developed a fluidic chip based on inertial impactions that incorporates electrochemical detection as shown in Figure 5.3 to achieve excellent collection efficiency and nanogram-level measurements of the bio accessible metal fraction. Aerosol copper dissolution, detection and collection rely on the microchannel's liquid/solid and air/solid interfaces and do not require a large setup; detection ranges impacted by collection time and airflow rate. The microsystem had a working flow rate of 3.1 L/min and a collection efficiency of 70%, and it was able to identify Cu concentrations above published atmospheric values of 53 ng/m$^3$, 32 ng/m$^3$ and 8 ng/m$^3$ during 3 h, 5 h and 20 h collecting periods, respectively.

Several other metals, including Fe, Ni, Pb, Cd and Cu may be collected and determined using the present or altered microsystem. Furthermore, the microsystem's

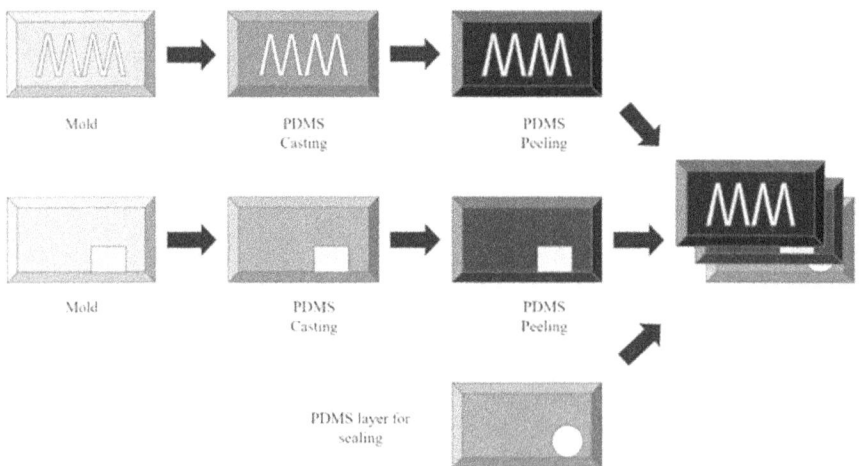

**FIGURE 5.3** Illustration of the composite aerodynamic and electrochemical microsystem for aerosol collection and electrochemical detection. Modified from (Zhao et al. 2022).

applicability may be appropriately expanded to collect and detect other elements for instance aerosol oxidative load and nitrate, providing a novel technique to construct an internet based, transportable, inexpensive and miniaturized measurement platform for various vapor-based elements (Zhao et al. 2022).

Cross et al. assessed the results of ARISense, a newly designed, integrated lower-cost electrochemical sensor system for simultaneous, real-time monitoring of a broad range of ambient-level air contaminants and associated meteorological variables. The ARISense system detects five gaseous pollutants (NO, $O_3$, $SO_2$, $NO_2$ and CO), atmospheric aerosol constituent part (diameter: 0.4–17 µm), and related environmental and meteorological factors. Data were collected using the ARISense integrated sensor package over a 4.5-month period, where the sensor system was collocated with a state-operated air quality monitoring station outfitted with reference instrumentation detecting the same pollutant species and validated electrochemical sensor readings of $NO_2$, CO, $O_3$ and NO at an urban neighborhood site with ranges of pollutant concentration (parts per billion by volume): [$NO_2$] = 11.7 ± 8.3 ppb, [CO] = 231 ± 116 ppb, [$O_3$] = 23.2 ± 12.5 ppb and [NO] = 6.1 ± 11.5 ppb (Cross et al. 2017).

### 5.4.3 MICROPLASTIC POLLUTANTS

Microplastics (MPs) are miniscule pieces of plastic (less than 5 mm in length) that contaminate the environment and are far more pervasive than plastics due to their persistent nature and non-biodegradability and are continually being introduced into the surroundings through multiple pathways (Veerasingam et al. 2020). Novel solutions are required to segregate and minimize their quantity in various environmental sectors, and several methods have been proposed and developed to screen and estimate microplastics.

Krishna et al. conceptualized a microfluidic device that in can sub-micron resolution divide any two varied sample of micron sized particles into two consistent samples. The apparatus is divided into two sections: the downstream section divides the concentrated microparticle stream into two samples, separated by size, while the upstream section concentrates on microparticles. With each electrode on top aligning with an electrode at the bottom to form a unit, each segment has a number of finite-sized planar electrodes that extend into the micron-sized channel from the bottom and top of each sidewall. This results in numerous units of electrodes on either side.

All of the microparticles are focused near one of the sides by the focusing component, which uses averting dielectrophoretic force from each electrode pair. To accomplish the necessary separation, this segment uses repulsive dielectrophoretic force to push the large microparticles inside the microchannel farther from the wall whilst minuscule particles are in motion unimpeded. The operating frequency of the separation section's set of electrodes is kept constant at the cut-off frequency of the microscopic micron sized particles.

The device's practical functionality is proved by distinguishing a varied combination of microparticles of polystyrene of varying sizes (radius: 2.25 and 2 µm) into two homogenous samples. The mathematical prototype is utilized for parametric analysis, and the achievement is measured with respect to the separation efficiency and purity; the factors taken into account include number of electrodes, electrode diameters, outlet widths, volumetric flow rate and applied electric voltages. For low volumetric flow rates, multiple electrode pairs, big electrode diameters, substantial voltage differences between sections, the purities and separation efficiencies for both microparticles are 100% (Krishna et al. 2020).

## 5.5   FUTURE PERSPECTIVES

The necessity for effective methods to rehabilitate ecosystems sparked interest in microfabricated electrochemical technology research and development for environmental applications. A thorough connectivity between the different analytical procedures, the assessment of user and environmental risk caused by the different chemicals and reagents utilized, temperature and voltage settings, emitted odors or noise levels must be appropriately balanced to deliver the best analytical characteristics while producing little or no user/environmental side effects. While microfluidics have traditionally been employed for analysis and detection, droplet-based microfluidics have lately focused on environmental remediation via the creation of functional microparticles. The most significant barrier to using microfluidic systems in effective water and air pollution treatment remains the technical and economic barriers to industrializing micro-systems or micro-technologies so they can be used or retrofitted into presently existing technologies on an industrial scale. Microfabricated device technology is, thus, still in its infancy and represents an emerging area of technological growth with tremendous promise, notably in the field of environmental care. The successful development of microwave coupling approaches to analytical facilities has significantly helped attempts to overcome on-chip detection difficulties, for instance low selectivity and sensitivity, allowing for the automation of integrated analytical systems (Yew et al. 2019). Regardless, there have been multiple incidents of commercialization of downsized analytical procedures, as evidenced by the number of patents published (Agrawal et al. 2021). Diebold et al. patented an electrochemical sensor capable of exact analyte concentration determination and a process for producing high-resolution, biocompatible electrodes for use in electrochemical sensors (Diebold et al. 1995). A patent regarding a compact electrochemical sensor system for field testing for metals was filed by Lin et al. (Lin et al. 2004). To detect at least one of hydrogen peroxide, potassium ferrocyanide, TNT or DNT in liquid or vapor phase, a CNT-based chemical sensor was patented. Underwater presence of substances such as heavy metals and explosives can be detected using textile-based sensors (Wang & Windmiller 2017). The majority of commercially available electrochemical sensors are now constrained to a "one design, one application" model, that necessitates additional customization effort to improve the functionality for different applications.

## 5.6   CONCLUSION

The addition of dependable and powerful electrical devices for pollution control or efficient process is enabled by the amalgamation of contemporary electrochemical practices by means of discoveries in miniaturization and microelectronics. Electrical sensors enable for remote deployment and near-real-time monitoring, as well as highly sensitive and selective detection. Nonetheless, electrochemical devices are constantly being offered based on cutting-edge technology and ongoing work in the fields of multiparameter sensor arrays, remote electrodes and "smart" sensors and molecular devices. "Smart" devices can do a variety of tasks "changing" between "screening/warning" and operate on "detailed" mode of analysis for diverse environmental conditions. Electrochemical research is moving forward in diverse directions which includes DNA-based miniaturized electrochemical sensor, paper-based electrochemical sensor, microfluidics-based electrochemical sensor, miniaturized capillary electrophoresis-based sensor and screen electrode-based microchip for various applications. Thus, electrochemical sensors are evolving based on newly emerging approaches that include active and passive label-free focusing, sorting and detection methods with enhanced selectivity and sensors that can determine multiple analytes at once based on application. The ability to work in varied environmental matrices will remain crucial in both of these fields, forcing researchers to overcome difficulties of specificity and stability. For example, incorporating electrochemical technology into microfluidic platforms would make in-situ and on-site environmental measurement operations even easier. Furthermore, developments in network communication and commercialization play a key role in extensive initiatives to meet today's environmental inspection needs.

## REFERENCES

Abu-Ali, Hisham, Alexei Nabok, and Thomas J. Smith. "Electrochemical inhibition bacterial sensor array for detection of water pollutants: Artificial neural network (ANN) approach." *Analytical and Bioanalytical Chemistry* 411, no. 29 (2019): 7659–7668.

Agrawal, Arpana, Rüstem Keçili, Fatemeh Ghorbani-Bidkorbeh, and Chaudhery Mustansar Hussain. "Green miniaturized technologies in analytical and bioanalytical chemistry." *TrAC Trends in Analytical Chemistry* 143 (2021): 116383.

Anshori, Isa, Ghani Faliq Mufiddin, Iqbal Fawwaz Ramadhan, Eduardus Ariasena, Suksmandhira Harimurti, Henke Yunkins, and Cepi Kurniawan. "Design of smartphone-controlled low-cost potentiostat for cyclic voltammetry analysis based on ESP32 microcontroller." *Sensing and Bio-Sensing Research* 36 (2022): 100490.

Bernalte, Elena, Sebastián Arévalo, Jaime Pérez-Taborda, Jannis Wenk, Pedro Estrela, Alba Avila, and Mirella Di Lorenzo. "Rapid and on-site simultaneous electrochemical detection of copper, lead and mercury in the Amazon river." *Sensors and Actuators B: Chemical* 307 (2020): 127620.

Bourgeois, Wilfrid, Anne-Claude Romain, Jacques Nicolas, and Richard M. Stuetz. "The use of sensor arrays for environmental monitoring: Interests and limitations." *Journal of Environmental Monitoring* 5, no. 6 (2003): 852–860.

Brett, Christopher MA. "Electrochemical sensors for environmental monitoring. Strategy and examples." *Pure and Applied Chemistry* 73, no. 12 (2001): 1969–1977.

Cross, Eben S., Leah R. Williams, David K. Lewis, Gregory R. Magoon, Timothy B. Onasch, Michael L. Kaminsky, Douglas R. Worsnop, and John T. Jayne. "Use of electrochemical sensors for measurement of air pollution: Correcting interference response and validating measurements." *Atmospheric Measurement Techniques* 10, no. 9 (2017): 3575–3588.

Das, Abhranila, Surajit Bose, Naresh Mandal, Bidhan Pramanick, and Chirasree RoyChaudhuri. "HOME-Stat: A handheld potentiostat with open-access mobile-interface and extended measurement ranges." *Proceedings of the Indian National Science Academy* 87, no. 1 (2021): 84–93.

Denchak, Melissa. "Water pollution: Everything you need to know." *Nat. Resour. Def. Counc. NY* (2018).

Diebold, Eric R., Richard J. Kordal, Nigel A. Surridge and Christopher D. Wilsey. "Electrochemical sensor." U.S. Patent 5,437,999, issued August 1, 1995.

Guo, Chun Xian, Zhi Song Lu, Yu Lei, and Chang Ming Li. "Ionic liquid–graphene composite for ultratrace explosive trinitrotoluene detection." *Electrochemistry Communications* 12, no. 9 (2010): 1237–1240.

Hanrahan, Grady, Deepa G. Patil, and Joseph Wang. "Electrochemical sensors for environmental monitoring: Design, development and applications." *Journal of Environmental Monitoring* 6, no. 8 (2004): 657–664.

Khairy, Mohamed, Rashid O. Kadara, and Craig E. Banks. "Electroanalytical sensing of nitrite at shallow recessed screen-printed microelectrode arrays." *Analytical Methods* 2, no. 7 (2010): 851–854.

Krishna, Salini, Fadi Alnaimat, Ali Hilal-Alnaqbi, Saud Khashan, and Bobby Mathew. "Dielectrophoretic microfluidic device for separating microparticles based on size with sub-micron resolution." *Micromachines* 11, no. 7 (2020): 653.

Lillehoj, Peter B., Ming-Chun Huang, Newton Truong, and Chih-Ming Ho. "Rapid electrochemical detection on a mobile phone." *Lab on a Chip* 13, no. 15 (2013): 2950–2955.

Lin, Yuehe, Wendy D. Bennett, Charles Timchalk, and Karla D. Thrall. "Compact electrochemical sensor system and method for field testing for metals in saliva or other fluids." U.S. Patent 6,699,384, issued March 2, 2004.

Mettakoonpitak, Jaruwan, Katherine Boehle, Siriwan Nantaphol, Prinjaporn Teengam, Jaclyn A. Adkins, Monpichar Srisa-Art, and Charles S. Henry. "Electrochemistry on paper-based analytical devices: A review." *Electroanalysis* 28, no. 7 (2016): 1420–1436.

Mohan, Jaligam Murali, Khairunnisa Amreen, Arshad Javed, Satish Kumar Dubey, and Sanket Goel. "Modified graphite paper based miniaturized electrochemically optimized hydrazine sensing platform." *ECS Journal of Solid State Science and Technology* 9, no. 11 (2020): 115001.

Nemiwal, Meena, and Dinesh Kumar. "Recent progress on electrochemical sensing strategies as comprehensive point-care method." *Monatshefte für Chemie-Chemical Monthly* 152, no. 1 (2021): 1–18.

Noviana, Eka, Cynthia P. McCord, Kaylee M. Clark, Ilhoon Jang, and Charles S. Henry. "Electrochemical paper-based devices: Sensing approaches and progress toward practical applications." *Lab on a Chip* 20, no. 1 (2020): 9–34.

Pang, Xiaobing, Marvin D. Shaw, Alastair C. Lewis, Lucy J. Carpenter, and Tanya Batchellier. "Electrochemical ozone sensors: A miniaturised alternative for ozone measurements in laboratory experiments and air-quality monitoring." *Sensors and Actuators B: Chemical* 240 (2017): 829–837.

Püntener, Martin, Tamás Vigassy, Ellen Baier, Alan Ceresa, and Ernö Pretsch. "Improving the lower detection limit of potentiometric sensors by covalently binding the ionophore to a polymer backbone." *Analytica chimica acta* 503, no. 2 (2004): 187–194.

Rodriguez-Mozaz, Sara, M-P. Marco, MJ Lopez De Alda, and Damià Barceló. "Biosensors for environmental applications: Future development trends." *Pure and Applied Chemistry* 76, no. 4 (2004): 723–752.

Tercier-Waeber, Mary-Lou, Fabio Confalonieri, Giuliano Riccardi, Antonio Sina, Stéphane Nöel, Jacques Buffle, and Flavio Graziottin. "Multi Physical–Chemical profiler for real-time in situ monitoring of trace metal speciation and master variables: Development, validation and field applications." *Marine Chemistry* 97, no. 3–4 (2005): 216–235.

Tetyana, Phumlani, Poslet Morgan Shumbula, and Zikhona Njengele-Tetyana. "Biosensors: Design, development and applications." In *Nanopores*. IntechOpen. (2021).

Tothill, Ibtisam E. "Biosensors developments and potential applications in the agricultural diagnosis sector." *Computers and Electronics in Agriculture* 30, no. 1–3 (2001): 205–218.

Veerasingam, S., M. Ranjani, R. Venkatachalapathy, Andrei Bagaev, Vladimir Mukhanov, Daria Litvinyuk, Liudmila Verzhevskaia, L. Guganathan, and P. Vethamony. "Microplastics in different environmental compartments in India: Analytical methods, distribution, associated contaminants and research needs." *TrAC Trends in Analytical Chemistry* 133 (2020): 116071.

Wang, Joseph, and Joshua Ray Windmiller. "Textile-based printable electrodes for electrochemical sensing." U.S. Patent 9,844,339, issued December 19, 2017.

Xu, Ke, Qiulin Chen, Yuanyuan Zhao, Chengjun Ge, Shiwei Lin, and Jianjun Liao. "Cost-effective, wireless, and portable smartphone-based electrochemical system for on-site monitoring and spatial mapping of the nitrite contamination in water." *Sensors and Actuators B: Chemical* 319 (2020): 128221.

Xuan, Xing, and Jae Y. Park. "A miniaturized and flexible cadmium and lead ion detection sensor based on micro-patterned reduced graphene oxide/carbon nanotube/bismuth composite electrodes." *Sensors and Actuators B: Chemical* 255 (2018): 1220–1227.

Yamada, Kentaro, Hiroyuki Shibata, Koji Suzuki, and Daniel Citterio. "Toward practical application of paper-based microfluidics for medical diagnostics: State-of-the-art and challenges." *Lab on a Chip* 17, no. 7 (2017): 1206–1249.

Yew, Maxine, Yong Ren, Kai Seng Koh, Chenggong Sun, and Colin Snape. "A review of state-of-the-art microfluidic technologies for environmental applications: Detection and remediation." *Global Challenges* 3, no. 1 (2019): 1800060.

Yu, Holly A., Junqiao Lee, Simon W. Lewis, and Debbie S. Silvester. "Detection of 2, 4, 6-trinitrotoluene using a miniaturized, disposable electrochemical sensor with an ionic liquid gel-polymer electrolyte film." *Analytical Chemistry* 89, no. 8 (2017): 4729–4736.

Yun, Kwang-Seok, Joonho Gil, Jinbong Kim, Hong-Jeong Kim, Kyunghyun Kim, Daesik Park, Myeung su Kim et al. "A miniaturized low-power wireless remote environmental monitoring system based on electrochemical analysis." *Sensors and Actuators B: Chemical* 102, no. 1 (2004): 27–34.

Zahid, Anum, Aref Lashin, Usman Ali Rana, Nassir Al-Arifi, Imdad Ullah, Dionysios D. Dionysiou, Rumana Qureshi, Amir Waseem, Heinz-Bernhard Kraatz, and Afzal Shah. "Development of surfactant based electrochemical sensor for the trace level detection of mercury." *Electrochimica Acta* 190 (2016): 1007–1014.

Zhao, Yi-Bo, Jiukai Tang, Tianyu Cen, Guangyu Qiu, Weidong He, Fuze Jiang, Ranxue Yu, Christian Ludwig, and Jing Wang. "Integrated aerodynamic/electrochemical microsystem for collection and detection of nanogram-level airborne bioaccessible metals." *Sensors and Actuators B: Chemical* 351 (2022): 130903.

Zhu, Xiaolin, Guanlan Wu, Chengzhi Wang, Dongmei Zhang, and Xing Yuan. "A miniature and low-cost electrochemical system for sensitive determination of rhodamine B." *Measurement* 120 (2018): 206–212.

Zhuiykov, Serge, Eugene Kats, and Donavan Marney. "Potentiometric sensor using submicron Cu2O-doped RuO2 sensing electrode with improved antifouling resistance." *Talanta* 82, no. 2 (2010): 502–507.

# 6 Flexible and Wearable Sensors for Health Monitoring Applications

*Navneet Gupta and Reva Teotia*

## CONTENTS

## 6.1 INTRODUCTION

Flexible electronics have a huge opportunity in the field of health care. With the advancement of Internet of Things (IoT), wearable devices for real-time health monitoring are growing at a rapid pace. The development of affordable and accurate health monitoring sensors can help patients in remote places and provide diagnosis and treatment without going to the hospital. The main advantage of flexible electronics is its ability to adjust onto a random curved surface like human skin and organs, keeping almost similar operational characteristics. This motivates developing wearable devices for non-invasive real-time human health monitoring [1,2].

Figure 6.1 shows the different types of flexible and wearable sensors that can be used to measure important parameters required to monitor human health. Using these sensors attached to the human skin, it is possible to monitor blood pressure [3], temperature [4,5], pulse and heart rate [6–8], and motion [2]. Recent developments in pressure sensors for blood pressure measurement show the possibility of a miniaturized real-time monitoring system for hypertension patients [9,10]. Non-implantable, as well as implantable, wearable sensors can also be used to study electrophysiological signals that

DOI: 10.1201/b23359-6

**93**

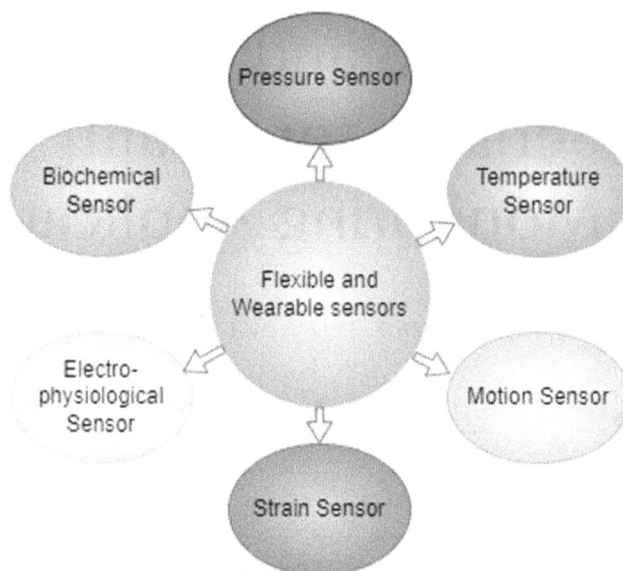

**FIGURE 6.1**  Various types of flexible and wearable sensors.

include electroencephalogram (EEG), electrocardiogram (ECG), and electromyogram (EMG), which help in providing more relevant information about a person's health [11].

Along with continuous monitoring of physical indicators, it is important to monitor molecular data from the body to have detailed health information. Non-invasive methods to acquire molecular biomarkers are through body fluids like sweat [12–14], tears [15], and saliva [16], etc.

Thus, flexible sensors are the potential devices for biomedical applications. In subsequent sections, we shall discuss the materials, structure, mechanism, and manufacturing processes of different flexible sensors with a special focus on temperature, pressure, and strain sensors.

## 6.2   DEVICE MATERIALS AND STRUCTURES

Flexible sensors mainly consist of two important layers: flexible substrate and an active material. This combination forms film-on-substrate structure. Substrate provide flexibility to the device and also decides its reliability of the device. In flexible devices, three types of substrates are mainly used: metal foils, glass sheets, and plastics (polymers).

The material chosen for a flexible substrate must possess the following characteristics:

  i. The substrate material must possess considerable low flexural rigidity.
 ii. It should be able to withstand high temperatures, must be lightweight, and easy to manufacture and handle.
iii. The substrate material should have very low surface roughness and also must be resistant to chemical reactions.

iv. In order to avoid internal stress, the substrate material must possess a low coefficient of thermal expansion (CTE). At the same time, the CTE of the film and the substrate should not be much different.

v. Low cost of these substrates is also required to scale up the device manufacturing.

Commonly used metal foils include steel, copper, aluminum, and kovar (nickel-cobalt-iron alloy). The major issue with metal foils is their surface roughness, which degrades the performance of the device. However, various techniques are available for reducing the surface roughness. One such technique is chemical mechanical polishing (CMP).

The most commonly used glass sheets include soda-lime glass and borosilicate glass. Glass substrates possess higher surface energy compared to metal foils and plastic substrates. Surface energy is the property that determines how well the thin film of other materials can be deposited and stick to it. The major drawback with glass substrates is their flexural rigidity.

Commonly used plastic substrates are poly(ethylene terephthalate) (PET), poly (ethylene napthalate) (PEN), Kapton (Polyimide), and polydimethylsiloxane (PDMS).

The choice of the substrate may vary depending on the specific application as well as the transduction method. The mechanical, electrical, thermal, and optical qualities of the substrate material determine its suitability for a given application. The material selection problem can be addressed using a well-established multi-criteria decision making (MCDM) framework [17–19].

## 6.3 TRANSDUCTION MECHANISMS

The main transduction mechanisms are capacitive, piezoresistive, and piezoelectric. These are discussed in the following section.

### 6.3.1 CAPACITANCE

The external stimuli cause a change in the capacitance, thereby changing the output response. The capacitance change is dictated by the equation of the parallel plate capacitor.

$$C = \varepsilon_0 \varepsilon_r A / d \qquad (6.1)$$

where $\varepsilon_r$ is the relative permittivity, $A$ is the area, and $d$ is the gap between plates. The change in capacitance can be due to the change in dielectric properties, plate area, or change in the plate gap as governed by the equation of parallel plate capacitance.

### 6.3.2 PIEZORESISTANCE

Piezoresistance is the change in the resistivity of a solid body due to a change in structure caused by an applied force or mechanical stress. The change in the

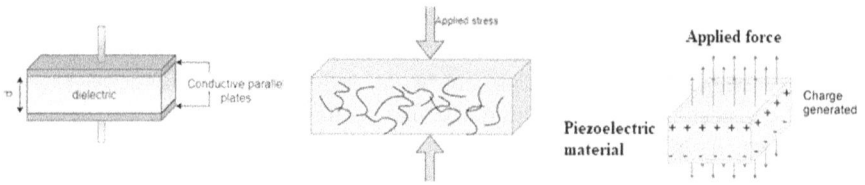

**FIGURE 6.2** Transduction mechanism: (a) capacitive, (b) piezoresistive, (c) piezoelectric.

structure causes a change in the electronic structure of the material, usually crystalline materials. The piezoresistive coefficient is given by:

$$P = (\Delta\rho/\rho)/\varepsilon \tag{6.2}$$

where $\Delta\rho$ is the change in resistivity, $\rho$ is original resistivity, and $\varepsilon$ is the strain.

### 6.3.3 PIEZOELECTRICITY

Piezoelectricity is the property of a certain material to generate an electric current in the presence of applied stress and vice versa. The direct piezoelectric effect is the generation of current due to applied force and the converse piezoelectric effect is the generation of stress or deformation due to applied electric current. Figure 6.2 illustrates the structure and corresponding transduction mechanism.

## 6.4  FLEXIBLE SENSORS

### 6.4.1 TEMPERATURE SENSORS

A temperature sensor is one of the important sensors in wearable devices for human health monitoring. The human body has a specific range of body temperature i.e 36.1°C to 37.2°C. The condition of elevated temperature, called hyperthermia, indicates the state of inflammation or fever, and lower body temperature called hypothermia indicates reduced blood flow or even organ failure. Continuous temperature monitoring is required by the doctors to monitor a patient's health condition. Temperature is closely related to inflammation and infection state of the wound; therefore, a smart wound dressing with temperature sensors is used to monitor the healing process of the wound [20].

The traditional methods of temperature sensors are thermistors and thermocouples. Thermistors work on the principle of change in resistance due to temperature change. This change in resistance is converted into an electrical signal. The temperature coefficient of resistance (TCR) is one of the main considerations for thermistors. TCR is defined as a relative change of the resistance per degree change in temperature. TCR and the sensitivity (S) is given by

$$TCR = (R_b - R_a)/R_a\Delta T \tag{6.3}$$

$$S = \Delta R/R_a \qquad (6.4)$$

where $R_a$ is initial resistance, $R_b$ is final resistance, and $\Delta T$ is the temperature change

Typical TCR values for metals fall in the range of 0.003–0.006 $°C^{-1}$, where higher values correspond to higher sensitivity. The most commonly used temperature-sensitive materials are metals like nickel (Ni), copper (Cu), gold (Au), platinum (Pt), etc., with Ni having the highest TCR. CNT combined with PEDOT:PSS exhibits TCR comparable to that of metals [21].

Conductive composites have been implemented in temperature sensing applications because conductive nanoparticles impart conductivity and the polymer matrix provides mechanical stability [22]. Temperature-sensitive conductive composites are formed by passing conductive fillers through a soft polymeric matrix. The active materials used as conductive fillers are carbon nanofibers, carbon black, carbon nanotubes (CNTs), single-walled carbon nanotubes (SWCNTs), multi-walled carbon nanotubes (MWCNTs) and CNT/PDMS composites [22–24].

One of the commonly used technique to sense temperature (change in temperature) is to monitor the electrical resistance (change in resistance) of a material. These types of temperature sensors are known as resistance temperature detectors (RTDs). Conventional RTDs are made up of platinum, copper, or nickel.

Flexible RTDs are made by depositing conductive metals or carbon derivatives like CNT, graphene, quantum dots, etc. on flexible substrates through different fabrication/printing processes. Ni-based flexible RTD [25] with sensitivity ~ 3.52 × $10^{-3}°C^{-1}$ is fabricated by laser digital printing (LDP), a non-photolithographic and cost-effective technique [26]. It was observed that up to a radius of curvature of 1.75 mm, the Ni-based RTDs possess superior mechanical and electrical stability without much change in the resistance.

Inkjet-printed silver interdigitated electrodes on Kapton (polyimide) is developed by Ali et al. [27]. The resulting device had good sensitivity of 0.00375 $°C^{-1}$ at a temperature range of 28 to 50°C, with a recovery time of 8.5 sec and a response time of 4 sec. An inkjet-printed sensor on Kapton with silver nanocomposites had a temperature range of 20–60°C and sensitivity was 2.23 × $10^{-3}°C^{-1}$[28] PEDOT:PSS[29]. Other fabrication techniques include sputtering [30], a spin-coating method, and lamination techniques [31–33].

## 6.4.2 Pressure Sensors

Pressure sensors are used to detect various body pressure like systolic pressure, pulse rate, and also the pressure generated on the skin to help us determine the contact with the external environment [34–36]. According to the method of transduction, flexible pressure sensors are divided into capacitive, piezoelectric, and piezoresistive pressure sensors [37].

The material consideration for capacitive pressure sensors can be divided into dielectric material and electrode material. The dielectric material should be deformable; therefore, microporous and microstructured PDMS is highly studied for the application. With improvement in capacitor sensitivity by modification of dielectric material, researchers are also working on the development of better electrodes.

Au, Ag, AgNW, conductive fabric, graphene, etc. are commonly used as electrodes. Ceramic dielectric barium-titanate is introduced into the dielectric layer to increase sensitivity and the dielectric constant of the pressure sensor in the work by Longquan et al. [38]. A flexible and wearable PDMS-DI water dielectric-based capacitive pressure sensor was fabricated and characterized for blood pressure measurement [39].

Piezoresistive pressure sensors work on the principle of change of resistance of the material due to deformation caused by applied pressure. Carbon-based materials like carbon nanotubes (CNTs), graphene, carbon black, polymers with metal nanowires, and metal nanoparticles have piezoresistive properties. Xu T et al. [40] have used MWCNT and PDMS composite networks to create a pressure sensor to detect 55.7 Pa of pressure.

Piezoelectric pressure sensors are based on the principle of conversion of mechanical energy into electrical energy. These sensors unlike capacitive and piezoresistive sensors can operate without an external power source. Commonly used piezoelectric materials are zinc oxide (ZnO), lead zirconium titanate (PZT), barium titanate (BaTiO$_3$), and polyvinylidene difluoro-trifluoroethylene (P(VDF-TrFe)). Chen et al. [41] presented a nanowire/graphene heterostructure for static pressure sensor with a sensitivity up to $9.4 \times 10^{-3}$kPa$^{-1}$.

### 6.4.3 STRAIN SENSORS

Strain is the relative change in the length of a structure under stress. Strain sensors convert mechanical stimuli into optical or electrical signals. Conventional strain gauges have been used in industrial applications for the measurement of strain in buildings, dams, and other architectural objects. With the advancement of soft sensing and wearable devices, flexible strain sensors gained importance. The commonly used transduction methods are piezoresistive, capacitive, and piezoelectric, each of which has been already discussed in section 6.3 of this chapter.

*Piezoresistive strain sensors* can detect compressive or tensile strain. The sensitivity for compressive strain is given by

$$S = (\Delta E/E_0)/\Delta P \qquad (6.5)$$

where $\Delta E$ represents the change in the electrical signals. The sensitivity for the tensile strain is given by the gauge factor (GF):

$$GF = (\Delta R/R_0)/\varepsilon, \qquad (6.6)$$

that is, the fractional change in the resistance, with $\varepsilon$ being the strain.

The piezoresistive strain sensors are low cost, with excellent sensing properties and high linearity. The substrate is implemented with microstructures like microdomes, micropillars, or micropyramids to increase the sensitivity of the sensor. Apart from this, paper- or textile-based substrates are also used for the implementation of the strain sensor.

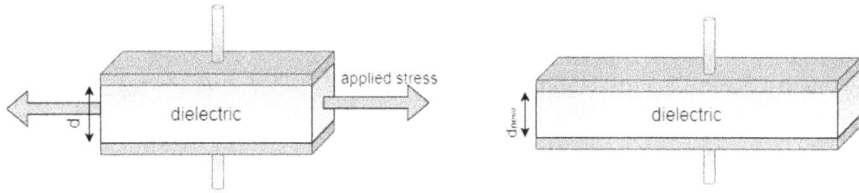

**FIGURE 6.3**   Change in plate area and gap due to applied stress.

The active materials generally used are carbon based like carbon black, which is a 0-D material, with a high surface-to-volume ratio and is conductive. Carbon nanotubes (CNTs) are also extensively studied for applications in strain sensors since they possess high flexibility and electrical conductivity. Metal nanowires can be sandwiched between flexible substrates like PDMS and can also be integrated with fibers and textiles. This results in a highly conductive active flexible material that can be easily incorporated into wearable devices.

In capacitance strain sensors, capacitance change can occur due to a change in the dielectric constant or the physical dimensions that are induced due to external stress. The gauge factor is

$$GF = (\Delta C/C_0)/\varepsilon \qquad (6.7)$$

where epsilon is the strain. The deformation due to applied stress can occur in the vertical and horizontal directions, changing the plate gap and the area, respectively, as illustrated in Figure 6.3. The capacitive sensors have low power input energy, low noise impact, and high dynamic response.

The electrodes in the capacitive sensor are high conductivity materials like metals, conductive polymers, or carbon-based materials. These materials are deposited on flexible substrates like PDMS using methods like spin coating, thermal evaporation, or sputtering. The active material in the capacitive sensor is the dielectric material. The dielectric material has to be both compressible and flexible. Elastomers are a great option for this application. These are polymers with high viscoelasticity and low Young's modulus like PDMS. Besides the material, the structural design also greatly affects the sensor's performance. Microstructures introduce air gaps that reduce viscoelastic damping and thereby reduce the response time. Also, the air gaps help in increasing the sensitivity by leading to enhanced variation in the physical dimensions.

### 6.4.4   Physiological Biochemical Sensors

These sensors are very important in understanding various aspects of human health. The biochemical sign includes blood glucose and body fluids (sweat, saliva, and tears) [42–44]. Flexible biochemical sensors work on the basis of chemical methods for detecting the composition and amount of a biological substance. The working principle of these sensors includes the change in electrical properties of the sensor due to chemical reaction between the active material and the target detection substance. By analyzing these electrical parameters, one can get the information about the physiological health of a human being.

For a diabetic person, it is important to continuously monitor the glucose level. The presently invasive technique is used to measure the glucose level that requires blood samples received using needles that lead to pain in the patient. On the other hand, flexible sensors provide a non-invasive method of measuring the glucose level.

Using electronic skin (e-skin) biosensor blood glucose monitoring is done by measuring electrochemical fluid from the blood vessel available on the skin surface and also by using sweat analysis [45,46].

## 6.5  CONCLUSION

This chapter explain the use of flexible and wearable sensors for human health montoring. This technique is non-invasive, low cost, and also provides continuos real-time monitoring of human health. Mainly four types of flexible sensors are used for this purpose-temperature sensor, pressure sensor, strain sensor, and physiological biochemical sensor. All of these sensors work on the basis of some mechanism. The three mechanisms are involved-capacitive, piezoresistive, and piezoelectric.

## REFERENCES

[1] K. Sakuma, and K. Iniewski, *Flexible, wearable, and Stretchable Electronics*. CRC Press, an imprint of Taylor & Francis Group, 2021.
[2] S. Z. Homayounfar, and T. L. Andrew, "Wearable sensors for monitoring human motion: A review on mechanisms, materials, and challenges," *Slas Technology: Translating Life Sciences Innovation*, vol. 25, pp. 9–24, Dec. 2019.
[3] Bijender, and A. Kumar, "Flexible and wearable capacitive pressure sensor for blood pressure monitoring," *Sensing and Bio-Sensing Research*, vol. 33, p. 100434, 2021.
[4] Y. Su et al., "Printable, highly sensitive flexible temperature sensors for human body temperature monitoring: A review," *Nanoscale Research Letters*, vol. 15, pp. 1–34. Oct. 2020.
[5] Y. Yu et al., "Wearable temperature sensors with enhanced sensitivity by engineering microcrack morphology in pedot:pss–pdms sensors," *ACS Applied Materials & Interfaces*, vol. 12, no. 32, pp. 36578–36588, 2020. PMID: 32667193.
[6] T. Yamaguchi et al., "Wrist flexible heart pulse sensor integrated with a soft pump and a pneumatic balloon membrane," *RSC Advances*, vol. 10, no. 29, pp. 17353–17358, 2020. 4.
[7] T. Sekine et al., "Fully printed wearable vital sensor for human pulse rate monitoring using ferroelectric polymer," *Scientific Reports*, vol. 8, pp. 1–10. Mar. 2018.
[8] Y. H. Kwak et al., "Flexible heartbeat sensor for wearable device," *Biosensors and Bioelectronics*, vol. 94, pp. 250–255, Aug. 2017.
[9] X. Wang, J. Yu, Y. Cui, and W. Li, "Research progress of flexible wearable pressure sensors," *Sensors and Actuators A: Physical*, vol. 330, p. 112838, 2021.
[10] K. Meng et al., "Flexible weaving constructed self-powered pressure sensor enabling continuous diagnosis of cardiovascular disease and measurement of cuffless blood pressure," *Advanced Functional Materials*, p. 1806388, Dec. 2018.
[11] J. Liu et al., "Recent progress in flexible wearable sensors for vital sign monitoring," *Sensors*, vol. 20, no. 14, p. 20, 2020.

[12] H. Lee et al., "A graphene-based electrochemical device with thermoresponsive microneedles for diabetes monitoring and therapy," *Nature Nanotechnology*, vol. 11, pp. 566–572, Mar. 2016.

[13] D.-H. Choi, Y. Li, G. R. Cutting, and P. C. Searson, "A wearable potentiometric sensor with integrated salt bridge for sweat chloride measurement," *Sensors and Actuators B: Chemical*, vol. 250, pp. 673–678, Oct. 2017.

[14] W. Gao et al., "Fully integrated wearable sensor arrays for multiplexed in situ perspiration analysis," *Nature*, vol. 529, pp. 509–514, Jan. 2016.

[15] J. Kim et al., "Wearable smart sensor systems integrated on soft contact lenses for wireless ocular diagnostics," *Nature Communications*, vol. 8, pp. 1–8, Apr. 2017.

[16] J. Kim et al., "Wearable salivary uric acid mouthguard biosensor with integrated wireless electronics," *Biosensors and Bioelectronics*, vol. 74, pp. 1061–1068, Dec. 2015.

[17] M. A. Hopcroft, W. D. Nix, and T. W. Kenny, "What is the young's modulus of silicon?," *Journal of Microelectromechanical Systems*, vol. 19, no. 2, pp. 229–238, 2010. 5

[18] P. Sharma, and N. Gupta, "Investigation on material selection for gate dielectric in nanocrystalline silicon (nc-Si) thin-film transistors (TFTs) using Ashby's, VIKOR and TOPSIS," *Journal of Material Science: Materials in Electronics, Springer-Verlag Berlin*, vol. 26, 9607–9613, 2015.

[19] B. N. Aditya, and N. Gupta, "Material selection methodology for gate dielectric material in metal-oxide-semiconductor devices," *Materials and Design-an Elsevier Journal*, vol. 35, 696–700, 2012.

[20] S. Hozumi et al., "Multimodal wearable sensor sheet for health-related chemical and physical monitoring," *ACS Sensors*, vol. 6, pp. 1918–1924, Apr. 2021.

[21] K.-Y. Chun, S. Seo, and C.-S. Han, "A wearable all-gel multimodal cutaneous sensor enabling simultaneous single-site monitoring of cardiac-related biophysical signals," *Advanced Materials*, vol. 34, p. 2110082, Mar. 2022.

[22] Q. Pang et al., "Smart flexible electronics-integrated wound dressing for real-time monitoring and on-demand treatment of infected wounds," *Advanced Science*, vol. 7, p. 1902673, Mar. 2020.

[23] P. Liu, Y. Huang, and C. Lian "Equation of resistance-temperature of flexible tactile sensor based on temperature-sensitive conductive rubber," *Polym. Mater. Sci. Eng.*, vol. 28, no. 6, 107–109, 2012.

[24] L. Wu, W. J. Qian, and J. Peng "Screen-printed flexible temperature sensor based on FG/CNT/PDMS Composite with constant TCR," *Mater. Elect.*, vol. 30, no. 10, 9593–9601, 2019.

[25] V. B. Nam, and D. Lee, "Evaluation of ni-based flexible resistance temperature detectors fabricated by laser digital pattering," *Nanomaterials*, vol. 11, p. 576, Feb. 2021.

[26] V. B. Nam, T. T. Giang, S. Koo, J. Rho, and D. Lee, "Laser digital patterning of conductive electrodes using metal oxide nanomaterials," *Nano Convergence*, vol. 7, pp. 1–17, July 2020.

[27] S. Ali, S. Khan, and A. Bermak, "Inkjet-printed human body temperature sensor for wearable electronics," *IEEE Access*, vol. 7, pp. 163981–163987, 2019.

[28] M. Dankoco, G. Tesfay, E. Benevent, and M. Bendahan, "Temperature sensor realized by inkjet printing process on flexible substrate," *Materials Science and Engineering: B*, vol. 205, pp. 1–5, Mar. 2016.

[29] C. Bali et al., "Fully inkjet-printed flexible temperature sensors based on carbon and PEDOT:PSS," *Materials Today: Proceedings*, vol. 3, no. 3, pp. 739–745, 2016. 6

[30] M. Seifi, S. Hamedi, and Z. Kordrostami, "Fabrication of a high-sensitive wearable temperature sensor with an improved response time based on PEDOT:PSS/rGO on a flexible kapton substrate," *Journal of Materials Science: Materials in Electronics*, vol. 33, pp. 6954–6968, Feb. 2022.

[31] T.-W. Lee et al., "Organic light-emitting diodes formed by soft contact lamination," *Proceedings of the National Academy of Sciences*, vol. 101, pp. 429–433, Jan. 2004.

[32] Y.-L. Loo et al., "Soft, conformable electrical contacts for organic semiconductors: High-resolution plastic circuits by lamination," *Proceedings of the National Academy of Sciences*, vol. 99, pp. 10252–10256, July 2002.

[33] T. Q. Trung, S. Ramasundaram, B.-U. Hwang, and N.-E. Lee, "An all-elastomeric transparent and stretchable temperature sensor for body-attachable wearable electronics," *Advanced Materials*, vol. 28, pp. 502–509, Nov. 2015.

[34] S. Baloda, Z. A. Ansari, S. Singh, and N. Gupta, "Development and analysis of graphene nanoplatelets (GNP) based flexible strain sensor for health monitoring applications," *IEEE Sensors Journal*, vol. 20, no. 22, 13302–13309, 2020.

[35] F. Xu et al., "Recent developments for flexible pressure sensors: A review," *Micromachines*, vol. 9, p. 580, 2018.

[36] Y. Yang et al., "Human skin based triboelectric nanogenerators for harvesting biomechanical energy and as self-powered active tactile sensor system," *ACS Nano*, vol. 7, pp. 9213–9222, 2013.

[37] W.-Y. Chang, C.-C. Chen, C.-C. Chang, and C.-L. Yang, "An enhanced sensing application based on a flexible projected capacitive-sensing mattress," *Sensors*, vol. 14, pp. 6922–6937, 2014.

[38] L. Ma et al., "Highly sensitive flexible capacitive pressure sensor with a broad linear response range and finite element analysis of micro-array electrode," *Journal of Materiomics*, vol. 6, pp. 321–329, June 2020.

[39] Bijender, and A. Kumar, "Flexible and wearable capacitive pressure sensor for blood pressure monitoring," *Sensing and Bio-Sensing Research*, vol. 33, p. 100434, Aug. 2021.

[40] T. Xu et al., "High resolution skin-like sensor capable of sensing and visualizing various sensations and three dimensional shape," *Scientific Reports*, vol. 5, pp. 1–9, Aug. 2015.

[41] Z. Chen et al., "Flexible piezoelectric-induced pressure sensors for static measurements based on nanowires/graphene heterostructures," *ACS Nano*, vol. 11, pp. 4507–4513, Apr. 2017.

[42] Y. Li et al., "Skin-like biosensor system via electrochemical channels for non-invasive blood glucose monitoring," *Sci Adv*, vol. 3, p. e1701629, 2017.

[43] A. J. Bandodkar et al., "Tattoo-based noninvasive glucose monitoring: A proof-of-concept study," *Anal Chem*, vol. 87, pp. 394–398, 2015.

[44] Huang X. et al., "Stretchable, wireless sensors and functional substrates for epidermal characterization of sweat," *Small*, vol. 10, pp. 3083–3090, 2014.

[45] Y. Li, X. Feng, Z. Qu et al, "Skin-like biosensor system via electrochemical channels for noninvasive blood glucose monitoring," *Sci Adv*, vol. 3, p. e1701629, 2017.

[46] W. Gao et al., "Fully integrated wearable sensor arrays for multiplexed in situ perspiration analysis," *Nature*, vol. 529, pp. 509–514, 2016.

# 7 Electrochemical Tools for the Prognosis of Skin Wounds

*Ramendra Kishor Pal, Debirupa Mitra, and Arnab Dutta*

Department of Chemical Engineering, BITS Pilani, Hyderabad Campus, Telangana, India

## CONTENTS

## 7.1 INTRODUCTION

Innovations in wound management are now moving towards the precision or perso-nalised medicine approaches requiring real-time monitoring, onboard prognosis, and on-demand medication administration (working in a closed loop). Such technologies can overcome the ongoing challenges of delayed treatment by accelerating prognosis and improving the administration of therapeutics in time and space. An envisioned wearable system having body-compliant biosensors for tracking the pathophysiological conditions of wounds along with data collection, wireless data transfer, artificial intelligence-based diagnosis, and treatment will enable tailored management of wounds according to the conditions of each patient and may revolutionise traditional medical practices.

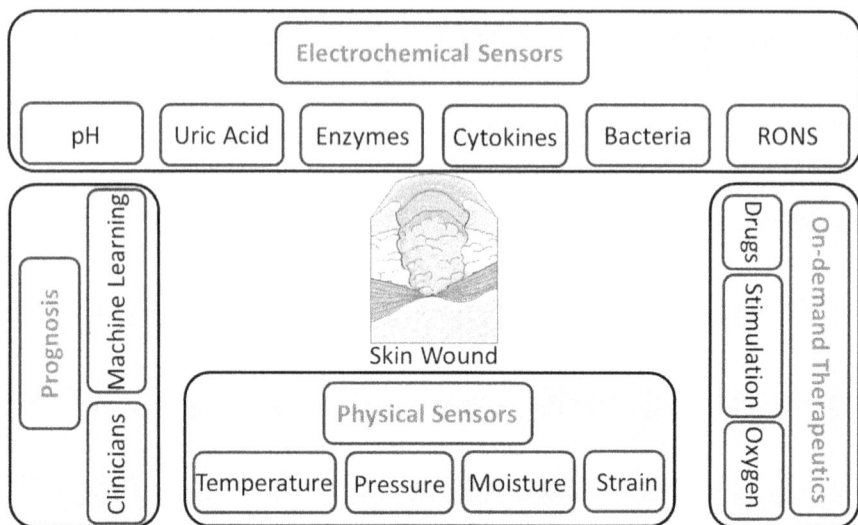

**FIGURE 7.1** Schematic of integrated wound care. Though crucial, physical sensors, on-demand therapeutics, and clinical prognosis are not in the scope of book chapter. The wound schematic in the figure is taken from the Servier Medical Art database (https://ssmart.servier.com/smart_image/ulcer-3/).

Wearable biosensors will enable real-time tracking of key biomarkers, which will help understand the dynamics, interplay between biomarkers, and the effects of their relative concentrations on the cellular repair process. Further, real-time wound data will help understand the effects of ongoing treatment and tweak the treatment to achieve fast recovery. Figure 7.1 shows an envisioned integrated system that aids wound healing. Such tools become important while treating patients suffering from chronic or slow-healing wounds, which are turning into an epidemic affecting ~8.2 million people in the United States alone, rising at 2% annually and affecting primarily elders, obese, and diabetics (Sen 2019).

This book chapter aims to provide a brief insight into wounds biomarkers that inform about the state of the wound, establish the requirements for real-time monitoring specific to skin wounds, highlight some recent developments, and discuss emerging trends in wound diagnostics and their value proposition to the precision medicine for wound management. The biosensors will play a key role in diagnosing and monitoring the wound condition to initial detection, progression, and effect of drug treatment. Though biosensors can have diverse detection strategies such as optical, electronic, and imaging, this book chapter will restrict the discussions on electrochemical biosensors to stay in line with the theme of this volume.

## 7.2 CHRONIC WOUNDS AND BIOMARKERS

Healing of skin wounds is a complex, dynamic, and sequential process comprising four overlapping phases: hemostasis, inflammation, proliferation, and remodelling. The healing time depends on multiple factors, including wound size, depth, location, patient age, and local and systemic disease. Acute wounds progress through these phases of healing in a typical and timely manner. In contrast, chronic wounds fail to proceed through a typical and timely repair sequence to restore normal anatomy and function.

Wound repair is a dynamic cellular-level process. The wound fluid contains several biochemical species playing a specific role in healing. These molecules may serve as biomarkers that indicate the progression of wound healing. The process of biomarker identification involves finding necessary molecules that interfere with healing by inhibiting or regulating it, followed by its validation in patients. These biomarkers must have predictive, diagnostic, or indicative information about the wound, allowing a personalised assessment of the wound healing and aiding in tuning therapeutics. Table 7.1 lists wound fluid biomarkers. Among these, pH and uric acid have been most commonly investigated for developing wound monitoring sensors. Enzymes also hold great promise as biomarkers (Mota et al. 2021), and their detection and monitoring tools are still under research.

## TABLE 7.1
## Biomarkers for Wound Healing Prognosis

| Type of Biomarker | Biomarker | Characteristics |
|---|---|---|
| Physicochemical markers | pH | Varies with stages of wound healing; differs between chronic and acute wounds [3] |
| | Oxygen | Decreased levels (hypoxic) in chronic wounds [4] |
| | Temperature | Differs between healing and nonhealing wounds [5] |
| | Moisture | Too dry or excess moisture associated with delayed healing [6] |
| Signalling molecules | C-reactive protein | Increased levels during acute phase response of inflammation [5] |
| | Nitric oxide | Decreased levels in nonhealing wounds [4] |
| | Hydrogen peroxide | Increased levels in nonhealing wounds |
| | Cytokines | [4], [7] |
| | Reactive Oxygen Species | |
| Enzymes | Myeloperoxidase | Increased levels in nonhealing wounds [7] |
| | Matrix metalloproteinases | |
| | Xanthine oxidase | |
| | Human-neutrophil elastase | Increased levels during acute phase |
| | Cathepsin G | response of inflammation [2] |
| | Lysozyme | Increased levels in infected wounds [8] |
| | Bacteria-secreted proteases | Increased levels in infected wounds [9] |
| Metabolites | Uric acid | Increased levels in nonhealing wounds [4] |
| | Lactic acid | |
| | Pyocyanin | Metabolite secreted by *P. aeruginosa* [4] |

## 7.3   WOUND INFECTIONS

Infection of wounds is the most common complication that hinders wound healing and poses a severe risk to wounded patients. The moist bed of wound provides an ideal site for bacterial colonisation leading to infection. Most infected wounds are polymicrobial in nature; the most frequently isolated pathogens are *Staphylococcus aureus* and *Pseudomonas aeruginosa* (Bessa et al. 2015; Serra et al. 2015). Some of the other wound pathogens are *Streptococcus* spp, *Escherichia coli*, *Klebsiella* spp, *Enterobacter* spp, *Proteus* spp, *Clostridium* spp, and *Corynebacterium* spp (Bowler 2002; Bessa et al. 2015). To diagnose infected wounds, clinicians typically look for physical signs like heat, pain, redness, swelling, discolouration, foul odour, and excess exudate formation (S. Li et al. 2021). These signs are still dubious and may not occur during the early stages of infection. For a reliable identification, two possible ways of diagnosing infected wounds are: direct detection of bacteria and detection of extracellular metabolites or toxins or enzymes secreted by the bacteria (indirect method). Direct detection of bacteria has a high diagnostic value but has low sensitivity and longer analysis times. The detection of specific metabolites or enzymes has high sensitivity. Bacteria-secreted enzymes, specifically proteases, are the first biomarkers appearing during colonisation and can help diagnose an early stage of infection. Increased pH, temperature, CRP, ROS, lysozyme, and decreased NO in wound fluid are also indicators of wound infection.

## 7.4   ELECTROCHEMICAL SENSORS FOR MONITORING BIOMARKERS

Electrochemical biosensors offer sensitive, quick, and continuous monitoring. They are compatible with rapid prototyping tools such as screen-printing, inkjet printing, photolithography, and 3D printing on flexible and stretchable substrates. They integrate well with flexible bioelectronics. These properties make them apt for monitoring wounds in wearable formats. Electrochemical biosensors consist of modified electrodes with responsive materials and specific biorecognition molecules, enabling labelled or label-free detection of bioanalytes in real time.

### 7.4.1   pH Sensors

Wound pH affects all the biochemical reactions during the healing processes. A high pH indicates bacterial colonisation, while a pH around or below 7 indicates a normal healing progression, optimised protease activities, fibroblast migration, and remodelling of ECM (Jones, Cochrane, and Percival 2015).

Potentiometric and voltammetric methods can measure pH in a liquid electrolyte. The potentiometric method measures the open circuit potential (OCP) between the working electrode (WE) and the reference electrode (RE) to track pH change in the electrolyte. The voltammetric method measures the current generated when a potential is applied in a three-electrode electrochemical cell having WE, RE, and counter electrode (CE). The potentiometric method is more popular than the voltammetric method for monitoring wound pH. They require simple electronics to measure OCP and do not require an external current supply like the voltammetric method (Mariani et al. 2021; Mostafalu et al. 2018a). The pH sensors integrated on flexible/stretchable substrates such as EcoFlex, bandage, polyethylene terephthalate (PET) sheet, polydimethylsiloxane (PDMS) film, paper, or textile display a linear sensitivity of ~50 mV/pH ($R^2 > 0.95$) with stable performance between 4 and 10 pH (Mostafalu et al. 2018b; Tang et al. 2022; Mariani et al. 2021; Zhang et al. 2022; Ghoneim et al. 2019). Nonetheless, the sensitivity of these sensors needs improvement to monitor minute pH variations (less than 0.05 units) (Manjakkal, Dervin, and Dahiya 2020).

### 7.4.2 URIC ACID SENSORS

Uric acid (UA) levels strongly correlate with the status of wound and infection. UA forms when a cell ruptures at the wound through the purine metabolic cycle. Elevated levels of UA (>500 μM) in wound fluids indicate the severity of chronic venous leg ulcers (VLU) (Fernandez et al. 2012). UA is an electroactive molecule meaning that its oxidation occurs within the electrochemical window, allowing direct detection; however, urate oxidase (UOx) enhances biosensors' sensitivity and detection limit by catalysing the oxidation reaction (Yan et al. 2020). A report also shows using a secondary enzyme such as horseradish peroxidase (HRP) to facilitate electron transport between the electrode surface and UOx (Bhushan et al. 2019). The detection limit of these sensors is in μM, and the linear range is between 10 to 800 μM (Bhushan et al. 2019).

### 7.4.3 CYTOKINE AND GROWTH FACTOR SENSORS

Cytokines and growth factors are signalling molecules that help regulate cell functions such as proliferation, migration, and matrix synthesis by communicating immune responses between cells and stimulating the cells to move towards inflammation, infection, and injury sites. They also play a crucial role in the redeposition of the extracellular matrix (Gharee-Kermani and Pham 2005). Important cytokines for wound healing are interleukin (IL-1, IL-6), chemokines (CC-chemokines ligand family, CXC-chemokine ligand family), and interferon (INF-γ). Important growth factor families for the wound healing include epidermal growth factor (EGF), transforming growth factor-beta (TGF-β), fibroblast growth factor (FGF), vascular endothelial growth factor (VEGF), granulocyte-macrophage colony-stimulating factor (GM-CSF), platelet-derived growth factor (PDGF), connective tissue growth factor (CTGF), and tumour necrosis factor-alpha (TNF-α) family (Barrientos et al. 2008). Concerted processes by cytokines, growth factors, and several cells lead to a successful wound healing process. Therefore, monitoring the levels of these biomarkers provide pathophysiological information about wounds. Cytokines and growth factors are electrochemically inactive molecules.

Electrochemical sensing of cytokines and growth factors involves a class of biorecognition molecules such as antibodies and aptamers with an affinity towards specific cytokines or growth factors. Antibodies are protein molecules that evolved naturally to bind target analytes with high specificity. Therefore, they are effective as capture probes for biosensing platforms. Covalent immobilisation of antibodies on the electrode surfaces may lead to randomly oriented antibodies losing effectiveness. However, the biotin-streptavidin chemistry provides an oriented immobilisation of antibodies on the electrode surface (Arshavsky-Graham et al. 2020). Aptamers are single-stranded (ss) DNA or RNA oligonucleotides with a binding affinity to nucleic acids, proteins, metal ions, and other small molecules with high selectivity and sensitivity. The chemical structure, size, and synthetic production of aptamers overcome some of the disadvantages of antibodies. Synthetic production allows conjugation of surface functional groups to the non-binding end; thus, aptamers do not suffer the orientation issues like antibodies. As they are ~10 times smaller than antibodies, immobilisation density is also greater than antibodies. Nonetheless, aptamer and oriented antibodies show similar sensing performance on a specific surface in practice (Arshavsky-Graham et al. 2020). Affinity-based sensors have several other limitations, such as binding affinity depends upon pH, temperature, and light, and binding is irreversible (sensors work only once). Further, the operation time of electrochemical sensors depends on the loading capacity of biorecognition molecules.

Electrochemical sensors can use a variety of electrochemical techniques to detect biorecognition events. The electrochemical methods include potentiostatic (current is measured for the applied controlled potential to the electrochemical cell); galvanostatic (potential is measured for the applied current); potentiometric (measurement of cell potential at near-zero current or open circuit potential); and impedance spectroscopy (current is measured for an applied alternating potential to the cell and analysed to obtain impedance – complex resistance) (Vogiazi et al. 2019). Further, amplification of the electrochemical response is possible using labels such as enzymes, redox-active molecules, nanoparticles, quantum dots, and low dimensional carbon materials (Koyappayil and Lee 2021).

### 7.4.4 Reactive Oxygen and Nitrogen Species (RONS) Sensors

RONS such as hydrogen peroxide ($H_2O_2$), singlet oxygen ($^1O_2$), and nitric oxide (NO) are signalling molecules to immunocytes and non-lymphoid cells involved in tissue repair. Electrochemical detection of $H_2O_2$ can take either enzymatic or non-enzymatic routes. Both routes show good analytical performance but have some limitations. Enzymatic sensors face issues such as enzyme stability and active site inhibition. Non-enzymatic sensors face problems such as high working potential, unpredicted redox reactions, slow electrokinetics, poisoning by intermediate, and weak sensing. As NO readily oxidises electrochemically, direct electrochemical detection of NO is possible.

Further, permselective polymers-coated (such as poly-eugenol) electrodes facilitate selective electrochemical measurement of NO in the presence of biological interferents (R. Li et al. 2020). Electrochemical detection of $^1O_2$ involves hydroquinone (HQ)/benzoquinone (BQ) redox reaction. $^1O_2$ first oxidises HQ to form BQ, which reduces back to HQ on the sensing electrode giving a reducing current corresponding to the BQ concentration (Ling et al. 2021).

## 7.4.5 ENZYME SENSORS

Proteolytic and antioxidative enzymes are essential in wound healing. Proteolytic enzymes such as matrix metalloproteinases (MMPs) degrade necrotic debris resulting from cellular breakdowns. Antioxidative enzymes such as superoxide dismutase (SOD), glutathione peroxidase (GPX), peroxiredoxin (PRDX), catalase, and hemeoxygenase (HO) regulate the RONS concentration in wounds. Electrochemical sensing of MMPs uses different approaches. The first approach exploits the proteolytic attributes of MMPs. MMPs act upon polypeptides having specific recognition and digestion sites. These polypeptides immobilised on electrode surfaces undergo degradation when interacting with specific MMP present in the wound fluid leading to improved electron transfer between the electrode and the electroactive species in the sample. Improved electron transfer increases the current and provides a label-free detection (Palomar et al. 2020). Other approaches involve the application of molecularly imprinted polymer (MIP) with selective receptors or aptamers as biorecognition elements (Bartold et al. 2022; Vladyslav Mishyn et al. 2022). An electrochemical sensor detects the binding events using electrochemical methods described in the previous section. Electrochemical sensors for the detection of antioxidant enzymes are still in their infancy.

## 7.5 ELECTROCHEMICAL SENSORS FOR PATHOGEN DETECTION

As mentioned earlier, detection of bacteria in wounds can be done by either detecting the bacteria or detecting any biomolecule secreted by the bacteria into the wound fluid (Figure 7.2). Direct detection of bacteria is usually done by culturing methods, nucleic acid-based assays like PCR (detection of bacteria-specific genetic material), or immunoassays (detection of bacteria-specific antibodies) like ELISA. While PCR and ELISA are the current "gold" standards for pathogen detection, the need for continuous monitoring, and rapid detection without using additional reagents or sample preparation steps has resulted in the development of biosensors.

Electrochemical biosensors for bacteria sensing employ an electrochemical cell comprising electrodes and pathogen-containing electrolytes. Based on differences in electrode configuration, applied signals (AC or DC), and measured signals, electrochemical sensors are broadly classified into potentiometric, amperometric, conductometric, and impedimetric sensors (Cesewski and Johnson 2020). One of the earliest sensors for direct bacteria detection was Bactometer®, which was developed using impedance microbiology (IM). The impedance at a single frequency is monitored at the electrodes immersed in a sample containing bacteria in the medium. Impedance changes result from ionic metabolites secreted by the actively growing bacteria into the solution, and beyond a threshold level, the sample tests positive for bacteria (Furst and Francis 2019). However, these sensors are non-specific and applicable for samples with high bacteria concentration (readout time of 2–3 h for $>10^5$ CFU/mL). The sensitivity of these has been improved by using interdigitated electrode arrays (IDA) and combining microfluidics with IDA (Furst and Francis 2019). The last decade has seen the development of electrochemical sensors with increased sensitivity as well as specificity. For this, the electrodes are usually modified with "biorecognition"

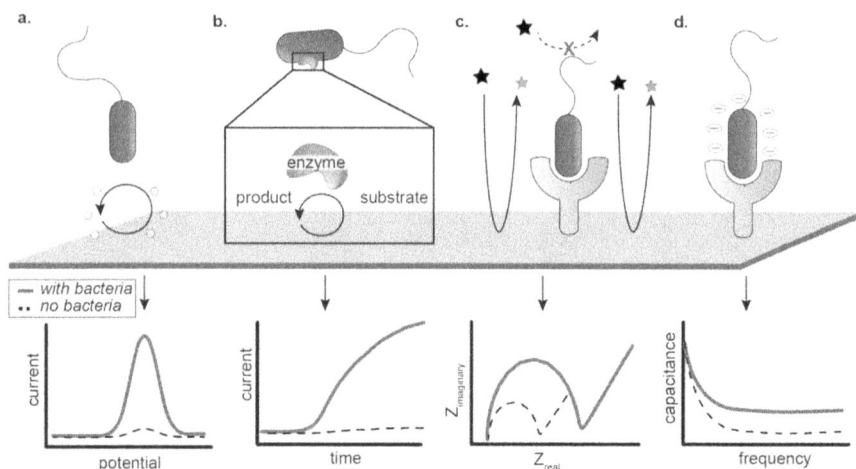

**FIGURE 7.2** Pathogen detection modes of electrochemical biosensors. Indirect detection by sensing (a) cell-secreted electroactive metabolites (hexagon, yellow = reduced, orange = oxidised) using square wave voltammetry (current vs potential) or (b) electroactive products formed by the exogenously secreted enzymes using chronoamperometry (current vs time). Direct detection by sensing cells binding to a biorecognition probe on the electrode is detected using impedimetric techniques. Binding may cause a change in (c) impedance due to reduced electron transfer activity of a mediator (star, dark blue = reduced, light blue = oxidised) as a result of surface passivation (faradaic mode) or a change in (d) surface dielectric properties such as capacitance in the absence of a redox mediator (nonfaradaic mode). Adapted with permission from [52]. Copyright © 2020, American Chemical Society.

elements that can recognise and bind to selective pathogens. These biorecognition elements can be antibodies, oligonucleotides (DNA/RNA), proteins/peptides, aptamers, phages, or even MIP (Cesewski and Johnson 2020). The chemical energy associated with this binding of pathogens to the electrodes is converted into electrical signals that give the readout. Antibodies have been the most used biorecognition element. For example, a paper-based electrochemical immunosensor was developed using antibody-immobilised SWCNT-coated carbon electrodes for the detection of *S. aureus* using differential pulse voltammetry. The limit of detection was found to be 13 CFU/mL with a detection time of 30 min (Bhardwaj et al. 2017). In another work, a fully automated microfluidic-based electrochemical biosensor with a detection limit of 50 CFU/mL was used to detect *E. coli* via cyclic voltammetry. A gold chip sensor functionalised with specific antibodies resulted in a high specificity in cross-reactivity studies with other bacterial species (Altintas et al. 2018). Simultaneous detection of multiple bacteria has also been studied, such as amperometric detection of *E. coli* and *S. aureus* using antibody-immobilised carbon nanotube-coated gold-tungsten microwire electrodes with a detection limit of 100 CFU/mL (Yamada et al. 2016) and impedimetric detection of *S. aureus* and *E. coli* using antibody-immobilised nanoporous alumina membranes integrated into a microfluidic device containing Pt wire electrodes with a detection limit of 100 CFU/mL (Tian et al. 2016). While the binding of the bacteria to the electrodes led to changes in current flowing through the circuit in the former case, in the latter case, the binding of the bacteria to the membrane blocked the pores, thereby changing the impedance of the device. The detection limit of these sensors can be further decreased to <10 CFU/mL by the use of aptamers as the biorecognition element (Cai et al. 2021; Cesewski and Johnson 2020). Aptamers can bind to their targets with high affinity and specificity, and they are more stable in a wide range of environmental conditions as compared to antibodies (F. Li et al. 2019). MIPs are another alternative that provides template-based sites for the binding of either whole microorganisms or even certain biomarkers like lysozyme and interleukins (Hasseb et al. 2022). Aside from the above, bacteriophages can also be used as the biorecognition element for the highly specific and selective detection of bacteria. There are two pathways of detection based on: (a) phage-induced lysis and (b) direct cell wall recognition (Hussain et al. 2021). In the first case, the phages induce bacterial lysis upon binding resulting in the release of cytoplasmic contents and progeny phases. Potential intracellular biomarkers such as enzymes and metabolites, if present in the released content, can be then detected electrochemically as discussed in the next paragraph. In the second case, the specific adsorption of bacteria onto phages immobilised on the transducer surface leads to changes in impedance, which can be monitored using EIS (Hussain et al. 2021).

Indirect detection relies on monitoring any exogenous metabolite or toxin (virulence factor) or enzymes that are secreted by the bacteria. If the secreted metabolite is electroactive, its detection via electrochemical tools is the most promising approach. A prominent example of this kind is pyocyanin (PYO), which is a phenazine redox-active metabolite secreted by almost all strains of *P. aeruginosa*. Detection of PYO is also clinically relevant because it is secreted in the early stages of infection (clinically detected concentration range 1–150 µM) and is a critical virulence factor for the generation and establishment of *P. aeruginosa* infection symptoms (Simoska and Stevenson 2022). Several studies have reported the detection of PYO with a linear detection range (LDR) of 1–100 µM and a limit of detection (LOD) of 0.1–1 µM. These studies have typically used cyclic voltammetry (CV), differential pulse voltammetry (DPV) or square wave voltammetry (SWV). Still, SWV appears to be particularly useful for the quantitative detection of low PYO concentrations, as it can suppress the background, non-Faradaic currents, providing high sensitivity in more complex biological samples (Simoska and Stevenson 2022). Amplification of PYO sensing can also be achieved by the addition of specific amino acids (proline, histidine, arginine, leucine, tyrosine, and valine) that upregulate PYO production or by the use of a bio-based redox capacitor (e.g. catechol-chitosan complex) that can amplify the electrochemical signals (Alatraktchi, Svendsen, and Molin 2020). Besides PYO, unique and redox-active metabolites secreted by other wound-relevant bacteria are yet to be identified.

Proteases are secreted by several bacteria during the early stages of infection and have already been identified as promising biomarkers for infected and nonhealing wounds. However, there exists no rapid diagnostic method for the detection of protease activity. Electrochemical detection using voltammetry or chronoamperometry may be a promising method where the electrodes can be functionalised with enzyme-specific substrates, and the enzymatic degradation of this substrate will lead to changes in the signal readout (Eissa and Zourob 2020). A similar concept can be applied for sensing bacteria-secreted toxins; for example, delta toxin and rhamnolipid secreted by *S. aureus* and *P. aeruginosa*, respectively, triggered the release of redox probes encapsulated in bilayer vesicles immobilised on electrodes enabling their detection via voltammetry (Thet and Jenkins 2015).

Most of the investigations for bacterial detection targeted foodborne pathogens. These can extend to detect wound bacteria; nonetheless, more studies are still needed.

## 7.6  MATERIALS FOR BIOSENSORS USED FOR MONITORING WOUNDS

Biosensors for monitoring wounds require integrating bandages, dressings, or gauge materials. Thus, in addition to providing constant and conformal contact between the electrodes and wounds, the biosensors must minimally hinder the properties of bandages such as absorption of wound fluids, maintaining a moist environment, breathable, biocompatible, non-adherence, and antibacterial properties. Electrodes fabricated on fabrics and Tegaderm® and using multifunctional hydrogels using different printing methods such as screenprinting, photolithography, and coating are under investigation (Mostafalu et al. 2018a; Liu et al. 2021; Y. Gao et al. 2021; Wang et al. 2022). Popular electrode materials are metals such as gold, platinum, and silver; metal oxides such as indium tin oxide (ITO), titanium dioxide ($TiO_2$), carbon-based materials such as carbon nanotubes (CNTs), graphene, and graphite; conducting polymers such as polyaniline (PANI), polypyrrole (PPy), poly(3,4-ethylenedioxythiophene), and polystyrene sulfonate (PEDOT:PSS) (Mostafalu et al. 2018a; Cesewski and Johnson 2020; Y. Gao et al. 2021).

The performance of an electrochemical biosensor also depends on the physical form of the electrodes. Nanoscale electrodes (e.g. Pt nanowires) or electrodes with micro and/or nanostructured topographical features lead to high sensitivity as these enhance the surface area without significantly increasing the electrode footprint. A different approach to achieving high sensitivity is to employ electrode arrays such as interdigitated array microelectrodes (IDAMs) consisting of alternating, parallel-electrode fingers assembled in an interdigitated pattern. These arrays exhibit rapid response and a high signal-to-noise ratio.

## 7.7  DEVELOPMENT OF MODERN-DAY BIOSENSORS FOR WOUND MANAGEMENT

### 7.7.1  MULTIANALYTE AND MULTIPLEXED BIOSENSORS

Simultaneous detection of multiple biomarkers and pathogens in a single biosensor is necessary for clinically relevant wound monitoring. Recent efforts are thus focused on multianalyte or multiplexed biosensors. For example, carbon ultramicroelectrode arrays (CUAs) fabricated on flexible polyethylene terephthalate (PET) substrate were used for sensing three different electroactive biomarkers: PYO, nitric oxide, and uric acid. The proof-of-concept wearable CUA sensor had mid-micromolar LOD for all the three analytes, and it exhibited different electrochemical signatures for each of them without significant interference (Simoska, Duay, and Stevenson 2020). Multiple biorecognition elements can also be immobilised to fabricate multiplexed sensors. In this case, specific antibodies or aptamers that bind to different targets with very high affinity are used (Y. Gao et al. 2021). Since multiple immobilisation sites are required, nanomaterials or nanostructured electrodes may be necessary to provide a high surface area.

## 7.7.2   REAL-TIME MONITORING

Real-time transmission/collection of wound biomarkers' data would enable sharing of information to medical professionals and clinicians to help them improve the quality of their medical treatments by enabling them to make informed decision. Monitoring wound biomarkers in real time can help realise personalised medicine where therapeutic interventions follow according to the obtained data and therapeutic outcomes predicated by the AI algorithms operating in a closed-loop (W. Gao and Yu 2021). One crucial aspect of real-time monitoring is the wireless transmission of data. A few leading technologies for wireless data transmission are near-field communication (NFC), Bluetooth, and radiofrequency (RF) (C. Xu, Yang, and Gao 2020). NFC allows simultaneous wireless data transmission and power between two electronic devices via inductive coupling when placed in proximity (~10 cm). NFC antenna can be in flexible formats. However, NFC requires the antenna in a particular orientation and proximity to the other electronic device, limiting its practical, long-term use. Bluetooth is an alternative wireless communication tool that allows a faster transmission rate to ~9 metres than NFC. Thus, Bluetooth suits multifunctional sensing and long-term use. Nonetheless, Bluetooth modules are rigid, which limits applicability on the skin. RF transmission operates at a fixed frequency of 13.56 MHz between a powered initiator and a passive target. RF antenna can be in flexible formats.

   A few recent efforts have reported real-time monitoring of wounds in wearable formats. An anomaly of wound temperature is an early indicator of infection. Electronic temperature sensors in flexible and wearable formats are sensitive, accurate, and easy to operate in clinical settings. These sensors can monitor wound temperature continuously over a period. Further, electronic temperature sensors integrate with Bluetooth, allowing real-time transmission of measured data (in this case, temperature) to a portable data-logging tool such as a smartphone (Pang et al. 2020). pH is a key indicator of bacterial infections. pH-measuring electrochemical sensors fabricated on flexible formats such as PET sheets, PDMS film, or textile can work for continuous monitoring of wound pH (Mariani et al. 2021; G. Xu et al. 2021). An onboard Bluetooth-enabled or NFC-enabled transceiver can transmit the pH data obtained by sensors and receive instructions to trigger/modulate the delivery of drugs loaded in the smart bandage (Liu et al. 2021; Mostafalu et al. 2018a; G. Xu et al. 2021). Integrated sensor modules may carry multiple sensor combinations that can track different indicators such as temperature, pressure, pH, uric acid, and blood glucose. Wireless data transmission tools can send multiple data sets from different sensors as well (Guo et al. 2021; G. Xu et al. 2021; Liu et al. 2021). The data repository in the cloud will enable statistical analysis, knowledge sharing, and the development of AI-based algorithms for personalised treatment.

## 7.8    ROLE OF MACHINE LEARNING IN WOUND MANAGEMENT

In the era of personalised medicine, the recent boom in artificial intelligence and data analytics holds potential for promising solutions in developing a reliable, intelligent wound management strategy (Wang et al. 2022). In this context, we must understand the philosophy behind developing such a smart strategy and the role of machine learning (ML) in it. Figure 7.3 illustrates an overview of an intelligent framework for wound monitoring and management.

To develop such a framework, it is essential to identify suitable biomarkers that can characterise the progression of chronic wounds, thus assisting in its diagnosis and treatment. Pre-processing of raw data reduce noise in data by handling missing information and outliers. Next, feature extraction techniques are used to obtain characteristic features representing the dataset. Feature representation techniques include principal component analysis (PCA), discrete Fourier transform (DCT), linear discriminant analysis (LDA), etc. Once suitable features are extracted from the data set, these are used as inputs to develop a data-driven model. Such models can be categorised as classification, prediction, and clustering. Wound healing classification problems include developing classifiers to identify the phases or the types of wounds. Some of the common algorithms that are deployed to develop classifiers are support vector machine (SVM), K-nearest neighbour (kNN), extreme gradient boosting (XGBoost), etc. They evaluate the performance of the classifiers various metrics like accuracy, precision, recall, and F1-score. The next category of models includes wound healing prediction by capturing relationship between sensor data and the target value(s). This approach regresses the target value using a set of features using algorithms like polynomial response surface models (PRSM), artificial neural network (ANN), convolutional neural network (CNN), etc. To assess the accuracy of these predictive models, root mean square error (RMSE), mean absolute percentage error (MAPE), etc. are used. The last category of models includes when no specific information about output or the target value(s) is explicitly available. Then, unsupervised algorithms like K-means clustering, density-based clustering, self-organising map (SOM), etc. are used to extract meaningful

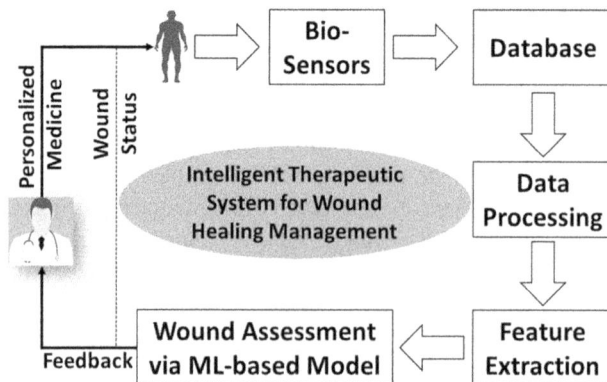

**FIGURE 7.3**    Overview of ML-assisted framework for wound management.

insights from the underlying patterns hidden in the input data set. The results obtained from different ML-based models can also be combined using effective decision fusion strategies, thus enhancing the final accuracy of the ML model. The ML framework should also have model agnostic methods to interpret the results obtained from the ML models, which will elucidate the importance of different features as well as provide valuable insights pertaining to the model. Based on the wound assessment obtained from the ML framework, an optimal strategy can be developed to accelerate the healing. Thus, integrating biosensors with an ML framework has the potential to expedite the recovery of any chronic wounds.

## 7.9 CHALLENGES AND THE WAY FORWARD

Integrated electrochemical sensing tools working in a closed loop aided with machine learning technologies possess a great promise to solve current challenges in wound management. Though a lot of progress has occurred in the last few years; nonetheless, several challenges remain for future works. One such issue is to develop sensors that can operate with a low volume of wound fluid. Efficient wound fluid collection, direction towards electrodes, and smaller electrodes requiring less volume of fluid may solve the low volume issue. Certain metabolites are unstable or short-lived or can undergo side reactions (e.g. PYO undergoes polymerisation). Therefore, the characterisation of sensor response time is essential. Detection of clinical isolates: Most studies were performed using wild-type strains. The expression of certain biomarkers varies among strains making it a challenging task. Complex biological matrix: Wound fluid is a very complex matrix. The use of simulated wound fluid is recommended to evaluate sensor performance. Affinity-based biosensors discussed in this chapter can only provide sensing once due to irreversible binding to biomarkers and bacteria.

Flexible, low-powered electronics and biocompatible power sources are still in their infancy for the continuous operation of wireless wearable systems. Finally, most of the efforts demonstrated the detection of biomarkers *in vitro* systems or on animal models for a short period. Human trials validating the utility of these devices are still unavailable. The long-term stable and reliable monitoring of wounds are crucial factors for ensuring successful application in humans. These unresolved issues still require continued efforts to develop reliable, stable multiplexed biosensing technology capable of monitoring a variety of biomarkers involved in wound healing in humans.

## REFERENCES

Alatraktchi, Fatima Alzahra'A, Winnie E. Svendsen, and Søren Molin. 2020. "Electrochemical Detection of Pyocyanin as a Biomarker for Pseudomonas Aeruginosa: A Focused Review." *Sensors (Switzerland)* 20 (18): 1–15. doi:10.3390/s20185218

Altintas, Zeynep, Mete Akgun, Guzin Kokturk, and Yildiz Uludag. 2018. "A Fully Automated Microfluidic-Based Electrochemical Sensor for Real-Time Bacteria Detection." *Biosensors and Bioelectronics* 100: 541–548. doi:10.1016/j.bios.2017.09.046

Arshavsky-Graham, Sofia, Katharina Urmann, Rachel Salama, Naama Massad-Ivanir, Johanna Gabriela Walter, Thomas Scheper, and Ester Segal. 2020. "Aptamers vs.

Antibodies as Capture Probes in Optical Porous Silicon Biosensors." *Analyst* 145 (14). The Royal Society of Chemistry: 4991–5003. doi:10.1039/D0AN00178C

Barrientos, Stephan, Olivera Stojadinovic, Michael S. Golinko, Harold Brem, and Marjana Tomic-Canic. 2008. "PERSPECTIVE ARTICLE: Growth Factors and Cytokines in Wound Healing." *Wound Repair and Regeneration* 16 (5). 585–601. doi:10.1111/J.1524-475X.2008.00410.X

Bartold, Katarzyna, Zofia Iskierko, Pawel Borowicz, Krzysztof Noworyta, Chu-Yun Lin, Jakub Kalecki, Piyush Sindhu Sharma, Hung-Yin Lin, and Wlodzimierz Kutner. 2022. "Molecularly Imprinted Polymer-Based Extended-Gate Field-Effect Transistor (EG-FET) Chemosensor for Selective Determination of Matrix Metalloproteinase-1 (MMP-1) Protein." *Biosensors and Bioelectronics* 208 (July). Elsevier: 114203. doi:10.1016/J.BIOS.2022.114203

Bessa, Lucinda J, Paolo Fazii, Mara Di Giulio, and Luigina Cellini. 2015. "Bacterial Isolates from Infected Wounds and Their Antibiotic Susceptibility Pattern: Some Remarks about Wound Infection." *International Wound Journal* 12 (1). John Wiley & Sons, Ltd: 47–52. doi:10.1111/iwj.12049

Bhardwaj, Jyoti, Sivaranjani Devarakonda, Suveen Kumar, and Jaesung Jang. 2017. "Development of a Paper-Based Electrochemical Immunosensor Using an Antibody-Single Walled Carbon Nanotubes Bio-Conjugate Modified Electrode for Label-Free Detection of Foodborne Pathogens." *Sensors and Actuators B: Chemical* 253: 115–123. doi:10.1016/j.snb.2017.06.108

Bhushan, Pulak, Yogeswaran Umasankar, Sohini Roy Choudhury, Penelope A. Hirt, Flor E. MacQuhaec, Luis J. Borda, Hadar A. Lev-Tov, Robert S. Kirsner, and Shekhar Bhansali. 2019. "Biosensor for Monitoring Uric Acid in Wound and Its Proximity: A Potential Wound Diagnostic Tool." *Journal of The Electrochemical Society* 166 (10). B830–B836. doi:10.1149/2.1441910JES/XML

Bowler, Philip G. 2002. "Wound Pathophysiology, Infection and Therapeutic Options." *Annals of Medicine* 34 (6): 419–427. doi:10.1080/078538902321012360

Cai, Rongfeng, Zhongwen Zhang, Haohan Chen, Yaping Tian, and Nandi Zhou. 2021. "A Versatile Signal-on Electrochemical Biosensor for Staphylococcus Aureus Based on Triple-Helix Molecular Switch." *Sensors and Actuators B: Chemical* 326: 128842. doi:10.1016/j.snb.2020.128842

Cesewski, Ellen, and Blake N. Johnson. 2020. "Electrochemical Biosensors for Pathogen Detection." *Biosensors and Bioelectronics*. 159. Elsevier Ltd. doi:10.1016/j.bios.2020.112214

Eissa, Shimaa, and Mohammed Zourob. 2020. "A Dual Electrochemical/Colorimetric Magnetic Nanoparticle/Peptide-Based Platform for the Detection of Staphylococcus Aureus." *Analyst* 145 (13). Royal Society of Chemistry: 4606–4614. doi:10.1039/d0an00673d

Fernandez, Melissa L., Zee Upton, Helen Edwards, Kathleen Finlayson, and Gary K. Shooter. 2012. "Elevated Uric Acid Correlates with Wound Severity." *International Wound Journal* 9 (2). Int Wound J: 139–149. doi:10.1111/J.1742-481X.2011.00870.X

Furst, Ariel L., and Matthew B. Francis. 2019. "Impedance-Based Detection of Bacteria." *Chemical Reviews*. American Chemical Society. doi:10.1021/acs.chemrev.8b00381

Gao, Wei, and Cunjiang Yu. 2021. "Wearable and Implantable Devices for Healthcare." *Advanced Healthcare Materials* 10 (17). John Wiley & Sons, Ltd: 2101548. doi:10.1002/ADHM.202101548

Gao, Yuji, Dat T. Nguyen, Trifanny Yeo, Su Bin Lim, Wei Xian Tan, Leigh Edward Madden, Lin Jin, et al. 2021. "A Flexible Multiplexed Immunosensor for Point-of-Care in Situ Wound Monitoring." *Science Advances* 7 (21). American Association for the Advancement of Science. doi:10.1126/SCIADV.ABG9614

Gharee-Kermani, Mehrnaz, and Sem Pham. 2005. "Role of Cytokines and Cytokine Therapy in Wound Healing and Fibrotic Diseases." *Current Pharmaceutical Design* 7 (11). Bentham Science Publishers Ltd.: 1083–1103. doi:10.2174/1381612013397573

Ghoneim, M. T., A. Nguyen, N. Dereje, J. Huang, G. C. Moore, P. J. Murzynowski, and C. Dagdeviren. 2019. "Recent Progress in Electrochemical PH-Sensing Materials and Configurations for Biomedical Applications." *Chemical Reviews* 119 (8). American Chemical Society: 5248–5297. doi:10.1021/ACS.CHEMREV.8B00655/ASSET/ IMAGES/LARGE/CR-2018-00655Y_0013.JPEG

Guo, Hongshuang, Ming Bai, Yingnan Zhu, Xinmeng Liu, Shu Tian, You Long, Yiming Ma, Chiyu Wen, Qingsi Li, and Jing Yang. 2021. "Pro-Healing Zwitterionic Skin Sensor Enables Multi-Indicator Distinction and Continuous Real-Time Monitoring." *Advanced Functional Materials* 31 (50). Wiley Online Library: 2106406.

Hasseb, Alaa A., Nourel din T. Abdel Ghani, Ola R. Shehab, and Rasha M. El Nashar. 2022. "Application of Molecularly Imprinted Polymers for Electrochemical Detection of Some Important Biomedical Markers and Pathogens." *Current Opinion in Electrochemistry* 31. Elsevier Ltd: 100848. doi:10.1016/j.coelec.2021.100848

Hussain, Wajid, Muhammad Wajid Ullah, Umer Farooq, Ayesha Aziz, and Shenqi Wang. 2021. "Bacteriophage-Based Advanced Bacterial Detection: Concept, Mechanisms, and Applications." *Biosensors and Bioelectronics* 177 (January). Elsevier B.V.: 112973. doi:10.1016/j.bios.2021.112973

Jones, Eleri M., Christine A. Cochrane, and Steven L. Percival. 2015. "The Effect of PH on the Extracellular Matrix and Biofilms." *Advances in Wound Care* 4 (7). Adv Wound Care (New Rochelle): 431–439. doi:10.1089/WOUND.2014.0538

Koyappayil, Aneesh, and Min Ho Lee. 2021. "Ultrasensitive Materials for Electrochemical Biosensor Labels." *Sensors (Basel, Switzerland)* 21 (1). Multidisciplinary Digital Publishing Institute (MDPI): 1–19. doi:10.3390/S21010089

Li, Fengqin, Zhigang Yu, Xianda Han, and Rebecca Y. Lai. 2019. "Electrochemical Aptamer-Based Sensors for Food and Water Analysis: A Review." *Analytica Chimica Acta* 1051. Elsevier Ltd: 1–23. doi:10.1016/j.aca.2018.10.058

Li, Rongfeng, Hui Qi, Yuan Ma, Yuping Deng, Shengnan Liu, Yongsheng Jie, Jinzhu Jing, et al. 2020. "A Flexible and Physically Transient Electrochemical Sensor for Real-Time Wireless Nitric Oxide Monitoring." *Nature Communications 2020 11:1* 11 (1). Nature Publishing Group: 1–11. doi:10.1038/s41467-020-17008-8

Li, Shuxin, Paul Renick, Jon Senkowsky, Ashwin Nair, and Liping Tang. 2021. "Diagnostics for Wound Infections." *Advances in Wound Care* 10 (6): 317–327. doi:10.1089/ wound.2019.1103

Ling, Pinghua, Xinyu Sun, Nuo Chen, Shan Cheng, Xianping Gao, and Feng Gao. 2021. "Electrochemical Biosensor Based on Singlet Oxygen Generated by Molecular Photosensitizers." *Analytica Chimica Acta* 1183 (October). Elsevier: 338970. doi:10. 1016/J.ACA.2021.338970

Liu, Ziqi, Junqing Liu, Tiancheng Sun, Deke Zeng, Chengduan Yang, Hao Wang, Cheng Yang, et al. 2021. "Integrated Multiplex Sensing Bandage for in Situ Monitoring of Early Infected Wounds." *ACS Sensors* 6 (8). American Chemical Society: 3112–3124. doi:10.1021/ACSSENSORS.1C01279/SUPPL_FILE/SE1C01279_SI_001.PDF

Manjakkal, Libu, Saoirse Dervin, and Ravinder Dahiya. 2020. "Flexible Potentiometric PH Sensors for Wearable Systems." *RSC Advances* 10 (15). Royal Society of Chemistry: 8594–8617. doi:10.1039/D0RA00016G

Mariani, Federica, Martina Serafini, Isacco Gualandi, Danilo Arcangeli, Francesco Decataldo, Luca Possanzini, Marta Tessarolo, Domenica Tonelli, Beatrice Fraboni, and Erika Scavetta. 2021. "Advanced Wound Dressing for Real-Time PH Monitoring." *ACS Sensors* 6 (6). American Chemical Society: 2366–2377. doi:10.1021/ACSSENSORS. 1C00552/SUPPL_FILE/SE1C00552_SI_001.PDF

Mostafalu, Pooria, Ali Tamayol, Rahim Rahimi, Manuel Ochoa, Akbar Khalilpour, Gita Kiaee, Iman K. Yazdi, et al. 2018a. "Smart Bandage for Monitoring and Treatment of Chronic Wounds." *Small* 14 (33). John Wiley & Sons, Ltd: 1703509. doi:10.1002/SMLL.201703509

Mostafalu, Pooria, Ali Tamayol, Rahim Rahimi, Manuel Ochoa, Akbar Khalilpour, Gita Kiaee, Iman K. Yazdi, et al. 2018b. "Smart Bandage for Monitoring and Treatment of Chronic Wounds." *Small* 14 (33). Wiley-VCH Verlag. doi:10.1002/SMLL.201703509

Mota, Fátima A.R., Sarah A.P. Pereira, André R.T.S. Araújo, Marieta L.C. Passos, and M. Lúcia M.F.S. Saraiva. 2021. "Biomarkers in the Diagnosis of Wounds Infection: An Analytical Perspective." *TrAC - Trends in Analytical Chemistry* 143. doi:10.1016/j.trac.2021.116405

Palomar, Quentin, Xing Xing Xu, Robert Selegård, Daniel Aili, and Zhen Zhang. 2020. "Peptide Decorated Gold Nanoparticle/Carbon Nanotube Electrochemical Sensor for Ultrasensitive Detection of Matrix Metalloproteinase-7." *Sensors and Actuators B: Chemical* 325 (December). Elsevier: 128789. doi:10.1016/J.SNB.2020.128789

Pang, Qian, Dong Lou, Shijian Li, Guangming Wang, Bianbian Qiao, Shurong Dong, Lie Ma, Changyou Gao, and Zhaohui Wu. 2020. "Smart Flexible Electronics-integrated Wound Dressing for Real-time Monitoring and On-demand Treatment of Infected Wounds." *Advanced Science* 7 (6). Wiley Online Library: 1902673.

Sen, Chandan K. 2019. "Human Wounds and Its Burden: An Updated Compendium of Estimates." *Advances in Wound Care* 8 (2). Mary Ann Liebert, Inc.: 39. doi:10.1089/WOUND.2019.0946

Serra, Raffaele, Raffaele Grande, Lucia Butrico, Alessio Rossi, Ugo Francesco Settimio, Benedetto Caroleo, Bruno Amato, Luca Gallelli, and Stefano de Franciscis. 2015. "Chronic Wound Infections: The Role of Pseudomonas Aeruginosa and Staphylococcus Aureus." *Expert Review of Anti-Infective Therapy* 13 (5). Taylor & Francis: 605–613. doi:10.1586/14787210.2015.1023291

Simoska, Olja, Jonathon Duay, and Keith J. Stevenson. 2020. "Electrochemical Detection of Multianalyte Biomarkers in Wound Healing Efficacy." *ACS Sensors* 5 (11): 3547–3557. doi:10.1021/acssensors.0c01697

Simoska, Olja, and Keith J. Stevenson. 2022. "Electrochemical Sensors for Detection of Pseudomonas Aeruginosa Virulence Biomarkers: Principles of Design and Characterization." *Sensors and Actuators Reports* 4 (December 2021). Elsevier B.V.: 100072. doi:10.1016/j.snr.2021.100072

Tang, Ning, Rongjun Zhang, Youbin Zheng, Jing Wang, Muhammad Khatib, Xue Jiang, Cheng Zhou, et al. 2022. "Highly Efficient Self-Healing Multifunctional Dressing with Antibacterial Activity for Sutureless Wound Closure and Infected Wound Monitoring." *Advanced Materials* 34 (3). John Wiley & Sons, Ltd: 2106842. doi:10.1002/ADMA.202106842

Thet, Naing Tun, and Toby A Jenkins. 2015. "An Electrochemical Sensor Concept for the Detection of Bacterial Virulence Factors from Staphylococcus Aureus and Pseudomonas Aeruginosa." *Electrochemistry Communications* 59: 104–108. doi:10.1016/j.elecom.2015.01.001

Tian, Feng, Jing Lyu, Jingyu Shi, Fei Tan, and Mo Yang. 2016. "A Polymeric Microfluidic Device Integrated with Nanoporous Alumina Membranes for Simultaneous Detection of Multiple Foodborne Pathogens." *Sensors and Actuators B: Chemical* 225: 312–318. doi:10.1016/j.snb.2015.11.059

Vladyslav Mishyn, Merve Aslan, Adrien Hugo, Teresa Rodrigues, Henri Happy, Rana Sanyal, Wolfgang Knoll, et al. 2022. "Catch and Release Strategy of Matrix Metalloprotease Aptamers via Thiol–Disulfide Exchange Reaction on a Graphene Based Electrochemical Sensor." *Sensors & Diagnostics*. 1. Royal Society of Chemistry. doi:10.1039/D2SD00070A

Vogiazi, Vasileia, Armah De La Cruz, Siddharth Mishra, Vesselin Shanov, William R. Heineman, and Dionysios D. Dionysiou. 2019. "A Comprehensive Review: Development of Electrochemical Biosensors for Detection of Cyanotoxins in Freshwater." *ACS Sensors* 4 (5). American Chemical Society: 1151–1173. doi:10.1021/ACSSENSORS.9B00376/ASSET/IMAGES/LARGE/SE-2019-003768_0004.JPEG

Wang, Lirong, Mengyun Zhou, Tailin Xu, and Xueji Zhang. 2022. "Multifunctional Hydrogel as Wound Dressing for Intelligent Wound Monitoring." *Chemical Engineering Journal* 433: 134625. doi:10.1016/j.cej.2022.134625

Xu, Changhao, Yiran Yang, and Wei Gao. 2020. "Skin-Interfaced Sensors in Digital Medicine: From Materials to Applications." *Matter* 2 (6). Cell Press: 1414–1445. doi:10.1016/J.MATT.2020.03.020

Xu, Gang, Lu, Yanli, Cheng, Chen, Li, Xin, Xu, Jie, Liu, Zhaoyang, Liu, Jinglong, Liu, Guang, Shi, Zhenghan, Chen, Zetao, Zhang, Fenni, Jia, Yixuan, Xu, Danfeng, Yuan, Wei, Cui, Zheng, Low, Sze Shin, & Liu, Qingjun (2021). Battery-Free and Wireless Smart Wound Dressing for Wound Infection Monitoring and Electrically Controlled On-Demand Drug Delivery. Advanced Functional Materials, 31 (26): 2100852. doi:10.1002/adfm.202100852.

Yamada, Kara, Won Choi, Inae Lee, Byoung-Kwan Cho, and Soojin Jun. 2016. "Rapid Detection of Multiple Foodborne Pathogens Using a Nanoparticle-Functionalised Multi-Junction Biosensor." *Biosensors and Bioelectronics* 77: 137–143. doi:10.1016/j.bios.2015.09.030

Yan, Qinghua, Na Zhi, Li Yang, Guangri Xu, Qigao Feng, Qiqing Zhang, and Shujuan Sun. 2020. "A Highly Sensitive Uric Acid Electrochemical Biosensor Based on a Nano-Cube Cuprous Oxide/Ferrocene/Uricase Modified Glassy Carbon Electrode." *Scientific Reports* 10 (1). Nature Publishing Group: 1–10. doi:10.1038/s41598-020-67394-8

Zhang, Zhiyang, Rui Su, Fei Han, Zhiqiang Zheng, Yuan Liu, Xiaomeng Zhou, Qingsong Li, et al. 2022. "A Soft Intelligent Dressing with PH and Temperature Sensors for Early Detection of Wound Infection." *RSC Advances* 12 (6). Royal Society of Chemistry: 3243–3252. doi:10.1039/D1RA08375A

# 8 Phosphorene-Based Electrochemical Systems

*Surendran Jyothis and Ravindran Sujith*
Mechanical Engineering Department, BITS Pilani Hyderabad Campus, Hyderabad, Telangana, India

*Sanket Goel*
Electrical and Electronics Engineering Department, BITS Pilani Hyderabad Campus, Hyderabad, Telangana, India

## CONTENTS

## 8.1 INTRODUCTION

The investigation of two-dimensional (2D) materials like graphene, BN, and $MoS_2$ has emerged as one of the most intriguing fields of nanoscience in the past few years [1–6]. However, a novel 2D material – phosphorene – has sparked substantial interest in the scholarly community recently. Phosphorene is a 2D allotrope of phosphorus with a single-layer sheet structure and is the building unit of black phosphorus (BP). Based on the number of layers, phosphorene may be classified as a monolayer, bilayer, and few-layer phosphorene. According to the morphology, it may be categorized as quantum dots, nanoribbons, nanowires, nanorods, microrods/platelets, nanotubes, nanoparticles, nanoflakes, and microscale flakes [7]. Phosphorene monolayers can be built to form black phosphorus crystal layers via van der Waals interactions, similar to how graphene layers stack to form graphite.

a)

b)

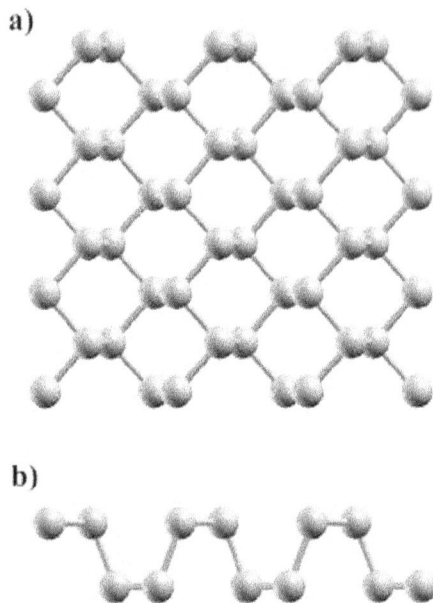

**FIGURE 8.1**   Crystal structure of single-layer phosphorene: (a) top view, (b) side view.

**Source: Adapted with permission from [10] Copyright (2020) Elsevier.**

Phosphorene has an in-plane bonding as a consequence of the $sp^3$-hybridization present in it, unlike in the case of graphene [8]. Since every phosphorus atom is bonded to three neighboring phosphorus atoms, each p-orbital maintains a lone pair of electrons. Due to the $sp^3$-hybridization, phosphorene does not form atomically flat sheets like graphene. Rather, they form honeycomb-structure puckered layers as seen from a side perspective normal to the armchair direction (Figure 8.1). Every such layer of the honeycomb network has two atomic layers with marginally separated two adjacent atoms ($d_1 = 2.224$ Å) and the top and bottom atoms ($d_2 = 2.244$ Å) due to this puckered structure. The values of $d_1$ and $d_2$ are quite close to each other due to covalent interactions between phosphorus 3p-orbitals. The stacking of phosphorene monolayers produces few layers of phosphorene. The bulk phosphorene lattice constants are $a_1 = 0.34$ nm, $a_2 = 0.45$ nm, and $a_3 = 1.12$ nm [9]. Between two phosphorene layers, the sheet-sheet gap is 0.53 nm, which is larger than the interlayer spacing of graphene (0.335 nm). The puckered structure along with the AB Bernal stacking of two phosphorene layers in the unit cell, account for the comparatively high interlayer separation. A weak interlayer contact (20 meV atom$^{-1}$) holds these phosphorene sheets together, although it can be easily disrupted. As a result, phosphorene is particularly appealing both for electronics and optoelectronics and has therefore been broadly studied in applications such as transistors, photodetectors, and solar cells. Owing to its novel structure and properties phosphorene is being researched for a variety of electrochemical applications. These applications are very much linked to the morphology of phosphorene and the synthesis technique adopted. This chapter aims to give a detailed description of various methods of phosphorene synthesis and its applications in electrochemical systems.

## 8.2   SYNTHESIS OF PHOSPHORENE FROM BLACK PHOSPHOROUS

There are mainly two methods that are adopted to synthesize phosphorene thin films. They are (i) the top-down approach involving exfoliation of black phosphorus, and (ii) the bottom-up approach, involving thermal depositions or assemblies from phosphorus precursors. Exfoliation methods rely on weak interlayer interactions between the two phosphorene sheets, which can be easily overcome by using external forces. The exfoliations may be classified into mechanical exfoliation, liquid-phase/sonication exfoliation, and microwave exfoliation based on the external forces. The sonication exfoliation is further categorized into organic liquid, aqueous surfactant, and water exfoliation based on the assisting agents.

### 8.2.1   MECHANICAL EXFOLIATION

The mechanical exfoliation or micromechanical cleavage is the process of removing nanoflakes from bulk crystals, as shown in Figure 8.2(a). Scotch tape is usually employed for this purpose and is hence known as the scotch-tape method [11].

Mechanical exfoliation is extensively used for the manufacture of 2D black phosphorus. Many 2D membranes, including graphene, $MoS_2$, $WS_2$, h-BN, and MXene, have been effectively isolated via the microcleavage method. This technique is advantageous since it is simple and reliable, as it does not require any chemical treatments or expensive instruments during the production process. Phosphorene is widely used in fundamental physics and electronic device research owing to its simplicity of isolation via scotch tape. Since the as-exfoliated phosphorene normally has a few layers, surface cleaning and layer thinning with argon or ozone plasma can be utilized to obtain monolayer phosphorene. Mechanical exfoliation with adhesive tape produces a low density of a few layers of black

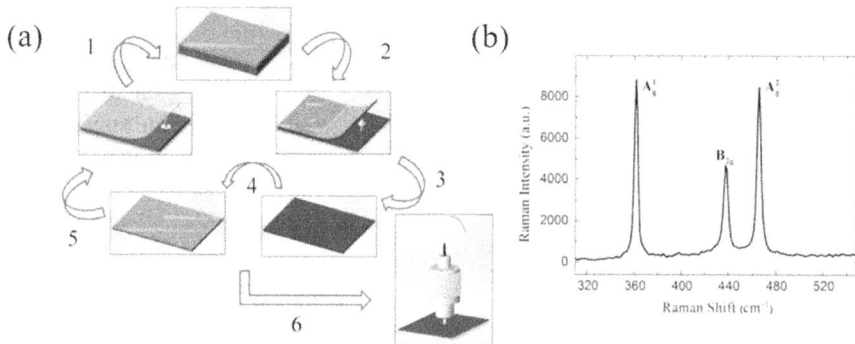

**FIGURE 8.2**   (a) Mechanical exfoliation of BP, (b) Raman spectra of BP.

**Source: Adapted with permission from [12] Copyright(2015) Optica Publishing Group.**

phosphorus flakes, leaving sticky traces on the surface. Exfoliating the flakes with an intermediate viscoelastic surface significantly boosts the yield and minimizes the contamination of the produced flakes [13]. Mechanical exfoliation requires repetitive steps of adhesion and splitting in order to obtain a 2D layer. Moreover, the layer thickness determination is to be carried out using microscopy and Raman spectroscopy techniques (Figure 8.2(b)) after each exfoliation. Hence, it is a sluggish process with low yield. As a result, this approach has limited mass-production potential. Nevertheless, mechanical exfoliation serves its purpose in university laboratories for basic research.

### 8.2.2 LIQUID-PHASE EXFOLIATION

Liquid phase exfoliation (LPE) via ultrasonication is suitable for large-scale production of black phosphorus nanosheets. Based on the solvent used for exfoliation, LPE may be classified as follows:

- Organic solvent-based exfoliation,
- Stabilizer-based exfoliation,
- Ionic liquid (IL)-based exfoliation,
- Salt-assisted exfoliation,
- Intercalant-assisted exfoliation,
- Ion exchange–based exfoliation [14].

The bath sonication or the probe sonication method can both be used for ultrasonication. Ultrasonication of the solvent results in the formation of bubbles which breaks to form high-energy jets, cleaving the layered 2D material [15]. The effectiveness of exfoliation depends on the surface energy of the solvent; the closer the surface energies of the solvent and the material, the better the exfoliation. The solvent with a Hansen solubility parameter comparable to that of the material is preferred [16]. Amide solvents such as N-cyclohexyl-2-pyrrolidone (CHP) or N-methyl-2-pyrrolidone (NMP) are often used for this technique, although isopropanol (IPA) has been found suitable in some cases. To obtain atomically thin BP nanoflakes, dimethylformamide (DMF) and dimethyl sulfoxide (DMSO) are suitable. They can yield uniform and stable dispersions after sonication. Initially, a lump of black phosphorous crystal ($0.02$ mg mL$^{-1}$) is added in a suitable solvent and is sonicated for 15 hours. The solution is then centrifuged and the supernatants are carefully collected by a pipette. A scanning electron microscope (SEM) image of a 3D layer of a phosphorene nanosheet on a silicon wafer is obtained after exfoliation in NMP, as shown in Figure 8.3. The translucent architecture of the phosphorene structure, with foldings at certain points, distinguishes it.

### 8.2.3 ELECTROCHEMICAL EXFOLIATION

When compared to liquid phase exfoliation through a sonication method, electrochemical exfoliation of BP is a more tractable, quicker, and less expensive approach [17]. Exfoliation of a broad range of 2D materials has been accomplished using this

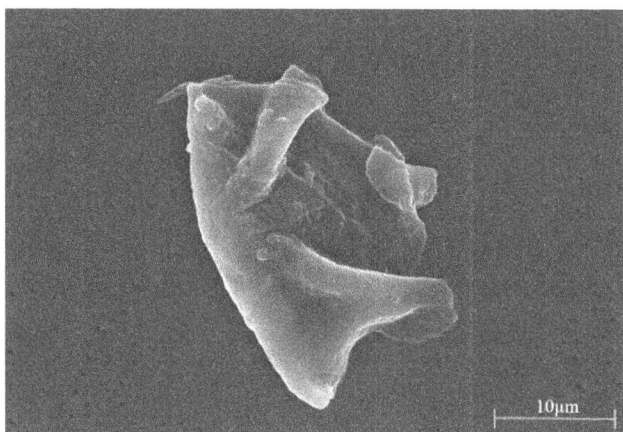

**FIGURE 8.3** SEM image of phosphorene nanosheet exfoliated in NMP solvent.

technique. At ambient conditions, electrochemical exfoliation of BP is a fairly simple procedure, and the reactions are completed in a short time, allowing control of the degradation of the generated phosphorene nanosheets. This approach employs the proper reduction and oxidation reactions to produce ideal nanosheet thicknesses in large quantities for practical applications [18–20]. A regular electrochemical cell for exfoliation consists of working and counter electrode connected to the power supply, and a reference electrode in an electrolyte (aqueous or non-aqueous solution) as shown in Figure 8.4. Electrolytes with appropriate surface tension

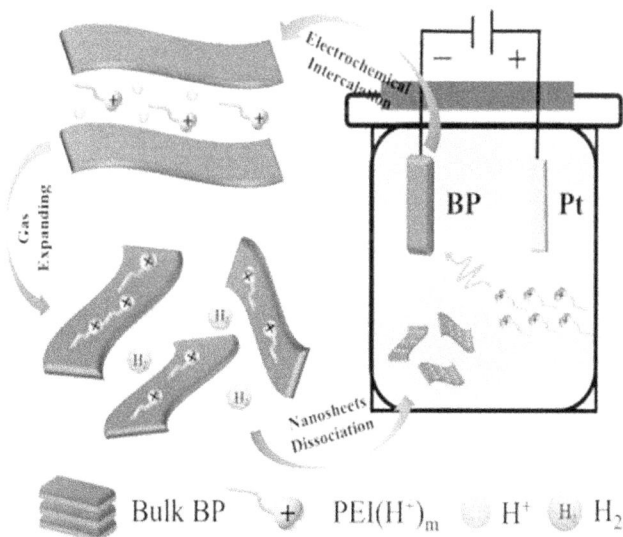

**FIGURE 8.4** Electrochemical anodic/cathodic exfoliation.

Source: Adapted with permission from [21] Copyright (2020) Elsevier.

promote anion and cation migration to the BP layers and prevent nanosheet re-stacking after exfoliation. The ionic species in the solution intercalate the bulk-layered BP in presence of an electric field and expand it into phosphorene nanosheets. The properties and the yield of the phosphorene nanosheets produced are controlled by the ionic intercalation and their interaction with electrodes during exfoliation.

## 8.3  APPLICATIONS

### 8.3.1  ANODE IN LITHIUM-ION BATTERIES

Rechargeable lithium-ion batteries (LIBs) are widely used in consumer electronics such as smartphones, laptops, tablets, digital cameras and so on. They are con-sidered superior energy storage units due to their stable cycling performance, high storage capacity, and high energy density [22]. Pure phosphorus has lately risen to prominence in lithium-ion batteries because of its high power density and long cycle life. Phosphorus in elemental form has a theoretical capacity of 2,596 mAhg$^{-1}$ (with Li$_3$P as the end compound) and has a low diffusion energy barrier of 0.08 eV for lithium ions [23]. The phosphorene anode has a larger surface area when compared to bulk BP which results in increased irreversible capacity at initial cycles. Figure 8.5 shows a Li-ion battery comprising of phosphorene anode and LiCoO$_2$ cathode. Theoretical studies indicate that due to its anisotropic nature Li-ion diffusion along the arm-chair direction is restricted by a high energy barrier of 0.68 eV [22]. Li and colleagues researched on LIBs with phosphorene anode and manifested that lithium-ion intercalation into phosphorene layers resulted in en-hanced electrical conductivity with a high open-circuit potential of 2.9 V, which is

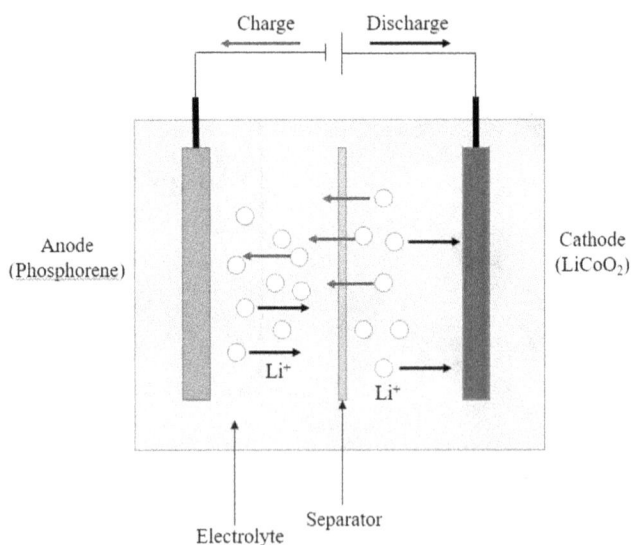

**FIGURE 8.5**  Schematic depicting lithium-ion battery with a phosphorene anode and LiCoO$_2$ cathode.

adequate for increased performance [24]. Theoretically, the phosphorene nanosheets have a larger specific capacity (432.79 mAhg$^{-1}$) than graphite anodes (372 mAh g$^{-1}$). Bringing in surface defects on the BP nanosheet can also enhance the binding energy of Li and phosphorene. In addition, the incorporation of a conductive material like graphene in phosphorene leads to hetero-structured anode enhancing the Li-ion mobility by minimizing the polarization effect.

Castillo et al. [25] studied the electrochemical performance of liquid phase exfoliated phosphorene nanosheets in acetone and CHP. In general, the electrochemical mechanism of BP nanosheets in the charging and discharging cycles can be described as follows:

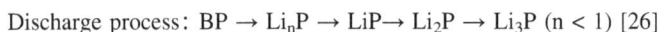

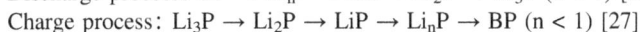

Discharge process: BP → Li$_n$P → LiP→ Li$_2$P → Li$_3$P (n < 1) [26]
Charge process: Li$_3$P → Li$_2$P → LiP → Li$_n$P → BP (n < 1) [27]

Initial capacities of 1,732 and 545 mAhg$^{-1}$ were obtained at 100 mAg$^{-1}$ for the exfoliated phosphorene in the solvents acetone and CHP respectively, from the galvanostatic charge/discharge profiles. However, after 20 cycles a significant capacity fading was observed (480 and 250 mAh g$^{-1}$ for acetone and CHP) and is due to the high volume expansion (300%) of phosphorene. Furthermore, the rate capability studies conducted on these anodes exhibited a lower specific capacity of 345 and 200 mAh g$^{-1}$ for the exfoliated phosphorene in acetone and CHP at 1 Ag$^{-1}$ [25]. The lower specific capacity and low Columbic efficiency in these systems were attributed to the irreversible Li$_3$P formed during the first discharge cycle [15].

Moreover, studies have shown that carbonaceous materials can be added to these systems to produce hybrid electrodes too improve the electrochemical performance and structural stability of black phosphorus nanosheets. Chen and co-workers [15] synthesized a 2D black phosphorus – 2D graphene (named BP-G) film by filtering black phosphorus nanosheets (80 wt.%) and graphene sheets. When these films were used as anodes, it leads to a specific capacity of 402 mAh/g and a Coulombic efficiency of ~100% at a current density of 500 mA/g.

In another study, Chen and colleagues investigated the electrochemical performance of exfoliated BP and phosphorene (80%) with graphene (20%) (P-G) hybrid as a LIB anode generated by vacuum filtration. The P/G nanosheets are stacked in parallel in the produced composite, which enhances the electrical conductivity of phosphorene and shortens the lithium-ion diffusion [28]. Graphene nanosheets serve as a preferential electrical pathway for the transit of electrons generated by the phosphorene redox process. The setting up of chemical P-C bonds increased the cyclic structural stability. This composite anode achieved a specific capacity of 920 mAh g$^{-1}$ at a current density of 100 mA g$^{-1}$, which is much higher compared to the pristine phosphorene (180 mAh g$^{-1}$) and graphene (435 mAh g$^{-1}$). The hybrid anode may still provide high specific capacity (501 mAh g$^{-1}$), reversibility, and rate capability by raising the current density to 500 mA g$^{-1}$ [29]. Figure 8.6 shows the rate capability performance of phosphorene-carbon composite anode at different current densities. Following this work, Zhang et al. investigated high-yield exfoliated black phosphorus in methanamide as an air-stable anode material for LIBs [30]. They found that phosphorene nanosheets react more readily

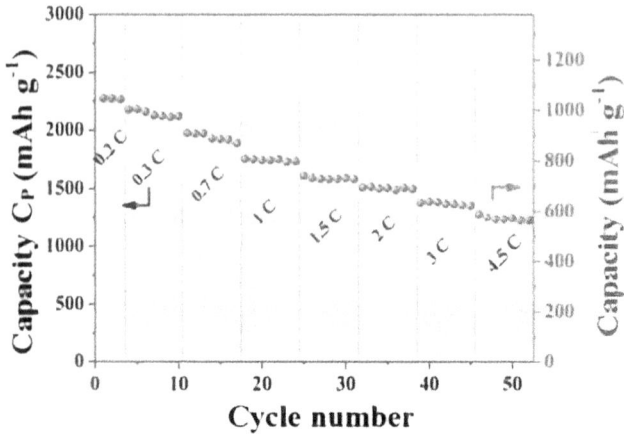

**FIGURE 8.6**   Rate capability performance of the phosphorene-carbon composite anode at different current densities.

**Source: Adapted with permission from [31] Copyright(2014) Elsevier.**

with graphene oxide than graphene. The fabricated anode revealed a high specific capacity of 1013 and 415 mAh g$^{-1}$ at the current density of 100 mA g$^{-1}$ and 10 Ag$^{-1}$. Hence, these studies show that the increased surface area of phosphorene has resulted in high irreversible capacities. Moreover, the addition of carbonaceous fillers in phosphorene and the resultant hetero-structures can improve the overall specific capacity of these systems due to its better structural stability and higher ionic diffusivity.

### 8.3.2   LITHIUM-SULFUR (LI-S) BATTERIES

Lithium-sulfur cathodes have gained increased attention and they stand out due to the high specific capacities of lithium and sulfur [32,33]. However, Li–S batteries have several limitations like the low electrical and ionic conductivities of sulfur, high volume change upon lithiation, sulfur loss due to dissolution in the electrolyte, and sulfur transfer between the anode and cathode. All of this leads to low reversibility and inefficient sulfur usage. As a solution to this problem, composites of sulfur were prepared with conductive matrices like graphene, carbon nanotubes, and porous carbonaceous materials. Phosphorene is preferred over nanocarbon materials due to its property of anchoring/immobilizing sulfur with strong P-S bonds to enable its efficient utilization [34]. Ren and his coworkers discovered that phosphorene works as a sulfur immobilizer and electro-catalyst, extending the cycle life lithium-sulfur batteries with a higher capacity [28]. In addition, when compared with conventional cathodes, phosphorene present in the cathode matrix remarkably decreases polarization and accelerates the redox reactions for quick charging/discharging and increases sulfur utilization. Furthermore, phosphorene can alter the separator to trap and activate the poly-sulfides, resulting in an enhanced capacity and increased cyclability.

Hence, the incorporation of phosphorene improves the performance of Li-S batteries by immobilizing the sulfur for its better utilization, reducing polarization and accelerating the redox reactions.

### 8.3.3 SODIUM-ION BATTERIES

Sodium-ion batteries look promising as a replacement to LIBs, thanks to their low cost and abundant supply of Na. Their charging and discharging processes are similar to those of Li-ion batteries. The first sodium-ion batteries with phosphorene/graphene (P/G) composite anode were fabricated by Cui and colleagues [35], which showed a reversible capacity of 2,440 mAhg$^{-1}$ at 0.02 C. The P/G layers were sandwiched together, which shortens the diffusion path of ions and electrons leading to increased rate performance. Additionally, this makes it possible to accommodate volume expansion via elastic buffer spacing. Graphene also permits electrons to be transported from the phosphorene redox process to the current collectors. The sodium-ion batteries also retain about 85% of their capacity after 100 cycles. Another promising strategy is to encapsulate phosphorene in h-BN nanosheets, which is found to result in high theoretical capacity and the least diffusion barrier. The sodium transport in phosphorene nanosheets is dependent on the surface and edge sites. Nie and coworkers studied the Na$^+$ ion transport in phosphorene using transmission electron microscopy (TEM) and the density functional theory (DFT) [36]. Their calculations revealed that the zig-zag direction of phosphorene allows faster diffusion path for Na$^+$ ion when compared to the armchair direction.

Due to this high capacity, electrical conductivity, Na+ ion diffusion and better stability phosphorene is considered to be a good candidate for sodium-ion batteries.

### 8.3.4 ELECTRODES IN SUPERCAPACITORS

Due to their excellent characteristics like high power density, long cycle life, and quick charge/discharge rate, supercapacitors are gaining a lot of interest [37–39]. Materials like 2D stacked graphene, with a large specific surface area, are frequently used in high-capacitance double-electrode capacitors which exhibited increased volumetric capacitance of about 1 F cm$^{-3}$. Wen and colleagues produced phosphorene films on PET (polyethylene terephthalate) substrates using phosphorene dispersions and drop coating processes, resulting in a supercapacitor with a capacitance of 13.75 F cm$^{-3}$ at a scanning rate of 0.01 V s$^{-1}$ [40]. These phosphorene double electrodes were extremely flexible and had negligible current density loss. They achieved a maximum current density and power density of 2.47 mWh cm$^{-3}$ and 8.83 Wcm$^{-3}$, respectively.

Based on the charge storage mechanism, supercapacitors are classified into the following types:

- Pseudo-capacitors
- Electrical double-layer capacitors (EDLCs) [7]

The main operating processes of pseudo-capacitors and EDLCs are rapidly reversible redox reactions at the electrode-electrolyte interfaces and charge storage at the electrode/electrolyte interfaces, respectively. On account of their high specific surface area, exfoliated phosphorene nanosheets were explored as electrode materials in EDLCs. A conventional sandwich model phosphorene electrode was fabricated by Hao et al. with the solid polyvinyl alcohol/phosphoric acid (PVA/$H_3PO_4$) electrolyte and demonstrated a double layer capacitance of 13.75 F $cm^{-3}$ at the scan rate of 0.01 V $s^{-1}$ with cyclability up to 30,000 cycles [40].

Phosphorene/carbon nanotube (CNT) composites are gaining increased attention in the production of all-solid-state supercapacitors [41]. Composites with phosphorene-CNT mass proportion 1:4 have revealed excellent electrical conductance and packing density. The electrodes showed a maximum volumetric capacity of 41 F $cm^{-3}$ and a maximum high power density of 821 W $cm^{-3}$. The superior mechanical flexibility of CNT contributes to the strength of the composite electrode.

Phosphorene is also incorporated into graphene as inserting spacers through redox reactions to produce hybrid electrodes to prevent agglomeration and improve mechanical strength [42]. The restacking issue of graphene is solved by this technique. Phosphorene nanosheets trapped in the graphene layers maintained rich pathways for electrolyte ion diffusion, resulting in improved electrochemical performance compared to pristine graphene. In order to further increase the specific capacitance of this electrode, pseudo-capacitive conductive agents like polyanlinine and polypyrrole were introduced, which resulted in an increased capacitance of 354 $Fg^{-1}$ [43]. The cyclic voltammetry (CV) curves at 100 $mVs^{-1}$ revealed a rectangular shape, verifying the double-layer charge storage mechanism of phosphorene.

The conventional sandwich type supercapacitors are incompatible when it comes to applications like wearable and portable microelectronics. The on-chip microsupercapacitors are suitable for these purposes. They are of three types:

- 3D micro-supercapacitors
- Fiber-shaped micro-supercapacitors
- In-plane micro-supercapacitors [44]

The in-plane micro-supercapacitors with micro-scale sizes exhibit high power density due to their planar structure, which promotes increased ionic transport. Recently, a simplified P/G micro-supercapacitor was developed by a single-step mask-assisted vacuum filtration method on a polyethylene terephthalate (PET) substrate [42]. The composite showed high mechanical flexibility and adhesion on to the PET substrate. It exhibited a typical EDLC-typed CV curve with a rectangular shape delivering an areal capacitance of 9.8 m F $cm^{-2}$ and volumetric capacitance of 37.5 F $cm^{-3}$.

Yang and coworkers used laser machining technology to develop solid-state microsupercapacitors using electrodes coated with thin film of BP. An ideal double-layer capacitor behavior is confirmed by the rectangular-shaped CV curves. Further, a linear capacitive performance with a negligible Ohmic drop was revealed by the galvanostatic charge/discharge at different current densities. The fabricated device displayed a high volumetric capacitance and energy density of 26.67 F $cm^{-3}$ and 3.63 mWh $cm^{-3}$ at 0.5 A $cm^{-3}$, respectively. A high energy density of 1.53 mW h $cm^{-3}$ persisted even at a very high power density of 10.1 Wcm$^{-3}$, which is higher when compared to other 2D-based micro-supercapacitors.

Phosphorene is found to enhance the capacitance, current density and power density in the case of conventional type supercapacitors. Combining phosphorene along with graphene improved the electrical conductivity due to improved ion diffusion. In addition, micro-supercapacitors with good mechanical flexibility and improved capacitance were developed, making phosphorene a possible contender for supercapacitor applications.

### 8.3.5 ELECTROCATALYST FOR OXYGEN EVOLUTION REACTION

The oxygen evolution reaction (OER) and oxygen reduction reaction (ORR) are critical as far as the electrochemical energy conversion and storage is concerned [6,27,45]. Both OER and ORR require efficient electrocatalysts to accelerate the reaction due to their sluggish reaction kinetics. The chemical reactions corresponding to OER and ORR are as shown below:

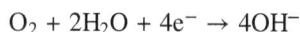

$$4OH^- \rightarrow 2H_2O + 4e^- + O_2.$$

$$O_2 + 2H_2O + 4e^- \rightarrow 4OH^-$$

For both reactions, the electrocatalyst should have good electrical charge conductivity, a large specific surface area, and high activity. Platinum, transitional metal oxides and nitrides are among the most advanced catalysts, but their high cost and scarcity prevent large manufacture. Hence, the efforts are towards development of novel electrocatalysts based on low-cost and commonly available elements.

Wang and colleagues made a recent breakthrough in electrocatalytic OER, where the bulk BP was produced for OER electrocatalysts [27]. An onset potential of ~1.49 V relative to the reversible hydrogen electrode (RHE) and a low Tafel slope of about 72.88 mV dec$^{-1}$ was observed in these systems, revealing an electrocatalytic performance comparable to that of commercial $RuO_2$ catalysts. Although the results are interesting and bulk BP is a promising OER electrocatalyst, the number of active sites are low in BP and hence there are challenges for achieving highly efficient OER catalytic performances. One strategy that is being adopted to increase the number of active sites in BP is by decreasing the number of layers of BP, thereby increasing the specific surface area [46]. Hence, it could be expected that these ultrathin lamellar structure may pave the way for developing electrocatalysts with enhanced catalytic activity.

Phosphorene and phosphorene-based hybrids are now being investigated as potential catalysts for a variety of electrocatalytic applications. Their exceptional electrochemical performance is comparable to those of transition-metal or noble-metal-based catalysts. The phosphorene synthesized by liquid phase exfoliation of black phosphorus displayed exceptional electrochemical OER performance exhibiting extended stability, exceeding that of the reported $RuO_2$ and $Co_3O_4$/N-graphene [46]. Owing to an onset potential of of 1.48 eV relative to the reversible hydrogen electrode and a current density of 10 mA $cm^{-2}$ at 1.6 V, black phosphorus has been proposed as a potential OER catalyst, with catalytic activity comparable to commercial $RuO_2$ [46].

Ren and co-workers utilized a conventional three-electrode cell consisting of potassium hydroxide (KOH) electrolyte with varying concentrations of $OH^-$ ($C[OH^-]$) to study the current variations during OER processes [46]. They took advantage of KOH solution as an electrolyte to improve the stability of as-prepared BP nanosheets and investigated the OER performance of BP nanosheets for different $OH^-$ concentrations. It was found that BP nanosheets have a higher current density and a lower onset potential (about 1.45 eV) than the previous reported 1.48 eV for bulk BP, in 1 M KOH. Further reduction of concentrations from 1 to 0.05 M can result in a fall in current density. In comparison to their pristine counterparts and other previously reported electrocatalysts, BP nanosheets demonstrate considerably increased OER activity, according to the computed Tafel slopes [47]. The superior electrochemical performance of BP nanosheets is ascribable to the additional active sites and significantly high electronic mobility benefited from ultrathin 2D-layered structures and intrinsic qualities of BP. Therefore, the low OER onset potential and the low Tafel slope of few-layered BP nanosheets show that they have considerable potential as OER electrocatalysts. Few-layered BP or phosphorene nanosheets have great potential in the applications of high-performance OER devices. Decreasing the thickness of BP nanosheets is recommended to be a good strategy to generate more active sites and improve the specific surface area in order to achieve superior electrochemical OER performance.

The ultrathin lamellar structure with increased specific surface area resulting in more active sites as well as high electronic mobility makes phosphorene a suitable material for OER electrocatalysis.

### 8.3.6 Electrochemical Sensors

Black phosphorus due to its good affinity for water molecules has been employed in the fabrication of humidity sensors based on the transduction principles [48]. Moreover, composites of BP with other materials, like indium oxide, CuO/pyrrole-BP nanocomposite, anthraquinone nanowire (AQNW), and chemically passivated phosphorene (CPP) with porous triazine-based two-dimensional polymer (T-2DP) or noble metals, have been used to detect gases like $NO_2$ or $H_2$. Silicon- or sulfur-doped phosphorene is found to have good sensitivity towards gases like NO, $NO_2$, CO, and $NH_3$ [49]. The doping results in the modification of the electronic structure of phosphorus atom thereby increase the binding energy. Noble metals, such as Au or Pt, are found to significantly improve the sensitivity of BP nanosheets to $H_2$. The field effect transistor (FET) is one of the most commonly used transducers modified by the dry transfer method. Degradation of BP while in contact with liquid or gaseous medium is a matter of concern. $Al_2O_3$ or Nafion, can be used as a passivation layer to improve BP stability against degradation.

BP is also employed in the detection of metal ions like Hg (II), As(III)/(V), Pb(II), and Ag(I) from liquid samples. Figure 8.7 shows the BP-based sensor for As(III)ion detection [50]. The deposition of antihuman IgG-conjugated gold nanoparticles on the

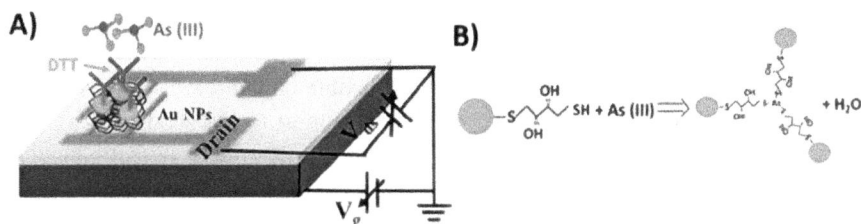

**FIGURE 8.7**  (a) Schematic of the BP/Au NPs/DTT sensing platform for the As ion detection;
(b) reaction between the dithiothreitol (DTT) and As(III) ion in the detection process.

**Source: Adapted with permission from [50] Copyright(2018) Elsevier.**

$Al_2O_3$ dielectric layer surface of a BP-based FET makes the sensor detect human
immunoglobulin G (IgG). BP is also used as substrates in nanocomposites for other
sensing devices. These rely on electrical impedance spectroscopy (EIS) or electrochemiluminescence for measurements (ECL). This method is commonly employed
to detect cancer cells, cell fragments, and proteins in human serum. It has been also
used in the detection of Pb(II)-ions in tap and river/lake water samples, with high
sensitivity.

Numerous sensors have been created, prototyped, and experimentally tested
using phosphorene/few-layer BP, including gas, humidity, light, biological molecules, and ion sensors. Their sensitivity performance is shown to be on par with, and
often better than, that of other 2D material platforms in competition, such as graphene and TMDs.

## 8.4  CONCLUSIONS AND FUTURE SCOPE

This chapter provides an overview of the various electrochemical applications of
phosphorene. Few layer/monolayer BP is an emerging material and is found to be
suitable for various electrochemical applications such as LIBs, supercapacitors,
catalysts, sensors, etc. However, the electrochemical performance of phosphorene is
strongly influenced by the synthesis procedures. Although liquid phase exfoliation
of BP through a sonication route is a scalable technique, it is time-consuming and is
susceptible to high defect concentration and usage of toxic solvents. Moreover, the
harsh environments and multi-stage preparation of phosphorene nanosheets make
this technique less appealing.

Another major challenge with the phosphorene nanosheets is its surface stability.
Among the various 2D materials, phosphorene nanosheets experience the most
severe degradation even in a few hours after exfoliation both in solvent media and
in the solid form. To obtain high-performance BP-based energy storage devices,
researchers must develop a new method to regulate the surface deterioration of
phosphorene nanosheets following exfoliation. Also, these phosphorene sheets are
prone to large volumetric expansions leading to rapid capacity weakening and low
Coulombic efficiency. Integration of carbonaceous materials and developing hetero-
structures seems to provide better structural stability and thereby achieving better

cyclic stability. However, the synthesis of a simple and elementary P/G hybrid material is still at an early stage.

Phosphorene coated on PET yields highly flexible double electrodes with minimal current density loss and high capacitance owing to its higher specific surface area. However, the morphology of the phosphorene nanosheets is detrimental in deciding the optimal behavior of CV curves, which is further dependent on the synthesis method. Efficient electrocatalysis of an oxygen evolution reaction requires a large number of active sites. Even though bulk BP lacks sufficient active sites, few layers of BP are found to be a superior candidate due to the reduced layer thickness which increases the surface area, along with high electron mobility. Improving the synthesis technique to produce monolayer BP in bulk quantity is an effective strategy. Degradation of phosphorene when in contact with water or other solutions is the major challenge for electrochemical sensors. Passivation layers that prevent the degradation of phosphorene nanosheets could improve the performance of the sensor. Integration of noble metals also help in improving the sensitivity.

Since the successful isolation of graphene, 2D materials have grown in importance across the world. Few layers of BP is being widely researched for various applications among the 2D material research enthusiasts' community. The structural features that dictate the useful properties that are of interest are to be explored in more depth, especially the effect of layer thickness. Very few research works are accessible on the parameters that govern layer thickness in the liquid phase exfoliation process, which demand further understanding. The probe sonication procedure can generate a lot of localized energy, which can be harmful for the BP layers. To avoid damage to the thin layers, proper optimization of the sonication energy is required. Moreover, the degradation of phosphorene when exposed to ambient conditions needs to be investigated in order to enhance the durability of the material following exfoliation.

Phosphorene/few-layer BP is hence found to be an upcoming versatile material with intriguing properties suitable for the aforementioned applications and could pave way for the manufacture of efficient and cost-effective electrochemical systems in the future.

## REFERENCES

[1] A.K. Geim, "Graphene: status and prospects", *Science*. 324 (5934) (2009) 1530–1534 10.1126/science.1158877

[2] Q.H. Wang, K. Kalantar-Zadeh, A. Kis, J.N. Coleman, M.S. Strano, "Electronics and optoelectronics of two-dimensional transition metal dichalcogenides", *Nat. Nanotechnol.* 7 (11) (2012) 699–712. 10.1038/nnano.2012.193

[3] K.S. Novoselov, V.I. Fal'Ko, L. Colombo, P.R. Gellert, M.G. Schwab, K. Kim, "A roadmap for graphene", *Nature*. 490 (2012) 192–200. 10.1038/nature11458

[4] A. C. Ferrari et al., "Science and technology roadmap for graphene, related two dimensional crystals, and hybrid systems", Nanoscale. 7 (2015) 4598–4810. 10.1039/c4nr01600a

[5] M. Chhowalla, H.S. Shin, G. Eda, L.J. Li, K.P. Loh, H. Zhang, "The chemistry of two-dimensional layered transition metal dichalcogenide nanosheets", *Nat. Chem.* 5 (2013) 263–275. 10.1038/nchem.1589

[6] W. T. Hong, M. Risch, K. A. Stoerzinger, A. Grimaud, J. Suntivich, Y. Shao-Horn, "Toward the rational design of non- precious transition metal oxides for oxygen electrocatalysis", *Energy Environ. Sci.* 8 (2015) 1404–1427 10.1039/C4EE03869J

[7] J. Pang, A. Bachmatiuk, Y. Yin, B. Trzebicka, L. Zhao, L. Fu, R.G. Mendes, T. Gemming, Z. Liu, M.H. Rummeli, "Applications of phosphorene and black phosphorus in energy conversion and storage devices", *Adv. Energy Mater.* 8 (8) (2018) 1702093. 10.1002/aenm.201702093

[8] C.R. Ryder, J.D. Wood, S.A. Wells, M.C. Hersam, "Chemically tailoring semi-conducting two-dimensional transition metal dichalcogenides and black phosphorus", *ACS Nano.* 10 (4) (2016) 3900–3917. 10.1021/acsnano.6b01091

[9] R.W. Keyes, "The electrical properties of black phosphorus", *Phys. Rev.* 92 (3) (1953) 580–584. 10.1103/PhysRev.92.580

[10] S. Ramachandran, K.V. Sai Srinivasan, R. Sujith, "Nickel-decorated single vacancy phosphorene – a favourable candidate for hydrogen storage", *Int. J. Hydrogen Energy.* 46 (54) (2021) 27597–27611. 10.1016/j.ijhydene.2021.05.206

[11] V. Eswaraiah, Q. Zeng, Y. Long, Z. Liu, "Black phosphorus nanosheets: synthesis, characterization and applications", *Small.* 12 (26) (2016) 3480–3502. 10.1002/smll.201600032

[12] Y. Chen, G. Jiang, S. Chen, Z. Guo, X. Yu, C. Zhao, H. Zhang, Q. Bao, S. Wen, D. Tang, D. Fan, "Mechanically exfoliated black phosphorus as a new saturable absorber for both Q-switching and Mode-locking laser operation", *Opt. Express.* 23 (10) (2015) 12823–12833. 10.1364/oe.23.012823

[13] S. Liang, M.N. Hasan, J.H. Seo, "Direct observation of raman spectra in black phosphorus under uniaxial strain conditions", *Nanomaterials.* 9 (4) (2019) 566. 10.3390/nano9040566

[14] A. Rabiei Baboukani, I. Khakpour, V. Drozd, C. Wang, "Liquid-based exfoliation of black phosphorus into phosphorene and its application for energy storage devices", *Small Struct.* 2 (5) (2021) 2000148. 10.1002/sstr.202000148

[15] L. Chen, G. Zhou, Z. Liu, X. Ma, J. Chen, Z. Zhang, H.M. Cheng, X. Ma, F. Li, W. Ren, "Scalable clean exfoliation of high-quality few-layer black phosphorus for a flexible lithiumion battery", *Adv. Mater.* 28 (3) (2016) 510–517. 10.1002/adma.201503678

[16] Y. Hernandez, M. Lotya, D. Rickard, S.D. Bergin, J.N. Coleman, "Measurement of multi- component solubility parameters for graphene facilitates solvent discovery", *Langmuir.* 26 (5) (2010), 3208–3213. 10.1021/la903188a

[17] P. Yasaei, B. Kumar, T. Foroozan, C. Wang, M. Asadi, D. Tuschel, J.E. Indacochea, R.F. Klie, A. Salehi-Khojin, "High-quality black phosphorus atomic layers by liquid-phase exfoliation", *Adv. Mater.* 27 (11) (2015) 1887–1892. 10.1002/adma.201405150

[18] Z. Song, Y. Ma, J Ye, "Preparation of stable black phosphorus nanosheets and their electrochemical catalytic study", *J. Electroanal. Chem.* 856 (2020) 113595. 10.1016/j.jelechem.2019.113595

[19] Y. Yang, H. Hou, G. Zou, W. Shi, H. Shuai, J. Li, X. Ji, "Electrochemical exfoliation of graphene-like two-dimensional nanomaterials", *Nanoscale.* 11 (2019) 16–33. 10.1039/C8NR08227H

[20] Z. Zeng, Z. Yin, X. Huang, H. Li, Q. He, G. Lu, F. Boey, H. Zhang, "Single layer semiconducting nanosheets: high yield preparation and device fabrication", *Angew. Chem.* 5 (50) (2011) 11093–11097. 10.1002/anie.201106004

[21] L. He, X. Zhou, W. Cai, Y. Xiao, F. Chu, X. Mu, X. Fu, Y. Hu, L. Song, "Electrochemical exfoliation and functionalization of black phosphorene to enhance mechanical properties and flame retardancy of waterborne polyurethane", *Compos. Part B Eng.* 202 (2020) 108446. 10.1016/j.compositesb.2020.108446

[22] P. Geng, S. Zheng, H. Tang, R. Zhu, L. Zhang, S. Cao, H. Xue, H. Pang, "Transition metal sulfides based on graphene for electrochemical energy storage", *Adv. Energy Mater.* 8 (15) (2017), 1703259. 10.1002/aenm.201703259

[23] J. Qian, X. Wu, Y. Cao, X. Ai, H. Yang, "High capacity and rate capability of amorphous phosphorus for sodium ion batteries", *Angew. Chemie - Int. Ed.* 52 (17) (2013) 4633–4636. 10.1002/anie.201209689

[24] K.J. Griffith, M. Mayo, A.J. Morris, C.P. Grey, "Phosphorus Electrodes for Li and Na-ion batteries: structural analysis and reaction intermediates", *ECS Meet. Abstr.* MA2016-03 (2016), 921. 10.1149/ma2016-03/2/921

[25] A. Capasso, A.E. Del Rio Castillo, H. Sun, A. Ansaldo, V. Pellegrini, F. Bonaccorso, "Ink-jet printing of graphene for flexible electronics- An environmentally friendly approach", *Chem. Mater.* 30 (50) (2018) 53–63. 10.1016/j.ssc.2015.08.011

[26] R. Geick, C. H. Perry, G. Rupprecht, "Normal modes in hexagonal boron nitride", *Phys. Rev.* 146 (2) (1966) 543. 10.1103/PhysRev.146.543

[27] X. Wang, S. Blechert, M. Antonietti, "Polymeric graphitic carbon nitride for heterogenous photocatalysis", *ACS Catal.* 2 (2012) 1596. 10.1088/2053-1591/1/4/045301

[28] L. Li, L. Chen, S. Mukherjee, J. Gao, H. Sun, Z. B. Liu, X. L. Ma, T. Gupta, C. V. Singh, W. C. Ren, H. M. Cheng, N. Koratkar, "Phosphorene as a polysulfide immobilizer and catalyst in high-performance lithium-sulfur batteries", *Adv. Mater.* 29 (2) (2017). 10.1002/adma.201602734

[29] J Kim, B. Park, "Fabricating and probing additive-free electrophoretic-deposited black phosphorus nanoflake anode for lithium-ion battery applications", *Mater. Lett.* 254 (2019) 367–370. 10.1016/j.matlet.2019.07.089

[30] Y. Zhang, H. Wang, Z. Luo, H. T. Tan, B. Li, S. Sun, Z. Li, Y.Yang, K.A. Khor, Z. J. Xu, Q. Yan, "An air-stable densely packed phosphorene-graphene composite toward advanced lithium storage properties", *Adv. Energy Mater.* 6 (12) (2016) 1600453. 10.1002/aenm.201600453

[31] J. Sun, G. Zheng, H.W. Lee, N. Liu, H. Wang, H. Yao, W. Yang, Y. Cui, "Formation of stable phosphorus-carbon bond for enhanced performance in black phosphorus nanoparticle-graphite composite battery anodes", *Nano Lett.* 14 (2014) 4573–4580. 10.1021/nl501617j

[32] W. Li, H. Yao, K. Yan, G. Zheng, Z. Liang, Y. M. Chiang, Y Cui, "The synergetic effect of lithium polysulfide and lithium nitrate to prevent lithium dendrite growth", *Nat. Commun.* 6 (7436) (2015) 26081242. 10.1038/ncomms8436

[33] Q. Pang, X. Liang, C. Y. Kwok, L.F. Nazar, "Advances in lithium-sulfur batteries based on multifunctional cathodes and electrolytes", *Nat. Energy.* 1 (2016) 1–11. 10.1038/nenergy.2016.132

[34] J. Zhao, Y. Yang, R. S. Katiyar, Z. Chen, "Phosphorene as a promising anchoring material for lithium sulfur batteries: a computational study", *J. Mater. Chem. A.* 4 (2016) 6124–6130. 10.1039/C6TA00871B

[35] J. Sun, H. W. Lee, M. Pasta, H. Yuan, G. Zheng, Y. Sun, Y. Li, Y. Cui, "A phosphorene-graphene hybrid material as a high-capacity anode for sodium-ion batteries", *Nat. Nanotechnol.* 10 (11) (2015) 980–985. 10.1038/nnano.2015.194

[36] A. Nie, Y. Cheng, S. Ning, T. Foroozan, P. Yasaei, W. Li, B. Song, Y. Yuan, L. Chen, A. Salehi-Khojin, "Selective ionic transport pathways in phosphorene", *Nano Letters.* 16 (2016) 2240–2247. 10.1021/acs.nanolett.5b04514

[37] R. Kotz, M. Carlen, "Principles and applications of electrochemical capacitors", *Electrochim. Acta.* 5 (2000) 2483. 10.1016/S0013-4686(00)00354-6

[38] P. Simon, Y.Gogotsi, "Materials for electrochemical capacitors", *Nat. Mater.* 7 (2008) 845. 10.1038/nmat2297

[39] X. Wang, Y. Chen, O. G. Schmidt, C.Yan, "Engineered nanomembranes for smart energy storage devices", *Chem. Soc. Rev.*, *Elsevier Ltd.* 45 (5) (2016) 1308–1330. 10.1016/j.csite.2017.08.002

[40] C. Hao, B. Yang, F. Wen, J. Xiang, L. Li, W. Wang, Z. Zeng, B. Xu, Z. Zhao, Z. Liu, Y.Tian, "Flexible all-solid-state supercapacitors based on liquid-exfoliated black-phosphorus nanoflakes", *Adv. Mater.* 28 (2016). 3194–3201. 10.1002/adma. 201505730

[41] B. Yang, C. Hao, F. Wen, B. Wang, C. Mu, J. Xiang, L. Li, B. Xu, Z. Zhao, Z. Liu, Y. Tian, "Flexible black-phosphorus nanoflake/carbon nanotube composite paper for high-performance all-solid-state supercapacitors", *ACS Appl. Mater. Interfaces.* 9 (2017) 44478–44484. 10.1021/acsami.7b13572

[42] J. Cao, P. He, J.R. Brent, H. Yilmaz, D.J. Lewis, I.A. Kinloch, B. Derby, "Supercapacitor electrodes from the in situ reaction between two-dimensional sheets of black phosphorus and graphene oxide", *ACS Appl. Mater. Interfaces.* 10 (2018) 10330–10338. 10.1021/ acsami.7b18853

[43] A. Sajedi-Moghaddam, C.C. Mayorga-Martinez, Z. Sofer, D. Bousa, E. Saievar-Iranizad, M. Pumera, "Black phosphorus nanoflakes/polyaniline hybrid material for high-performance pseudocapacitors", *J. Phys. Chem. C.* 121 (2017) 20532–20538. 10.1021/acs.jpcc.7b06958

[44] F. Wang, X. Wu, X. Yuan, Z. Liu, Y. Zhang, L. Fu, Y. Zhu, Q. Zhou, Y. Wu, W. Huang, "Latest advances in supercapacitors: from new electrode materials to novel device designs", *Chem. Soc. Rev.* 46 (2017) 6816–6854. 10.1039/C7CS00205J

[45] K.J. Katsounaros, I. Cherevko, S. Zeradjanin, A. R. Mayrhofer, "Oxygen electrochemistry as a cornerstone for sustainable energy conversion", *Angew. Chem.* 53 (1) (2014) 102–121. 10.1002/anie.201306588

[46] X. Ren, J. Zhou, X. Qi, Y. Liu, Z. Huang, Z. Li, Y. Ge, S.C. Dhanabalan, J.S. Ponraj, S. Wang, J. Zhong, H. Zhang, "Few-layer black phosphorus nanosheets as electrocatalysts for highly efficient oxygen evolution reaction", *Adv. Energy Mater.* 7(19) (2017) 1700396. 10.1002/aenm.201700396

[47] Z. Ma, Y. Zhang, W. Xu, Y.-C. Hsieh, P. Liu, L. Wu, Y. Zhu, K. Sasaki, R.R. Adzic, J.X. Wang, "Facile synthesis and reaction mechanism of oxygen evolution on RuO 2, IrO 2, and RuO 2 @IrO 2 core-shell nanocatalysts", *ECS Meet. Abstr.* MA2016-01 (2016) 1725. 10.1149/ma2016-01/35/1725

[48] J. Du Y. Yao, H. Zhang, J. Sun, J. Ma, L. Li, W. Li, "Novel QCM humidity sensors using stacked black phosphorus nanosheets as sensing film", *Sensor. Actuator. B Chem.* 244 (2017) 259–264. 10.1016/j.snb.2017.01.010

[49] J. Prasongkit, V. Shukla, A. Grigoriev, R. Ahuja, V. Amornkitbamrung, "Ultrahigh-sensitive gas sensors based on doped phosphorene: A first-principles investigation", *Appl. Surf. Sci.* MA2016-01 (2019). 10.1016/j.apsusc.2019.143660

[50] G. Zhou, H. Pu, J. Chang, X. Sui, S. Mao, J.Chen, "Real-time electronic sensor based on black phosphorus/Au NPs/DTT hybrid structure: application in arsenic detection", *Sensor. Actuator. B Chem.* 257 (2018) 214–219. 10.1016/j.snb.2017. 10.132

# 9 Microfluidic/ Miniaturized Electrochemical Devices for Bacteria Sensing Application

*Kumar Shivesh and Sanket Goel*

MEMS, Microfluidics and Nanoelectronics (MMNE) Lab, Department of Electrical and Electronics Engineering, Birla Institute of Technology and Science, Pilani, Hyderabad, India

## CONTENTS

## 9.1 INTRODUCTION

Bacterial infections remain the root reason for mortality worldwide. However, delivering a low-cost, portable, rapid, and durable detection technique for bacteria is still difficult (Simoska and Stevenson 2019). In the past few years, different methods, like conventional bacterial culture, molecular biology assays, staining, mass spectrometry

DOI: 10.1201/b23359-9

techniques, and microscopy based have been used for detecting pathogenic bacteria (McEachern et al. 2020).

Healthcare institutions generally make use of the cell culture method, which consists of the culture of specimens with well-defined growth media by altering the antibiotics and nutrients. Visual observation of growth patterns is a method of pathogens detection. Due to the implication of different steps like biochemical screening, selective enrichment, and bodily fluid confirmation, a culture-based method needs more labor as well as time. For a starting result of pathogen detection, it requires 1 to 3 days (Hudu et al. 2016; Paniel et al. 2013; Goluch 2017). In addition, this method also has problems of low selectivity and sensitivity, and inhibition of pathogen growth rate because of the high possibility of contamination of the culture (Paniel et al. 2013).

An alternative to the cell culture–based technique is molecular-based techniques like polymerase chain reaction (PCR) and nucleic acid sequencing. This method provides results in very less time with more sensitivity and selectivity. However, sample manipulation and preparation as well as the requirement of high-cost imaging equipment are disadvantages of the biomolecular-based approach. Moreover, there is obstruction of the amplification reaction because of simple matrix cellular components, inaccurate results because of unintentional contamination, and difficulties in distinguishing the dead and living cells (Buchan and Ledeboer, n.d.; Douterelo et al. 2016; Cho 2001). In addition, the requirement for specialized laboratory personnel and long incubation duration make researchers find another way (Pourakbari et al. 2019).

Mass spectrometry (MS) is another method preferred for the analysis of bacteria detection. It gives the final results in a few minutes via the identification of a wide scale of different types of molecules generated by pathogens for a range of concentrations (Ho and Muralidhar Reddy 2010). Matrix-assisted laser desorption time-of-flight (MALDI-TOF), desorption electrospray ionization (DESI), and electrospray ionization (ESI) are different types of MS. MALDI-TOF delivers more rapid and highly accurate, whilst less than molecular-based techniques, and does not need skilled laboratory personnel (Grieshaber et al. 2008). Nevertheless, the expensive initial cost of MALDI-TOF gives a motivation to find a cost-effective option for better healthcare.

An electrochemical biosensor is composed of a bioreceptor element as well as a transduction element to identify specific entities under a given environment. It is highly used in healthcare because of its high sensitivity and lucidity. Microbial electrochemical systems (MESs) are one of the most reliable, rapid, and precise for ascertainment of pathogens, the viability of cells in the medium, and interpretation of the rate of mechanism and pathway. The basis of MESs is based on two studies, namely mediated electron transfer (MET) and direct electron transfer (DET). In direct study, the oxidation current produced by biomolecules is the most vital part in the determination of the cell viability of examined bacteria. However, MET has a potential use in the detection of bacteria; it is not preferred due to its cytotoxicity (Sedki et al. 2017).

Recently, many reviews have been presented related to the sensing of bacteria using electrochemicals. Simoska *et al.* explained the merits of miniaturized array sensors for the electrochemical sensing of pathogens. In the same mini-review, graphene-based structures, carbon nanotubes, and multiplexed electrochemical sensing platforms have been elaborated on (Simoska and Stevenson 2019). McEachern et al. (2020) showed the indirect and direct methods for electrochemical detection of *Pseudomonas aeruginosa* based on pyocyanin detection. Khalilzadeh group presented a detailed article on application of organic and inorganic nano-materials for bacteria pathogens detection (Pourakbari et al. 2019). In this book chapter review, different types of electroanalytical methods viz., voltampero-metric techniques (cyclic voltammetry, chronoamperometry), electrochemical impedance spectroscopy, pulsed techniques (square wave voltammetry, differential pulse voltammetry), were eloborated in detail for the application of bacteria detection. In addition, the use of microfluidics electrochemical devices and lab-on-a-chip-based electrochemical sensors are elucidated.

## 9.2   ELECTROCHEMICAL ANALYTICAL METHODS

### 9.2.1   ELECTROCHEMICAL IMPEDANCE SPECTROSCOPY

Electrochemical impedance spectroscopy (EIS) is one of the most promising techniques for quantification of bacteria cells because the compilation of the output signal is accomplished without damaging the cells or altering the surrounding conditions. A study of a board range of frequencies is possible in this technique, as shown in Figure 9.1 (A). On the other hand, an analysis desired signal at a fixed frequency can be conducted (Kim et al. 2012). In this method, impedance in a circuit is calculated against a wide range of frequencies in ohms, which is similar to a unit of resistance (Magar et al. 2021).

A study organized by Tian et al. (2016) detected two different kinds of bacteria, namely *E. coli O157:H7* and *S. aureus,* simultaneously using a polyethylene glycol (PEG) microfluidic device consolidated with an alumina membrane setup via an EIS electrochemical analytical technique in the blend sample. The outcomes of the study revealed the selectivity toward the desired bacteria and shallow cross-attachment to the non-desired bacteria. The impedance was measured between 0.1 Hz, $10^4$ Hz in the frequency range. The prepared sensor possessed an LDR of between $10^2$ and $10^5$ cfu/ml along with a detection limit of about $10^2$ cfu/ml (Tian et al. 2016).

Recently in 2019, Saucedo and research group implemented EIS for monitoring and conformation of electrode fabrication; however, generally, researchers applied EIS for the detection of bacteria. So, this work showed that the application of EIS is not only restricted to the sensing bacteria but also used as a characterization tool for a fabricated electrode. In this study, Nyquist plots plotted for four different conditions viz., bare gold disk working electrode, modified gold electrode with poly 4-(3-pyrrolyl) butyric acid film, functionalized film of Concanavalin A on a polymerized gold electrode, and after incubation with *E. coli* cells. The results demonstrated that genuine functionalization of film attributed to enhancement in charge transfer resistance because of the insulating properties of Concanavalin A and bacteria layers. The LDR of the biosensor was calculated between $6.0 \times 10^3$ and $9.2 \times 10^7$ cfu/ml (Saucedo et al. 2019).

In another example, Wan et al. (2016) utilized EIS as a characterization tool for a modified gold electrode of a signal-off impedimetric immunosensor for sensing *E. coli O157:H7*. The Nyquist plots illustrated that attachment of gold nanoparticles

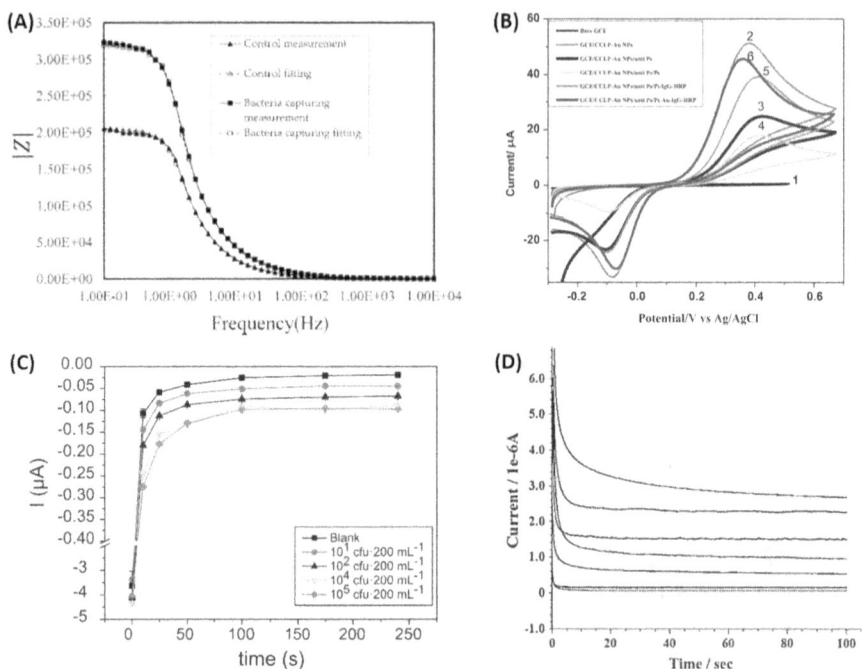

**FIGURE 9.1** (A) A typical impedance versus frequency plot for bacteria sensing (adapted with copyright permission taken from (Tian et al. 2016) Copyright 2016 Elsevier). (B) CV curves for different modifications on GCE (adapted the permission from (Krithiga et al. 2016) Copyright 2016 Elsevier). (C) Current versus time response for different range of *L. pneumophilla* (reprinted on receiving permission from (Ezenarro et al. 2020) Copyright 2020 MDPI). (D) Amperometric results for the sensing of target DNA at different concentrations (reprinted the image by taking copyright permission from (Li et al. 2011) Copyright 2011 Elsevier).

to the captured *E. coli* cells on the Au electrode drastically dropped the electron-transfer resistance ($R_{et}$). In this work, the value of $R_{et}$ reduced with increasing the concentration of bacteria, which gave the idea of using $R_{et}$ changed for the detection of bacteria. These types of biosensors would be useful for the rapid and sensitive sensing of many pathogens (Wan et al. 2016).

Oliveira Jr. and co-workers (2019) reported a sensitive, fast, cost-effective sensor for sensing three bacteria viz. *S. typhi*, *S. aureus*, and *E. coli* in water and apple juice. In this study, EIS was used for quantification of pathogen cells on utilizing homemade IDE-based screen-printed silver electordes along magnetic nanoparticles, modified using melittin. Capacitance spectrum plotted from 1 Hz to $10^6$ Hz frequency range for seven standard solutions of concentration between 1 and $10^6$ cfu/ml for bacteria separately. This method detected *E. coli* concentrations of 1 cfu/ml in portable water and 3.5 cfu/ml in apple juice in a time duration of less than half an hour. This procedure may provide a rapid, cost-effective way to sense bacteria in foodborne illnesses (Wilson et al. 2019).

## 9.2.2 Voltamperometric Techniques

Voltamperometric techniques are a type of electrochemical analytical method where current is measured while applying a specific voltage. There are various kinds of voltamperometric techniques namely, cyclic voltammetry (CV), open-circuit voltage (OCV), linear sweep voltammetry (LSV), chronopotentiometry (CP), chronoamperometry (CA), large-amplitude sinusoidal voltammetry (LASV), etc. These techniques have a wide linear dynamic range (LDR) and low limit of detection (LOD). However, CV and CP are generally used for pathogen detection. In the review, only these two techniques will be explained in detail.

### 9.2.2.1 Cyclic Voltammetry

In this method, at a fixed scan rate, voltage is changed from one point to another. The current is gauged between auxiliary and working electrode; however, voltage is gauged between reference and working electrode. On the received data plot, the Y axis represents current and the X axis depicts voltage, which is known as a voltammogram, as illustrated in Figure 9.1 (B). There are two convections used for representing the voltammogram, namely U.S. convection and IUPAC convection. In the former, voltage is plotted from a higher to lower value and in the latter, voltage is plotted from a lower value to higher value. The pattern of voltammogram is limited to scan rate and electrode surface, whilst the concentration of the catalyst also plays a vital role. For instance, enhancement in a concentration of particular enzymes gives a higher current compared to a non-catalyzed reaction. CV is not only restricted to sensing purposes but also used as a tool for the analysing surface morphology of electrodes (Grieshaber et al. 2008).

Zhou et al. (2016) utilized the *Pseudomonas aeruginosa* strain CP1 to enhance the catalytic reduction of nitric oxide. Bacterial presence was claimed to have acted as a catalyst, which resulted in an enhancement in the overall current generation by 4.36%. An experimental setup using cyclic voltammetry (CV) resulted and revealed the mechanism wherein the bacteria strain helped the electrochemical indirect electron transfer with a glassy-carbon electrode for reduction of NO. Subsequent detection of nitrous oxide was used to confirm the NO reduction in the presence and control environment. This model paves the way toward a bioelectrochemical approach to NO removal using biocatalysts (Zhou et al. 2016).

Do et al. (2019) also exploited cyclic voltammetry in combination with surface-enhanced Raman spectroscopy (SERS) to study pyocyanin (PYO), which is among the toxicants released by the pathogenic bacteria *P. aeruginosa*. This work focused on studying the molecular redox behavior of PYO under various conditions as pH and potential variation. The study with the help of SERS and EC discloses strong pH-dependent bands that can produce knowledge on the bacterial environment. The authors studied pellicle biofilms to find a similar behavior with different applied potential. An electrochemical and SERS combination can be an effective strategy for diagnostics and also studying the progress of infection (Do et al. 2019).

Another example, Sedki *et al.* used reduced graphene oxide (rGO) and hyper-branched chitosan (HBCs) composite as an electrode surface modifier to analyze the microbiological responses applying cyclic voltammetry. CV is used for measuring the amount of secretion of metabolites and cytotoxicity testing of cell suspensions. This revealed the high electro-catalytic function of the electrodes. Furthermore, the bacterial assay activity assay was studied on *E. coli* and *Pseudomonas aeruginosa* as pathogens to target. This modification on rGO exhibited ten times lower sensitivity (OD60 = 0.0025) than the other reports. Additionally, cell viability and susceptibility of *Pseudomonas aeruginosa* at various concentrations of ciprofloxacin, simvastatin kanamycin, and antibiotics were identified (Sedki et al. 2017).

Alatraktchi et al. (2016) used cyclic voltammetry to find a solution for the detection and quantification of Pyocyanin, which reveals crucial data about the existence of microorganisms in the human body. The study paves a way for selective sensing of Pyocyanin in the observence of an electroactive surrounding via CV. A detection window of 0.58–0.82 V showed a stable response uninfluenced by dissimilar redox species as an interferent. Pyocyanin was quantified in a wide range of concentrations of 2–100 μM along $R^2$ value of 0.991. Additionally, tests were performed on human saliva, resulting in an excellent standard deviation (SD) of 2.5%. This work using electrochemistry offers a cost-effective and reliable determination of *Pseudomonas aeruginosa* infections (Alatraktchi et al. 2016).

### 9.2.2.2 Chronoamperometry

In this technique, a steady-state current is gauged against time while supplying a square-wave potential to a working electrode, as shown in Figure 9.1 (C) and (D). The change in current is attributed to the electrochemical oxidation and reduction of an analyte. In this electrochemical technique, current is gauged at a fixed potential, unlike cyclic voltammetry, where potential is scanned over a fixed range. As the concentration of detection specie increases, current also increases. Due to its lucidity and low limit of detection, this method is preferred in affinity and bio-catalytic sensors. The reason behind the minimum background current in chron-oamperometry is insignificant charging current because of the supply of a constant potential (Ronkainen et al. 2010).

Ezenarro et al. (2020) worked on the determination of Legionella, which is a pathogenic bacterium. It can greatly impact humans since it is widely found in freshwaters and can easily spread in water management systems created by humans. A portable, low-cost, and rapid detection of Legionella was done using chron-oamperometry. To achieve this, a nitrocellulose-based membrane was used that not only performs sample concentration but also immunodetection of Legionella. Simple fabrication methods i.e., screen printing on paper were done to prepare the device. This work significantly decreases the sensing duration from 10 days to 2–3 hours. Additionally, the LDR was reported between $10^1$ and $10^4$ cfu/ml with a LOD of 4 cfu/ml (Ezenarro et al. 2020).

Diouani et al. (2021) exploited modified metallic nanoparticles to segregate and detect strains of *Bacillus cereus* and *Shigella flexneri*. Gold nanoparticle-based contri-chronoamperometry was applied for fast sensing of pathogenic bacteria. The sensing mechanism was based on the hydrogen reduction reaction of the complex formed by the gold nanoparticles; thus, the current generated was taken as a response to the bacteria. Such work can be useful in food quality and safety monitoring applications, thus helping society on a global scale by efficiently detecting the pathogens causing food contamination (Diouani et al. 2021).

Kuss et al. (2019) also employed modified screen-printed electrodes (SPEs) for bacteria sensing in the relevant concentration range. A specific detection of a wide range of bacteria could be determined with the help of thiol-based chemistry and an antibody binding mechanism with a simple SPE as a base electrode. Not only is this approach highly sensitive and selective but also cost-effective, which takes only a few seconds to determine bacteria in a wide range. This work reported the detection of *E. coli* and *N. gonorrheae*. Such approaches are helping to develop portable devices that can find applications in the food industry, bio-industry, etc. to bring down the mortality caused due to bacterial infections (Kuss et al. 2019).

Electrochemical behavior of a redox liposome was used for bacterial virulence factor sensing proposed by Luy et al. (2022). Rhamnolipid could be determined at the lowest of 500 nm rapidly within 30 minutes. The mechanism involved here was based on the deteriorating liposomes lipid membrane as they interact with toxins such as Rhamnolipid. This toxicity leads to the breakdown of the liposome to eventually release the encapsulated redox probe. Chronoamperometry responses were used to measure the concentration of the Rhamnolipid. The work presented exhibited a proof of concept with the given mechanism. However, this study was being performed with various other toxin identifications and their quantification (Luy et al. 2022).

### 9.2.3    PULSED TECHNIQUES

Out of all the electroanalytical techniques, pulse techniques are the most sensitive. The square wave voltammetry (SWV), differential pulse voltammetry (DPV), and normal pulse voltammetry (NPV) are the frequently used pulse strategies. In the case of cyclic voltammetry, voltage is swept from one to another, while in the case of pulse methods, a change of potential takes place by pulsing from one to another. The difference from decay rates of charging to Faradic currents is the basic principle of pulse procedures. The Faradic current is inverse to the half of power of time and charging current decays exponentially. Due to this fact, charging current decays quite faster than the Faradic current (Chen and Shah 2013). Herein, SWV and DPV will be discussed in detail for pathogenic bacteria detection.

#### 9.2.3.1    Square Wave Voltammetry

SWV is one of the prominent methods in pulsed techniques. It is an amplitude differential technique in which a waveform is composed of a square wave with a fixed amplitude, superposed on a pair of stairs waveform, is implemented to the working electrode of the sensor. The peak altitude in SWV is directly corresponding to the concentration of the bio-analyte in the specimen. It is one of the most sensitive techniques because it can reduce the background non-faradic current and has extensive usage in biosensors, chemical analysis, detecting various analytes, the study of kinematics, etc. SWV can also be called modified staircase voltammetry. The merits of this method are its high susceptibility, allowing one to sense the concentrations in order of ppb, less sweep time in the order of a few seconds, and understanding of kinematic reaction rate either slow or fast reaction.

Recently Wang and co-authors worked on the sensing of bacteria named *Streptococcus pneumonia*. In this work, DNA tetrahedron is anchored onto gold electrodes for rapid and selective sensing of pneumococcal surface protein A peptide and *S. pneumonia* from synthetic as well as an actual human specimen. A LOD of 0.218 ng/mL and a linearity range from 0 to 8 ng/mL for pneumococcal surface protein A peptide is achieved. However, a linearity range of 5 to 100 cfu/ml and a LOD of 0.093 cfu/ml are attained in the case of *S. pneumonia* (J. Wang et al. 2017).

Elliott and research group proposed an electrochemical sensing platform by modifying Au nanoparticles on a carbon-based transparent ultra-microelectrode for sensing of a unique biological compound Pyocyanin. An appreciable LOD of 0.75 µM and a wide linearity range from 0.75 µM to 25 µM was obtained (Elliott et al. 2017).

A novel work was done on sensing live bacteria in the *E. coli* cultures at a wide concentration range of $10^3$ cfu/ml in a span of 4 hours with the use of the SWV technique. This proposed sensor uses a carbon fiber mat and works on the principle that the sensing peak reduces with the effect of pH and the peak of riboflavin changes to less potential (Bigham et al. 2019).

In the year 2022, Gunasekaran *et al.* utilized the SWV electroanalytical technique for the sensing of *Escherichia coli* via bi-functional magnetic nanoparticle (MNP) conjugates in the range of $10^1$–$10^7$ cfu/ml, as illustrated in Figure 9.2 (A).

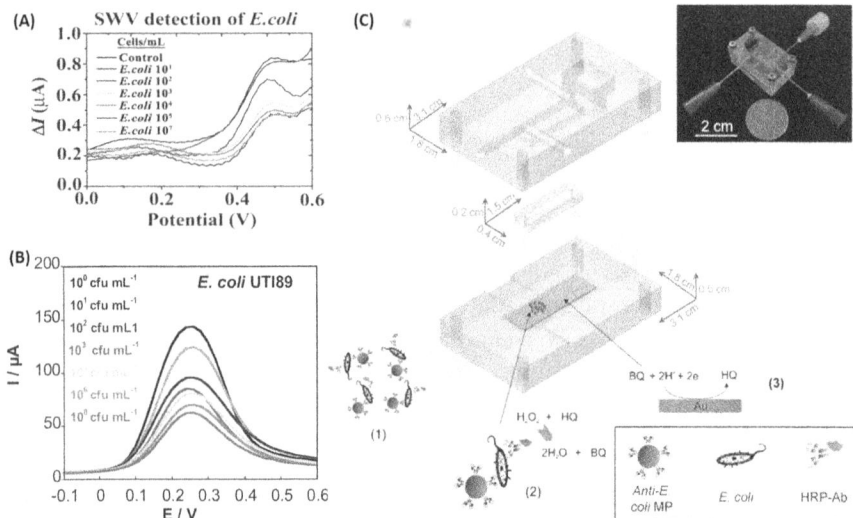

**FIGURE 9.2** (A) SWV of Bi-Functional MNPs for different *E. coli* concentrations (adapted with permission from (Gunasekaran et al. 2022) Copyright 2022 MDPI). (B) DPV output using modified Au electrodes in *E. coli* UTI89 of different concentrations ranging from $10^1$ to $10^8$ cfu/ml (Reprinted with permission from (Jijie et al. 2018) Copyright 2017 Elsevier). (C) Explains about the lab on a chip using microfluidic platform to sense bacterial on an electrode surface. This chip is made up of polycarbonate (PC), containing a well for holding the two magnets to provide confinement of magnetic particles upstream, the upper enclosure is fabricated using polymethylmethacrylate (PMMA), the interconnects were developed using polydimethylsiloxane (PDMS). (Reprinted by taking permission from (Laczka et al. 2011) Copyright 2011 Elsevier).

The presented method achieved a LOD of 10 cfu/ml in 1 hour. This approach of bifunctional conjugates will provide an excellent platform for the sensing of other pathogens in contaminated water (Gunasekaran et al. 2022).

### 9.2.3.2 Differential Pulse Voltammetry

Differential pulse voltammetry is used in electrochemical analysis where the series of constant voltage pulses is superimposed on the linear sweep voltammetry. In DPV, a reference value is selected at which there is no Faradic reaction and is implemented to the electrode. The reference potential is elevated between pulses with uniform addition. The effect of the charging current can be changed by sampling the current waveform just before the potential change. This pulsed technique can be used to analyze the redox reactions at very low concentrations since the charging current can be reduced and only Faradaic current is taken, which leads to high selectivity and sensitivity.

Raoof *et al.* developed the DNA sensor using an electrochemical approach where the MWCNT-based carbon paste electrodes are used for sensing. The hybridization of DNA is observed using DPV and EIS techniques. Further, the effects of adsorption and immobilization of the probe are studied using the DPV method (Raoof et al. 2009).

Xu et al. (2020) used the DPV technique in determining the viability of bacteria named T4 bacteriophage as an essential parameter in water quality monitoring. Systematic research was conducted in order to understand the combined effect on phage immobilization due to an outwardly implemented electric field and chemical modification. This study revealed that the condition for attaining the utmost capture efficiency of the deposited phages, it is similar to Debye length and phage size (Xu et al. 2020).

Jijie et al. (2018) utilized a modification of rGO by electrochemical deposition and polyethyleneimine on the gold electrodes for sensing uropathogenic *Escherichia coli* using an electrochemical approach for detecting in the range of 1 to $10^8$ cfu/ml, as reflected in Figure 9.2 (B). A LOD of 10 cfu/ml and a wide linearity range of $10 \times 10^1$ to $1 \times 10^4$ is achieved. Further, real-time specimens like blood serum and urine were tested by the sensor (Jijie et al. 2018).

## 9.3 DETECTION OF PATHOGENS USING VARIOUS TECHNIQUES

Diseases in human beings are mainly because of different microbes like protozoa, prions, viruses, parasitic worms, fungi, and bacteria. Of these microbial organisms, health issues caused by bacteria are more challenging for a human being to survive. In comparison to outbreaks caused by viruses or chemicals, bacteria outbreaks result in the most hospitalizations and deaths worldwide. The advancement of accurate, specific, and early detection methods of pathogenic diseases is the solution for controlling communicable diseases. Alongside the standard electrochemical approaches have arisen, allowing for unique results for pathogen detection techniques and fundamental ideas. The newest and noteworthy achievements are highlighted in the following.

### 9.3.1 ELECTROCHEMICAL DETECTION OF PATHOGENIC BACTERIA IN A MICROFLUIDIC CHAMBER

Integrated electrochemical sensors are smoothly adapted with both microfluidic technology and integrated circuits, which offer a promising solution to detect bacteria and other pathogenic microorganisms. Looking at microfluidics technology in point-of-care applications is highly emerging, automated, and miniaturized and could be used to provide speedy, economical, efficient, and on-site diagnoses instead of existing procedures. The most successful approaches for pathogen detection are molecular diagnostics, which are commonly utilized in microfluidic devices. Several pathogenic functions can be integrated and programmed using microfluidic chips. Over the past few years, many researchers have been working on integrating sensing electrodes into the microfluidic chip to perform on-chip electrochemical analysis for the measurement and sensing of bacteria and many other microorganisms.

In 2018, Altintas and co-authors published the design of a fully automated and combined microfluidic biosensor for real-time sensing of *E. coli* with a LOD of 50 cfu/ml. Herein, detection is done using cyclic voltammetry and amperometric measurement. The microfluidic channel which is incorporated for sensor insertion for electronics and fluidic interaction is approximately 7 µL. This custom-designed sensor is highly automated and portable with 8 Au electrode arrays. In the proposed system, microbes are concentrated on the Au electrodes. Especially, a biosensor is fabricated as a sandwich model using horseradish peroxidase-sensing. Gold nano-material is used to improve the sensor results (Altintas et al. 2018).

In another example, Dastider et al. (2018) published the development and fabrication of a MEMS-based biosensor integrated with the microfluidic chamber for sensing *Escherichia coli* O157:H7. The electrodes, which act as sensing regions, are interdigitated electrodes array and are treated with negative *E. coli* antibodies before testing, which results in adequate impedance changes while monitoring of *E. coli*. Within a total detection time of 2 hours, there was successful sensing of a pathogen with a concentration of 39 cfu/ml (Dastider et al. 2018).

In the same year 2018, Wang and research group proposed a novel chip based detection consisting of interdigitated electrodes with Tesla structure junction for micro-mixing useful for the sensing of *E. coli O157:H7*. At first, *E. coli* is coated with diallyl dimethyl ammonium chloride (PDDA) polymer to make the *E. coli* positively charged, followed by gold nanoparticles (AuNPs). The research group mainly focused on proving the silver enhancement reaction is a powerful analysis to ascend the workings of microfluidic chip-based for impedimetric bacteria detection (IBD). Within an hour, the LOD was brought to 500 cfu/ml. The experiment is validated by sensing *E. coli* in the tap water and eggshell wash (R. Wang et al. 2018).

### 9.3.2 ELECTROCHEMICAL DETECTION OF PATHOGENIC BACTERIA VIA LAB-ON-A-CHIP

Over the past few years, integrated and flexible lab-on-a-chip (LOC) biosensors have been enlightening the research community in the field of disease diagnosis and personnel healthcare (Yang et al. 2017). Various sensing techniques using micro-fluidic channels for bacteria include particle counter, on-chip microfluidic culture-based, biosensors, PCR, and isothermal amplification methods analyzed using electrochemical detection techniques. Among them, impedance-based biosensors play a promising place in bacteria detection for their simple, adaptive nature for lab-on-a-chip applications (Escamilla-Gómez et al. 2009).

Primiceri et al. (2016) used a lab-on-a-chip arrangement of gold electrodes integrated on a glass substrate arranged in four different sensing areas incorporated with microfluidic for the movement of the solution on the electrodes. The impedance spectroscopy was carried out using a potentiostat on applying the frequency range of $0.1-10^5$ Hz using a sinusoidal amplitude of 15 mV AC potential. The work demonstrated using a meat specimen drop cast onto the sensor array, thereby sensing *S. aureus* and *L. monocytogenes* was done. Hence, the proposed work concluded by inferring the detection limit is 1.2 cfu/ml and 5 cfu/ml for *L. monocytogenes* and *S. aureus*. The sensing is done using this biochip in less time. Further improvement in the device is by incorporating it with the sample pretreatment module to make it fully integrated for an on-chip assay (Primiceri et al. 2016).

In referring to the work done by Shaik and co-authors it was explained about the sensing of *E. coli 723 (coli 723)* in various water specimens. Most of them were reverse osmosis water (ROW), recycled water (RW), and standard treated tap water (STW). The impedance growth curve was analyzed by using the potentiostat in the electrochemical impedance spectroscopy study. These impedance spectra were analyzed by applying the frequency range of 1 Hz to 1 MHz for a potential window of 10 mV. Drop in impedance was analyzed at different time zones. ROW takes 60 minutes with an impedance drop of 60%. This drop was analysed corresponding to $1.04 \times 10^8$ cells/ml. The main outcome of the paper was this entire study was done using interdigitated (IDE) screen-printed electrodes (SPE), which were highly sophisticated to reuse it on thorough washing of sensors with deionized water (Shaik et al. 2021).

In another work, Laczka *et al.* explained about *E. coli* detection using a microfluidic chip where electrochemical and enzymatic reactions take place as illustrated in Figure 9.2 (C). Further, the results were analyzed using amperometric analysis in a single-step format, which shows the identification of *E. coli* in a wide range of $10^2$ to $10^8$ cfu/ml. Hence, whole assay detection takes about 1 hour, showing 55 cfu/ml as LOD (Laczka et al. 2011).

## 9.4    CONCLUSION

The ceaseless growth in the field of electrochemical biosensors for bacteria sensing clearly shows its significance and provides a rapid, reliable, and cost-effective method. The electrochemical method outweighs the cell culture methods, molecular-based techniques like nucleic acid sequencing and polymerase chain reaction, and mass spectroscopy. The low cost, along with rapid detection, is the most advantageous part of it. Furthermore, electrochemical sensors provide a wearable, implantable, and flexible design, which is desirable for point-of-care testing. There are many electro-analytical techniques; however, voltamperometric techniques, pulse techniques, and

**TABLE 9.1**

**Summary of Bacteria Detection Discussed in This Review**

| Pathogen Type | Sensing Technique | Electrochemical Techniques | LOD | Detection Time | Ref. |
|---|---|---|---|---|---|
| *P. aeruginosa* | Electrochemical sensing of nitric oxide by utilising catalyzed *P. aeruginosa* strain by CP1 | CV | N/A | N/A | (Zhou et al. 2016) |
| *P. aeruginosa* | Direct detection of Pyocyanin by applying spectroelectrochemical setup | CV | N/A | N/A | (Do et al. 2019) |
| *E. coli*<br>*P. aeruginosa* | Rapid, accurate, and real-time cell viability assessment | CV | $OD_{600}$ of 0.025 | N/A | (Sedki et al. 2017) |
| *P. aeruginosa* | Selective detection of Pyocyanin among interfering redox-active compounds | CV | 2 µM | 2–100 µM | (Alatraktchi et al. 2016) |
| *Legionella* | Combination of sample of concentration and immunoassay detection | CA | 4 cfu/ml | $10^1$–$10^4$ CFU/ml | (Ezenarro et al. 2020) |
| *B. cereus*<br>*S. flexneri* | Detection of pathogenic bacteria using metallic nanoparticles | CA | 3 cfu/ml<br>12 cfu/ml | N/A | (Diouani et al. 2021) |
| *E. coli*<br>*N. gonorrheae* | Selective electrochemical detection using cytochrome c oxidase | CA | $10^6$ cfu/ml<br>$10^7$ cfu/ml | N/A | (Kuss et al. 2019) |
| *P. aeruginosa* | Detection of Rhamnolipid virulence factor using Redox Liposome | CA | 500 nM | N/A | (Luy et al. 2022) |
| *S. pneumoniae* | Detection of *S. pneumoniae* using gold electrodes anchored with DNA tetrahedron | SWV | 0.093 cfu/ml | 5–100 cfu/ml | (J. Wang et al. 2017) |
| *P. aeruginosa* | Electrochemical sensing of Pyocyanin using carbon based ultramicroelectrode | SWV | 1 µM | 1–250 µM | (Elliott et al. 2017) |
| *E. coli* | Detection of coliforms using riboflavin–ferrocyanide redox couples | SWV | $10^3$ cfu/ml | $10^3$–$10^7$ cfu/ml | (Bigham et al. 2019) |

| Bacteria | Description | Technique | LOD | Range | Reference |
|---|---|---|---|---|---|
| E. coli | Development of T4 bacteriophage micro electrochemical sensor for detection of viability of pathogenic bacteria | DPV | 14 cfu/ml | $1.9 \times 10^1$–$1.9 \times 10^8$ cfu/ml | (Xu et al. 2020) |
| E. coli | Detection of uropathogenic E. coli using modified rGO/ /polyethylenimine gold electrode | DPV | 10 cfu/ml | $10^1$–$10^4$ CFU/ml | (Jijie et al. 2018) |
| E. coli | Development of electrochemical biosensor based on Rolling Circle Amplification (RCA) coupled peroxidase-mimicking DNAzyme amplification | DPV | 8 cfu/ml | $9.4$–$9.4 \times 10^4$ cfu/ml | (Guo et al. 2016) |
| E. coli | Sensing of pathogen in microfluidics chip using electrochemical biosensor | CV | 50 cfu/ml | $10$–$3.97 \times 10^7$ cfu/ml | (Altintas et al. 2018) |
| E. coli | Detection of low concentration bacteria using microfluidic MEMS biosensor modified with gold | EIS | 39 cfu/ml | N/A | (Dastider et al. 2018) |
| E. coli | IDB sensing via modified silver microfluidic chip | EIS | 500 cfu/ml | $2 \times 10^3$–$2 \times 10^5$ cfu/ml | (R. Wang et al. 2018) |
| E. coli | Sensing of pathogens on chip-based chamber combined with modified nanoporous membrane | EIS | 100 cfu/ml | $10^2$–$10^5$ cfu/ml | (Tian et al. 2016) |
| L. monocytogenes S. aureus | Label-free impedimetric detection of food pathogen using multipurpose biochip | EIS | 5 cfu/ml 1.26 cfu/ml | $5$–$7.5 \times 10^8$ CFU/ml $1.2$–$10^9$ CFU/ml | (Primiceri et al. 2016) |
| E. coli | Detection of bacteria via coupling magneto-immunocapture and electrochemistry | CA | 55 CFU/ml | $10^2$–$10^8$ CFU/ml | (Laczka et al. 2011) |

electrochemical impedance spectroscopy are the most favored methods for the sensing of bacteria. The recent advances are a combination of microfluidic devices and LOC. These types of combined electrochemical sensing platforms increase the sensitivity of the detection of bacteria and a summary of bacteria detection is listed in Table 9.1. However, microfluidic devices may be used for keeping an eye on catheters and wound dressings to uninterrupted tracking of infection. The combined electrochemical platform will solve many existing problems in upcoming years due to its low cost and accuracy.

## REFERENCES

Alatraktchi, Fatima Alzahra'a, Sandra Breum Andersen, Helle Krogh Johansen, Søren Molin, and Winnie E. Svendsen. 2016. "Fast Selective Detection of Pyocyanin Using Cyclic Voltammetry." *Sensors (Switzerland)* 16 (3). doi:10.3390/s16030408

Altintas, Zeynep, Mete Akgun, Guzin Kokturk, and Yildiz Uludag. 2018. "A Fully Automated Microfluidic-Based Electrochemical Sensor for Real-Time Bacteria Detection." *Biosensors and Bioelectronics* 100 (July 2017) : 541–548. doi:10.1016/j.bios.2017.09.046

Bigham, Teri, Charnete Casimero, James S.G. Dooley, Nigel G. Ternan, William J. Snelling, and James Davis. 2019. "Microbial Water Quality: Voltammetric Detection of Coliforms Based on Riboflavin–Ferrocyanide Redox Couples." *Electrochemistry Communications* 101 (February 2019) : 99–103. doi:10.1016/j.elecom.2019.02.022

Buchan, Blake W., and Nathan A. Ledeboer. n.d. "Emerging Technologies for the Clinical Microbiology Laboratory INTRODUCTION." *Clinical Microbiology Reviews*. doi:10.1128/CMR.00003-14

Chen, Aicheng, and Badal Shah. 2013. "Electrochemical Sensing and Biosensing Based on Square Wave Voltammetry." *Analytical Methods* 5 (9): 2158–2173. doi:10.1039/c3ay40155c

Cho, Jae-chang. 2001. "Bacterial Species Determination from DNA-DNA Hybridization by Using Genome Fragments and DNA Microarrays" *Applied and Environmental Microbiology* 67 (8): 3677–3682. doi:10.1128/AEM.67.8.3677

Dastider, Shibajyoti Ghosh, Amjed Abdullah, Ibrahem Jasim, Nuh S. Yuksek, Majed Dweik, and Mahmoud Almasri. 2018. "Low Concentration E. Coli O157:H7 Bacteria Sensing Using Microfluidic MEMS Biosensor." *Review of Scientific Instruments* 89 (12). doi:10.1063/1.5043424

Diouani, Mohamed Fethi, Maher Sayhi, Zehaira Romeissa Djafar, Samir Ben Jomaa, and Dhafer Laouini. 2021. "Magnetic Separation and Centri-Chronoamperometric Detection of Foodborne Bacteria Using Antibiotic-Coated Metallic Nanoparticles."Biosensors 11.

Do, Hyein, Seung Ryong Kwon, Kaiyu Fu, Nydia Morales-Soto, Joshua D. Shrout, and Paul W. Bohn. 2019. "Electrochemical Surface-Enhanced Raman Spectroscopy of Pyocyanin Secreted by Pseudomonas Aeruginosa Communities." *Langmuir* 35 (21). doi:10.1021/acs.langmuir.9b00184

Douterelo, Isabel, M. Jackson, C. Solomon, and J. Boxall. 2016. "Microbial Analysis of in Situ Biofilm Formation in Drinking Water Distribution Systems: Implications for Monitoring and Control of Drinking Water Quality," 100: 3301–3311. doi:10.1007/s00253-015-7155-3

Elliott, Janine, Olja Simoska, Scott Karasik, Jason B. Shear, and Keith J. Stevenson. 2017. "Transparent Carbon Ultramicroelectrode Arrays for the Electrochemical Detection of a Bacterial Warfare Toxin, Pyocyanin." *Analytical Chemistry* 89 (12): 6285–6289. doi:10.1021/acs.analchem.7b00876

Escamilla-Gómez, Vanessa, Susana Campuzano, María Pedrero, and José M. Pingarrón. 2009. "Gold Screen-Printed-Based Impedimetric Immunobiosensors for Direct and Sensitive Escherichia Coli Quantisation." *Biosensors and Bioelectronics* 24 (11): 3365–3371. doi:10.1016/j.bios.2009.04.047

Ezenarro, Josune J., Noemí Párraga-Niño, Miquel Sabrià, Fancisco Javier Del Campo, Francesc Xavier Muñoz-Pascual, Jordi Mas, and Naroa Uria. 2020. "Rapid Detection of Legionella Pneumophila in Drinking Water, Based on Filter Immunoassay and Chronoamperometric Measurement." *Biosensors* 10 (9). doi:10.3390/bios10090102

Goluch, Edgar D. 2017. "Microbial Identification Using Electrochemical Detection of Metabolites." *Trends in Biotechnology* 35 (12): 1125–1128. doi:10.1016/j.tibtech.2017.08.001

Grieshaber, Dorothee, Robert MacKenzie, Janos Vörös, and Erik Reimhult. 2008. "Electrochemical Biosensors - Sensor Principles and Architectures." *Sensors* 8 (3): 1400–1458. doi:10.3390/s8031400

Gunasekaran, Dharanivasan, Yoram Gerchman, and Sefi Vernick. 2022. "Electrochemical Detection of Waterborne Bacteria Using Bi-Functional Magnetic Nanoparticle Conjugates." *Biosensors* 12 (1). doi:10.3390/bios12010036

Guo, Yuna, Yu Wang, Su Liu, Jinghua Yu, Hongzhi Wang, Yalin Wang, and Jiadong Huang. 2016. "Label-Free and Highly Sensitive Electrochemical Detection of E. Coli Based on Rolling Circle Amplifications Coupled Peroxidase-Mimicking DNAzyme Amplification." *Biosensors and Bioelectronics* 75 (January): 315–319. doi:10.1016/J.BIOS.2015.08.031

Ho, Yen Peng, and P. Muralidhar Reddy. 2010. "Identification of Pathogens by Mass Spectrometry." *Clinical Chemistry* 56 (4): 525–536. doi:10.1373/clinchem.2009.138867

Hudu, Shuaibu Abdullahi, Ahmed Subeh Alshrari, Ahmad Syahida, and Zamberi Sekawi. 2016. "Cell Culture, Technology: Enhancing the Culture of Diagnosing Human Diseases." *Journal of Clinical and Diagnostic Research* 10 (3): DE01–DE05. doi:10.7860/JCDR/2016/15837.7460

Jijie, Roxana, Karima Kahlouche, Alexandre Barras, Nao Yamakawa, Julie Bouckaert, Tijani Gharbi, Sabine Szunerits, and Rabah Boukherroub. 2018. "Reduced Graphene Oxide/Polyethylenimine Based Immunosensor for the Selective and Sensitive Electrochemical Detection of Uropathogenic Escherichia Coli." *Sensors and Actuators, B: Chemical* 260: 255–263. doi:10.1016/j.snb.2017.12.169

Kim, Seonghwan, Guiduk Yu, Taeyoung Kim, Kyusoon Shin, and Jeyong Yoon. 2012. "Rapid Bacterial Detection with an Interdigitated Array Electrode by Electrochemical Impedance Spectroscopy." *Electrochimica Acta* 82: 126–131. doi:10.1016/j.electacta.2012.05.131

Krithiga, N., K. Balaji Viswanath, V.S. Vasantha, and A. Jayachitra. 2016. "Specific and Selective Electrochemical Immunoassay for Pseudomonas Aeruginosa Based on Pectin-Gold Nano Composite." *Biosensors and Bioelectronics* 79: 121–129. doi:10.1016/j.bios.2015.12.006

Kuss, Sabine, Rosa A.S. Couto, Rhiannon M. Evans, Hayley Lavender, Christoph C. Tang, and Richard G. Compton. 2019. "Versatile Electrochemical Sensing Platform for Bacteria." *Analytical Chemistry* 91 (7): 4317–4322. doi:10.1021/acs.analchem.9b00326

Laczka, Olivier, José María Maesa, Neus Godino, Javier del Campo, Mikkel Fougt-Hansen, Jorg P. Kutter, Detlef Snakenborg, Francesc Xavier Muñoz-Pascual, and Eva Baldrich. 2011. "Improved Bacteria Detection by Coupling Magneto-Immunocapture and Amperometry at Flow-Channel Microband Electrodes." *Biosensors and Bioelectronics* 26 (8): 3633–3640. doi:10.1016/j.bios.2011.02.019

Li, Kang, Yanjun Lai, Wen Zhang, and Litong Jin. 2011. "Fe2O3@Au Core/Shell Nanoparticle-Based Electrochemical DNA Biosensor for Escherichia Coli Detection." *Talanta* 84 (3): 607–613. doi:10.1016/j.talanta.2010.12.042

Luy, Justine, Dorine Ameline, Christine Thobie-gautier, Mohammed Boujtita, and Estelle Lebègue. 2022. "Detection of Bacterial Rhamnolipid Toxin by Redox Liposome Single Impact Electrochemistry" 61: 202111416. doi:10.1002/ange.202111416

Magar, Hend S., Rabeay Y.A. Hassan, and Ashok Mulchandani. 2021. "Electrochemical Impedance Spectroscopy (Eis): Principles, Construction, and Biosensing Applications." *Sensors* 21 (19). doi:10.3390/s21196578

McEachern, Francis, Edward Harvey, and Geraldine Merle. 2020. "Emerging Technologies for the Electrochemical Detection of Bacteria." *Biotechnology Journal* 15 (9). doi:10.1002/biot.202000140

Paniel, N., J. Baudart, A. Hayat, and L. Barthelmebs. 2013. "Aptasensor and Genosensor Methods for Detection of Microbes in Real World Samples." *Methods* 64 (3): 229–240. doi:10.1016/j.ymeth.2013.07.001

Pourakbari, Ramin, Nasrin Shadjou, Hadi Yousefi, Ibrahim Isildak, Mehdi Yousefi, Mohammad Reza Rashidi, and Balal Khalilzadeh. 2019. "Recent Progress in Nanomaterial-Based Electrochemical Biosensors for Pathogenic Bacteria." *Microchimica Acta* 186 (12). doi: 10.1007/s00604-019-3966-8

Primiceri, Elisabetta, Maria Serena Chiriacò, Francesco De Feo, Elisa Santovito, Vincenzina Fusco, and Giuseppe Maruccio. 2016. "A Multipurpose Biochip for Food Pathogen Detection." *Analytical Methods* 8 (15): 3055–3060. doi:10.1039/c5ay03295d

Raoof, Jahan Bakhsh, Mohammad Saeid Hejazi, Reza Ojani, and Ezat Hamidi Asl. 2009. "A Comparative Study of Carbon Nanotube Paste Electrode for Development of Indicator-Free DNA Sensors Using DPV and EIS: Human Interleukin-2 Oligonucleotide as a Model." *International Journal of Electrochemical Science* 4: 1436–1451.

Ronkainen, Niina J., H. Brian Halsall, and William R. Heineman. 2010. "Electrochemical Biosensors." *Chemical Society Reviews* 39 (5): 1747–1763. doi:10.1039/b714449k

Saucedo, Nuvia Maria, Sira Srinives, and Ashok Mulchandani. 2019. "Electrochemical Biosensor for Rapid Detection of Viable Bacteria and Antibiotic Screening." *Journal of Analysis and Testing* 3 (1): 117–122. doi:10.1007/s41664-019-00091-2

Sedki, Mohammed, Rabeay Y.A. Hassan, Amr Hefnawy, and Ibrahim M. El-Sherbiny. 2017. "Sensing of Bacterial Cell Viability Using Nanostructured Bioelectrochemical System: RGO-Hyperbranched Chitosan Nanocomposite as a Novel Microbial Sensor Platform." *Sensors and Actuators, B: Chemical* 252: 191–200. doi:10.1016/j.snb.2017.05.163

Shaik, Subhan, Aarthi Saminathan, Deepak Sharma, Jagdish A. Krishnaswamy, and D. Roy Mahapatra. 2021. "Monitoring Microbial Growth on a Microfluidic Lab-on-Chip with Electrochemical Impedance Spectroscopic Technique." *Biomedical Microdevices* 23 (2): 1–12. doi:10.1007/s10544-021-00564-1

Simoska, Olja, and Keith J. Stevenson. 2019. "Electrochemical Sensors for Rapid Diagnosis of Pathogens in Real Time." *Analyst* 144 (22): 6461–6478. doi:10.1039/c9an01747j

Tian, Feng, Jing Lyu, Jingyu Shi, Fei Tan, and Mo Yang. 2016. "A Polymeric Microfluidic Device Integrated with Nanoporous Alumina Membranes for Simultaneous Detection of Multiple Foodborne Pathogens." *Sensors and Actuators, B: Chemical* 225: 312–318. doi:10.1016/j.snb.2015.11.059

Wan, Jingzhuan, Junjie Ai, Yonghua Zhang, Xiaohui Geng, Qiang Gao, and Zhiliang Cheng. 2016. "Signal-off Impedimetric Immunosensor for the Detection of Escherichia Coli O157:H7." *Scientific Reports* 6: 2–7. doi:10.1038/srep19806

Wang, Jinping, May Ching Leong, Eric Zhe Wei Leong, Win Sen Kuan, and David Tai Leong. 2017. "Clinically Relevant Detection of Streptococcus Pneumoniae with DNA-Antibody Nanostructures." *Analytical Chemistry* 89 (12): 6900–6906. doi:10.1021/acs.analchem.7b01508

Wang, Renjie, Yi Xu, Thomas Sors, Joseph Irudayaraj, Wen Ren, and Rong Wang. 2018. "Impedimetric Detection of Bacteria by Using a Microfluidic Chip and Silver Nanoparticle Based Signal Enhancement." *Microchimica Acta* 185 (3). doi:10.1007/s00604-017-2645-x

Wilson, Deivy, Elsa M. Materón, Gisela Ibáñez-Redín, Ronaldo C. Faria, Daniel S. Correa, and Osvaldo N. Oliveira. 2019. "Electrical Detection of Pathogenic Bacteria in Food Samples Using Information Visualization Methods with a Sensor Based on Magnetic Nanoparticles Functionalized with Antimicrobial Peptides." *Talanta* 194: 611–618. doi:10.1016/j.talanta.2018.10.089

Xu, Jingting, Cong Zhao, Ying Chau, and Yi Kuen Lee. 2020. "The Synergy of Chemical Immobilization and Electrical Orientation of T4 Bacteriophage on a Micro Electrochemical Sensor for Low-Level Viable Bacteria Detection via Differential Pulse Voltammetry." *Biosensors and Bioelectronics* 151. doi:10.1016/j.bios.2019.111914

Yang, Yanbing, Xiangdong Yang, Yaning Tan, and Quan Yuan. 2017. "Recent Progress in Flexible and Wearable Bio-Electronics Based on Nanomaterials." *Nano Research* 10 (5): 1560–1583. doi:10.1007/s12274-017-1476-8

Zhou, Shaofeng, Shaobin Huang, Jiaxin He, Han Li, and Yongqing Zhang. 2016. "Electron Transfer of Pseudomonas Aeruginosa CP1 in Electrochemical Reduction of Nitric Oxide." *Bioresource Technology* 218: 1271–1274. doi:10.1016/j.biortech.2016.07.010

# 10 Cloth and Paper-Based Miniaturized Electrochemiluminesce-nce Platforms

*Manish Bhaiyya, Prasant Kumar Pattnaik, and Sanket Goel*

MEMS, Microfluidics and Nanoelectronics (MMNE) Lab, Department of Electrical and Electronics Engineering, Birla Institute of Technology and Science, Pilani, Hyderabad, India

## CONTENTS

## 10.1 INTRODUCTION

Technological advancements and the development of point-of-care diagnostic testing have played an essential role in early disease identification, analysis and maintenance. Since the last decade, point-of-care testing (POCT) devices have significantly impacted our lives and have been a focus for researchers. POCT is also known as near-beside patient testing or on-site testing. The primary purpose of POCT is to offer a quick response with a limited sample volume so that proper treatment can be given, resulting in a better clinical or economic outcome (Willmott and Arrowsmith 2013; Kost et al. 1999). Even today, most diagnoses are performed using a laboratory-based technique with expensive equipment that requires skilled individuals to operate, which leads to a time-consuming approach in which the patient must wait a long time for their results and has to pay a higher test fee. As a result, there has been a critical need for rapid diagnosis at or near the patient with no trained personnel to encourage acute disease monitoring of various biomolecules (Jung et al. 2015). Before being deployed in real-time applications, each POCT device should meet the following essential requirements: (1) give a speedy test report to allow the patient to undergo follow-up treatment; (2) deliver selective, accurate, and quantitative results that are comparable to benchmark analyzers used in laboratories; (3) easy-to-use devices that require minimal user participation (Sachdeva et al. 2021; Christodouleas et al. 2018).

Paper and cloth-based devices have opened up many doors to be used in the health care sector as they meet all of the above mentioned requirements of POCT devices. Paper and cloth-based POCT devices have replaced the conventional laboratory-based based diagnosis because of it being low-cost with simple fabrication techniques. Further, there is enormous potential to do multiple diagnosis using a single platform driven ban and excellent capillary mechanism (no need to use external pumps). Such devices are rapidly prototyped, easy to dispose, easy to form channels (hydrophilic and hydrophobic), highly selective with excellent sensitivity, and have mass fabrication capability and excellent biological compatibility (He et al. 2015; Yetisen et al. 2013; Songjaroen et al. 2011). Hence, paper and cloth-based POCT devices are broadly used in environmental detection, food industry, molecular analysis, and in POCT diagnosis (Sher et al. 2017; Yamada et al. 2017).

Chemiluminescence, electrochemistry, colorimetry, fluorescence, electrochemiluminescence (ECL), and other sensing methods have been widely reported and are suitable with paper and cloth-based devices. However, ECL has become the method of choice due to its great selectivity, superior sensitivity, low background signal, larger linear range, and ease of instrumentation (R. Liu et al. 2015; Miao 2008). When an adequate voltage is given to electrochemical reagents, electron transfer occurs, resulting in the production of light known as ECL (Richter 2004; Sakura 1992). The rate of photon production is measured in luminescence measurements, and the light intensity is consequently dependent on the rate of the luminescent reaction. This leads to making the intensity of light in a luminescence process proportionate to the concentration of a limiting reactant. Luminol and $Ru(bpy)_3^{2+}$-based ECL reactions have been reported and are utilized in a variety of disciplines such as enzyme-based sensing, target bimolecular sensing, DNA biosensors, and POCT applications (Marquette and Blum 2008). Generally, bipolar electrode (BPE) (open and closed) and single electrode (SE) these two most suitable ECL systems have been reported which are predominantly used with Luminol-based and $Ru(bpy)_3^{2+}$-based ECL reactions (Bhaiyya et al. 2021b).

A BPE electrode is a conductive substance that exhibits electrochemical processes (oxidation and reduction) at its poles (anode and cathode) without being in direct contact with external power supply. More explicitly, when adequate external voltage is applied to driving electrodes, a strong electric filed is created across channel, leading to initiates redox process (oxidation and reduction) at the anode and cathode poles of the BPE (Fosdick et al. 2013; Wu et al. 2014). As there is no direct contact needed in between BPE and the external power supply to initiate a redox process, a large number of BPE electrode arrays can be organized by just a single external DC power supply or even a battery. The working concept of a BPE-ECL system is illustrated Figure 10.1(a). The U-shaped open BPE-ECL and closed type of BPE-ECL device is revealed in Figure 10.1(c) and (d), and in-depth explanation related to these is given in Section 10.2.

In case of SE ECL system, when sufficient positive and negative power supply from external DC source is applied to the anode and cathode terminal respectively, high-intensity electric field is generated across the channel from anode to cathode.

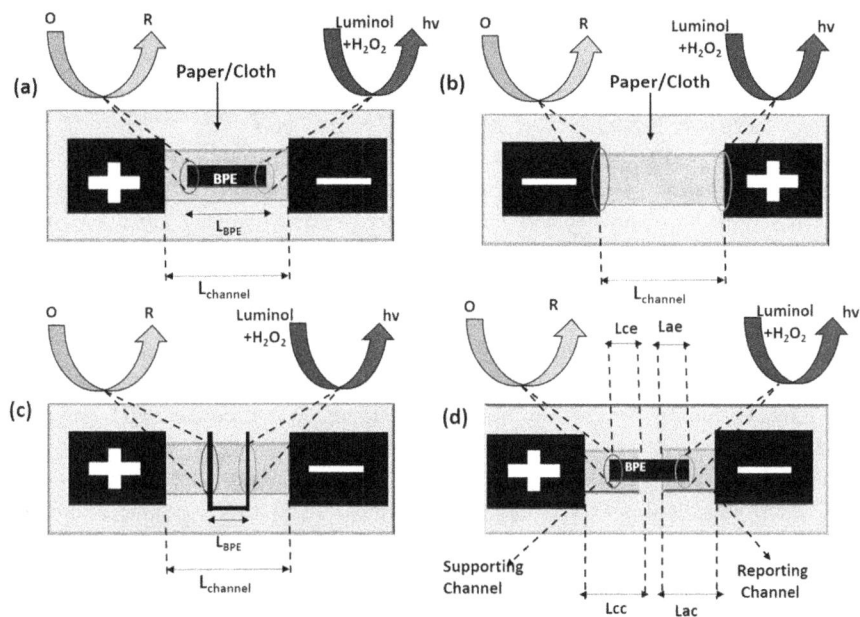

**FIGURE 10.1**   Different paper-based ECL systems: (a) BPE electrode–based ECL system; (b) single electrode–based ECL system; (c) U-Shaped open bipolar ECL system; (d) closed bipolar ECL system.

This leads to producing an ECL signal at the edge of anode of the SE-ECL device. SE-ECL with two driving electrodes and related oxidation and reduction process is illustrated in Figure 10.1(b).

## 10.2   PAPER AND CLOTH-BASED ECL SYSTEMS

### 10.2.1   Open Bipolar ECL Systems

Chen et al. (2016) fabricated a handheld open bipolar paper-based ECL device to sense $H_2O_2$ and glucose. Herein, they have utilized Whatman chromatography paper #1 as a substrate for a paper-based ECL device. The screen printing method utilizing carbon ink was used to developed the driving and BPE electrodes. Hydrophilic and hydrophobic zones were created over substrate using wax-screen printing. The performance validation for a fabricated device was accomplished using luminol/$H_2O_2$-based electrochemistry and detection of $H_2O_2$ and glucose were performed. A real sample analysis of glucose was performed in PBS and in artificial urine and obtained the limits of detection (LoD) of 0.0017 mM and 0.030 mM respectively. A smartphone was used as an ECL detector.

H. Liu et al. (2016) fabricated a portable open bipolar paper-based ECL system to detect the genetic detection of pathogenic bacteria. Conductive carbon ink was utilized to print BPE and driving over filter paper and hydrophilic and hydrophobic zones were created over substrate using wax-screen printing. $[Ru(phen)_2dppz]^{2+}$-based chemistry was used to detect pathogenic bacteria by using all optimized parameters DNA of *Listeria monocytogenes* was detected. Last, selectivity of fabricated device for *Listeria monocytogenes* was proved with respect to other interfering compounds such as *Staphylococcus aureus, Escherichia coli*, and *Salmonella*.

R. Liu et al. (2015) utilized paper as a substrate to fabricate the BPE-ECL device for visual detection. Herein, they have validated Luminol/$H_2O_2$ and Ru(bpy)$_3^{2+}$/TPrA-based electrochemistry and carried out the sensing of $H_2O_2$ and TPrA for the linear range (LR) 10 to 1,000 μM and 50 to 5,000 μM having LoDs of 8.7 μM and 46 μM, respectfully. Furthermore, the authors designed a second prototyping platform that compasses 11 BPEs across the two arms of a U-shaped paper channel. The screen printing method utilizing carbon ink was used to developed the driving and BPE electrodes and hydrophilic and hydrophobic zones were created using wax-screen printing. A high-resolution CCD camera was used for ECL imaging.

M. Liu et al. (2017) developed the cloth-based U-shaped open BPE-ECL device to detect $H_2O_2$ and glucose. Carbon ink was used to screen-print the driving and BPE electrodes, and wax-screen printing was utilized to form a microfluidics channel across the substrate. To detect glucose, luminol/$H_2O_2$ based chemistry was utilized, yielding a LR of 25 μM to 10,000 μM with a LoD of 23 μM. Finally, a real sample study of $H_2O_2$ and glucose in milk and blood serum was performed to demonstrate the device's performance.

R. Liu et al. (2017) introduced a simple, cost-effective thread-based open bipolar ECL device to detect TPrA and $H_2O_2$. A carbon ink–based screen printing approach was utilized to form a driving and bipolar electrodes over thread. The most usable two chemistries, i.e. Luminol/$H_2O_2$ and Ru(bpy)$_3^{2+}$/TPrA, were validated by doing the sensing of the TPrA and $H_2O_2$, and LoD of of 4.3 μM and 6.3 μM, respectively, were obtained. Furthermore, the functioning ability of the thread-based ECL device was proven by attaining a LoD 20.5 μM for glucose sensing.

Wang et al. (2018) fabricated a U-shaped paper-based BPE-ECL device to sense various biomarkers. They have used a different oxidase to detect choline (choline oxidase), lactate (lactate oxidase), and cholesterol (cholesterol oxidase), and achieved a LR of 10 to 5,000 μM for all three biomarkers, with distinct LoD values of 0.53 μM, 3.12 μM, and 7.418 μM for each. Carbon ink–based driving electrodes with a U-shaped open BPE electrode were fabricated, and a microfluidic channel was formed over a paper-based substrate utilizing a wax-printing method. Finally, to prove the applicability of the developed device, the real sample analysis was carried out using blood serum. A CCD camera was used successfully as an ECL detector.

## 10.2.2 Closed Bipolar ECL Systems

Although the open bipolar ECL devices have been used widely in diverse POCT applications, they lead to several disadvantages such as background noise produced by driving electrodes and poor current efficiency (X. Zhang et al. 2017). To overcome the above-mentioned drawbacks, M. Liu et al. (2016) developed a low-cost cloth-based closed BPE-ECL system and for the sensing of $H_2O_2$ and glucose. In a closed BPE-ECL system, the anode and cathode of BPE are independently placed into the reporting (RC) and supporting channel (SC), respectively (see Figure 10.1(d)).

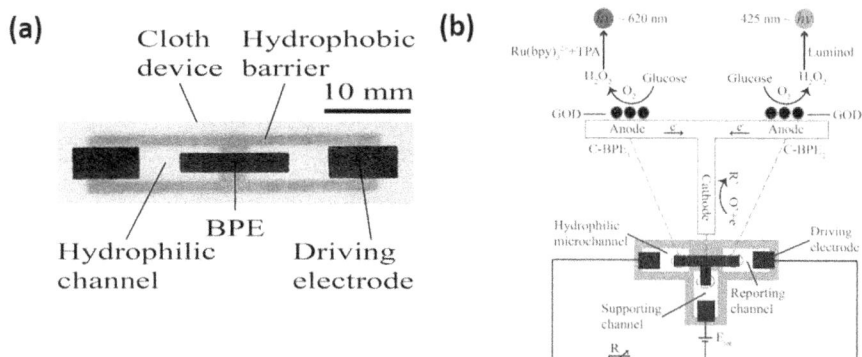

**FIGURE 10.2** Closed BPE-ECL systems: (a) cloth-based BPE-ECL device having screen-printed driving and closed BPE electrodes, taken from (M. Liu et al. 2016) Royal Society of Chemistry (open access); (b) multiplex detection of various biomolecules using shared cathode closed BPE-ECL device, taken from (Lai et al. 2022) with the prior approval of Elsevier.

Carbon ink–based driving and closed BPE electrodes were fabricated, and a microfluidic channel (supporting and reporting) was formed over a cloth-based substrate utilizing a wax-printing method (see Figure 10.2(a)). To ECL signal imaging, a CCD camera was used. The SC is filled with PBS solution, while RC is filled with target analytes with suitable luminol or ruthenium chemistry. Herein, the two most validated Luminol/$H_2O_2$ and Ru(bpy)$_3^{2+}$/TPrA-based ECL reactions were harnessed to detect $H_2O_2$ and TPrA and obtained LoD values of 24 μM and 85 μM, respectively.

R. Zhang et al. (2021) developed the cloth-based closed BPE-ECL device and demonstrated its application by sensing the *Salmonella enteritis* invA gene. Ru(bpy)$_3^{2+}$/TPrA was used to detect the*Salmonella enteritis* invA gene and obtained a LR 0.001 pM to 1,000 pM with LoD 0.05 fM. To demonstrate the feasibility of the fabricated device, selectivity analysis was performed against various interfering compounds such as *E. coli, S. aureus,* and *L. monocytogenes*. A repeatability analysis was conducted over a period of 26 days to demonstrate long-term capability of the device. Finally, real sample analysis of the *Salmonella enteritis* invA gene was investigated using, milk, grated tomato, and egg white.

Lai et al. (2022) fabricated cloth-based shared cathode closed bipolar ECL chip to sense multiple analytes over a single chip. A carbon ink–based screen-printing method was utilised to fabricated shared cathode electrodes, shown in Figure 10.2(b). Herein, both Ru(bpy)$_3^{2+}$/TPrA and Luminol/$H_2O_2$-based chemistry were used to detect glucose in real time, yielding LR of 0.005 to 1 mM and 0.05 to 10 mM, respectively, with LoD values of 0.0382 mM and 0.0422 mM, respectively. The simultaneous detection of glucose for seven different samples using a single chip was carried out. Furthermore, simultaneous detection of three targets such as uric acid, $H_2O_2$, and glucose have been accomplished to prove the potential of a fabricated device in multiple detections.

Feng et al. (2014) demonstrated a paper-based disposable closed BPE-ECL device to sense a prostate specific antigen (PSA). Wax printing was used to make the supporting and reporting channels, while carbon ink was used to make the closed BPE and driving electrodes. The performance of the fabricated device was validated using optimized parameters for a different concentration of PSA by obtaining a linear range 1 pg mL-1 to 100 pg mL$^{-1}$ having a limit of detection of 1 pg mL$^{-1}$.

C. Liu et al (2018) manufactured a paper-based cross channel paper-based closed BPE-ECL device for the multiple sensing. A cross channel BPE and driving electrodes were screen-printed using carbon ink over filter paper, and a microfluidic channel was formed using the wax-screen printing process. Multiple sensing and analytical performance of the fabricated device under optimal conditions were carried out. $H_2O_2$ and glucose sensing were performed over LR of 0.075 mM to 10 mM and 0.08 mM to 5 mM, respectively, with LoD values of 0.04 mM and 0.03 mM, respectively. Finally, interference study, stability, and real sample study were accomplished to demonstrate the feasibility of the developed cross channel paper-based closed BPE-ECL device.

Guan et al. (2016) showed a cloth-based three electrode system for ECL detection. Carbon ink was used to screen print the three electrodes (working, counter, and reference) and wax-screen printing was used to create microfluidics channel over the cloth-based substrate. Herein, both $Ru(bpy)_3^{2+}$/TPrA and Luminol/$H_2O_2$-based chemistries were validated by detecting TPrA and $H_2O_2$ by obtaining linear ranges of 2.5 µM to 2,500 µM and 0.05 µM to 2,000 µM with detection limits of 1.265 µM and 27 µM, respectively. A CCD camera was successfully used as an ECL signal detector. Finally, glucose sensing in PBS and artificial urine was carried out to validate the applicability of the device, by obtaining a LoD of 0.032 mM and 0.038 mM, respectively.

### 10.2.3 OTHER MATERIAL USED TO DEVELOP ECL SYSTEMS

Apart from paper and cloth, ECL systems have been developed using a variety of substrates, such as indium tin oxide (ITO), laser induced graphene on polyimide, 3D-printed material, etc. Different material-based bipolar and single electrode ECL systems have been thoroughly discussed in this section.

In a BPE ECL system, for multiplexed detection, a multiplex electrode will be required, which is a very time-consuming and expensive approach that is realistically unaffordable. Furthermore, BPE-ECL systems have been suffering from two major issues: background signal generated by driving electrodes that provide very poor current efficiency. Hence, to avoid those lacunas, Gao et al. (2018) for the first time, a low-cost single electrode-based ECL (SEES) system was developed and successfully validated by sensing $H_2O_2$, glucose, and uric acid. In this study, ITO was used with commercially available polyethylene terephthalate labels to fabricate SEES. Luminol/$H_2O_2$-based chemistry was used to sense all analytes. Herein, a photomultiplier tube (PMT) was used as an ECL detector.

Du et al. (2022) fabricated a SEES ECL device to detect cardiac troponin (cTnI). To detect cTnI, Luminol/$H_2O_2$-based electrochemistry was utilized, and it was found that when the concentration of cTnI increases, the ECL intensity decreases. A smartphone-based approach was used as a detector, and cTnI was detected by obtaining LR from 1 to 1,000 ng mL$^{-1}$ with a LoD 0.94 ng mL$^{-1}$. Herein, to fabricate

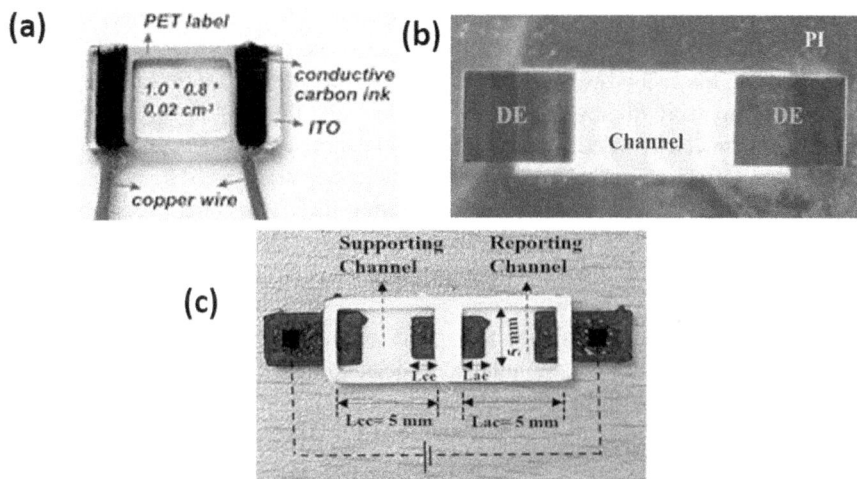

**FIGURE 10.3** Single electrode ECL systems: (a) fabricated ITO electrode-based SEES ECL device, adapted from (Gao et al. 2018) Royal Society of Chemistry (open access); (b) final fabricated LIG-based SE-ECL device, taken from (Manish Bhaiyya, Prasant Pattanaik 2021) with the prior permission of IEEE; (c) 3D-printed closed bipolar ECL device for multiple analytical sensing, taken from (Bhaiyya et al. 2021) with the prior approval of Wiley.

SEES, an array of electrodes was screen printed using carbon ink. The fabrication flow for SEES is illustrated in Figure 10.3(a).

Manish Bhaiyya (2021) reported the laser-induced graphene-based SE-ECL device for multiple biosensing. The LIG-SE-ECL device was fabricated using a basic low-cost polyamide substrate. The LIG-based electrodes were formed over the polyamide using optimized $CO_2$ laser parameters (speed = 80% and power = 50%). When a $CO_2$ laser with a specified speed and power is directed on a polyamide, the non-conductive zone becomes conductive. Herein, non-enzymatic detection of glucose, dopamine, and xanthine was carried out to confirm the performance of LIG-SE-ECL device, by obtaining LoD values of 3.76 μM, 1.25 μM, and 3.40 μM, respectively.

The smartphone-based technique was employed to obtain ECL images. The LIG-SE-ECL device was powered by a buck boost converter that was driven by a smartphone, eliminating the need for an external power supply. The final manufactured LIG-SE-ECL device is illustrated in Figure 10.3(b).

Further, Bhaiyya et al. (2021) developed a 3D-printed closed BPE-ECL device to sense the dopamine and choline. Conductive graphene filament was successfully used to develope a closed BPE and driving electrodes. $Ru(bpy)_3^{2+}$/TPrA chemistry with a quenching mechanism was effectively used to detect dopamine and obtained the LR 0.5 to 100 μM with a LoD value of 0.33 μM. Furthermore, Luminol/$H_2O_2$-based chemistry was used to detect choline and obtained the LR from 5 to 700 μM with LoD of 1.25 μM. To detect the ECL signal, they utilized two different detectors: a smartphone with a mini 3D-printed platform and a photo multiplier tube embedded with the Internet of Things. The final fabricated 3D-printed closed BPE-ECL device is demonstrated in Figure 10.3(c).

H. R. Zhang et al. (2016) developed two channel closed bipolar electrode ECL hip to detect cancer cell. The chip has three reservoirs, one of which was filled with PBS solution, the second with $Ru(bpy)_3^{2+}$/TPrA, and the third with Luminol/$H_2O_2$. Both reservoirs were filled with $Ru(bpy)_3^{2+}$/TPrA and Luminol/$H_2O_2$ acting as ECL reporting channels. The detection of the HL-60 cancer cell was accomplished using a quenching effect of $Ru(bpy)_3^{2+}$/TPrA. As the concentration of HL-60 cancer cell increases, the ECL signal starts reducing. Herein, ITO glass was successfully utilized to develop a closed bipolar electrode ECL chip. The validation of fabricated device was done by obtaining two linear ranges for HL-60 cancer cell with LoD value of 80 cells/mL.

Bhaiyya et al. (2022) continued their work by developing a six-well 3D-printed closed BPE-ECL device for simultaneous detection of glucose and choline. The analytical performance of the fabricated device was corroborated by performing one-step detection of glucose and choline, obtaining LR of 0.1 to 10 mM and 0.1 to 5 mM, respectively, with LoD values of 24 µM and 10 µM, respectively. The detection of glucose and choline for three distinct concentrations of each was accomplished on a single device to demonstrate the constructed device's multimodal sensing capability. A custom-made 3D-printed black box assembly integrated with a smartphone was used for ECL imaging. Furthermore, they have developed a mobile app to measure the ECL image intensity. Finally, an interference study, stability, and real sample study were done to validate the applicability of the device.

Salve et al. (2021) proposed a 3D-printed BPE-ECL system for multiple sensing. Graphite pencil–based electrodes were used as a BPE and DEs. $H_2O_2$, $O_2$, and $CO_2$ were detected using Luminol/$H_2O_2$ electrochemistry with LR values of 0.08 µM to 5,000 µM, 0.3 mg/L to 9 mg/L, and 0.6 mg/L to 9 mg/L, respectively, with LoDs of 0.069 M, 0.15 mg/L, and 0.45 mg/L. Furthermore, the enzymatic detection of glucose was performed using glucose oxidase and obtained the LR from 1 µM to 10,000 µM with LoD 0.31 µM. To eliminate the use of darkroom environment, a 3D-printed black box integrated with a smartphone and 9 V battery was used to detect the ECL signal. Finally, the real sample study was accomplished to strengthen the applicability of developed portable device.

In an another separate, Bhaiyya et al. (2021) proposed a LIG-based bipolar ECL device to detect multiple biomolecules such as glucose and $H_2O_2$. A polyimide sheet was used as substrate material and by directing a laser over polyamide with optimized speed (80%) and power (50%) of universal $CO_2$ laser system, a LIG was formed. Enzymatic detection of glucose was carried out using glucose oxidase by utilizing Luminol/$H_2O_2$ electrochemistry and achieved LR from 1 to 100 µM having a LoD of 0.138 µM. Furthermore, sensing of $H_2O_2$ was carried out and obtained the LR 1 µM to 100 µM with LoD of 5.87 µM. In addition, the same group developed a LIG-based multichannel closed BPE-ECL for simultaneous detection of vitamin C and vitamin $B_{12}$ (Bhaiyya et al. 2021c). With all optimized parameters, Luminol/$H_2O_2$-based chemistry was used to detect vitamin C and vitamin $B_{12}$ and obtained LR 1 to 1,000 µM and 0.5 to 1,000 nM, respectively, with LoD values of 0.96 µM and 0.109 µM, respectively.

Kwon et al. (2020) developed a carbon ink screen-printed three electrode ECL device to detect the concentration of dopamine using $Ru(bpy)_3^{2+}$/TPrA electrochemistry. The detection of dopamine was achieved using a quenching process, which reveals that when dopamine concentrations increase, the ECL signal intensity decreases. Herein, the sensing capability of the three electrode devices was validated and compared by changing the working electrode with gold, carbon and platinum and it was experimentally observed that carbon electrode–based three electrode system provides better performance compared to gold and platinum. With optimized conditions, dopamine was sensed and obtained the LR from 1 to 50 µM having a LoD of 0.5 µM. To detect as well as to calculate the intensity of the ECL signal, a smartphone was used.

Rivera et al. (2021) proposed a disposable screen-printed carbon ink–based three electrode ECL approach to sense phenolic compounds i.e. vanillic and coumaric acids. $Ru(bpy)_3^{2+}$/TPrA electrochemistry with a quenching mechanism was effectively used to detect vanillic and coumaric acids and obtained the limits of detection of 0.26 µM and 0.68 µM, respectively. ECL imaging was done using a smartphone. A mobile app was developed to control the phone camera and potentiostat parameters in the sensor design. For sensing, thre commercially available (Drop Sens, DRP-110) screen-printed electrodes were utilized.

Diverse effort has been done in the field of ECL to fabricate low-cost, portable, miniaturized devices by harnessing the capabilities of paper and cloth as a substrate. Table 10.1 summarizes the study findings for the POCT application using paper and cloth-based mini ECL devices.

## 10.3  EXISTING RESEARCH GAPS AND POTENTIAL SOLUTIONS

In recent years, many researchers have worked hard to build paper and cloth-based ECL devices that can be used for a variety of POCT applications. Despite much research into paper and cloth-based ECL devices, there is still a major gap in making such devices commercially feasible. After doing extensive research, two key research gaps were identified, and possible solutions were proposed.

- To develop electrodes and microfluidic channels on paper and cloth-based ECL devices, screen-printing and wax-printing methods have been widely used. As both screen-printing and wax-printing methods require several fabrication steps (usually nine to ten steps), this leads to time consuming and complex procedure.
- It was observed, in most of the cases, high-cost CCD and PMT cameras were used to detect ECL signals, leading to increasing the overall ECL system cost. Furthermore, PMT and CCD camera require their own stepup, which prohibits the ECL system from being used as a portable system.

The following suggested methods are proposed to overcome the aforementioned challenges:

- 3D printing and stereolithography (SLA) 3D printing based on ECL platforms with conductive filaments are the best suitable alternative that fabricate ECL devices with great accuracy in a single step.

**TABLE 10.1**

**Research Summary Using Paper and Cloth-Based Substrate in ECL**

| Material Used | Electrode Type | Chemistry Used | Application | Linear Range (µM) | LOD (µM) | Ref. |
|---|---|---|---|---|---|---|
| Paper | Open BPE | Luminol/$H_2O_2$ and Ru(bpy)$_3$$^{2+}$/TPrA | $H_2O_2$ and TPrA | 10 to 1,000 and 50 to 5,000 | 8.7 and 46 | (R. Liu et al. 2015) |
| Paper | Open BPE | Luminol/$H_2O_2$ | $H_2O_2$ and glucose | 5 to 5,000 and 0.1 to 5,000 | 1.75 and 30 | (Chen et al. 2016) |
| Paper | Open BPE | [Ru(phen)$_2$dppz]$^{2+}$ | pathogenic bacteria | - | 10 copies/µL | (H. Liu et al. 2016) |
| Cloth | U-shaped open BPE | Luminol/$H_2O_2$ | $H_2O_2$ and glucose | 25 to 10,000 and 25 to 10,000 | 24 and 23 | (M. Liu et al. 2017) |
| Cotton threads | Open BPE | Luminol/$H_2O_2$ and Ru(bpy)$_3$$^{2+}$/TPrA | $H_2O_2$ and TPrA | 10 to 1,000 and 10 to 1,000 | 6.3 and 4.3 | (R. Liu et al. 2017) |
| Paper | U-shaped open BPE | Luminol/$H_2O_2$ | Choline, lactate, and cholesterol | 10 to 5,000 | 0.53, 3.12 and 7.418 | (Wang et al. 2018) |
| Cloth | Closed BPE | Luminol/$H_2O_2$ and Ru(bpy)$_3$$^{2+}$/TPrA | $H_2O_2$ and glucose | 25 to 2,500 and 0 to 1,000 | 24 and 85 | (M. Liu et al. 2016) |
| Cloth | Closed BPE | Ru(bpy)$_3$$^{2+}$/TPrA | *Salmonella enteritis* invA gene | 0.001 pM to 1,000 pM | 0.05 fM | (R. Zhang et al. 2021) |
| Cloth | Closed BPE | Luminol/$H_2O_2$ and Ru(bpy)$_3$$^{2+}$/TPrA | Glucose | 50 to 1,000 and 50 to 1,000 | 38 and 42 | (Lai et al. 2022) |
| Paper | Closed BPE | Ru(bpy)$_3$$^{2+}$/TPrA | Prostate specific antigen | 1 pg mL$^{-1}$ pg mL$^{-1}$1 to 100 pg mL$^{-1}$ | 1 pg mL$^{-1}$ | (Feng et al. 2014) |
| Paper | Closed BPE | Luminol/$H_2O_2$ | $H_2O_2$ and glucose | 75 to 10,000 and 80 to 5,000 | 40 and 30 | (C. Liu et al. 2018) |
| Cloth | Three electrodes | Luminol/$H_2O_2$ and Ru(bpy)$_3$$^{2+}$/TPrA | $H_2O_2$ and TPrA | 50 to 2,000 and 2.5 to 2,500 | 27 and 1.26 | (Guan et al. 2016) |
| ITO | Single electrode | Luminol/$H_2O_2$ | $H_2O_2$ | 1 to 100 | 0.27 | (Gao et al. 2018) |
| ITO | Single electrode | Luminol/$H_2O_2$ | Cardiac troponin (cTnI). | 1 to 1,000 ng mL$^{-1}$ | 0.94 ng mL$^{-1}$ | (Du et al. 2022) |

| Platform | Electrode | ECL system | Analyte | Linear range | LOD | Reference |
|---|---|---|---|---|---|---|
| Laser-induced graphene | Single electrode | Luminol/$H_2O_2$ | Glucose, xanthine and dopamine | 0.1 to 70, 0.1 to 100 and 0.1 to 100 | 3.76 µM, 1.25 µM, and 3.40 µM | (Bhaiyya et al., 2021) |
| 3D Printed | Closed BPE | $Ru(bpy)_3^{2+}$/TPrA and Luminol/$H_2O_2$ | Dopamine and choline | 0.5 to 100 and 5 to 700 | 0.33 and 1.25 | (Bhaiyya, Kulkarni, et al. 2021) |
| ITO | Closed BPE | $Ru(bpy)_3^{2+}$/TPrA | HL-60 cancer cell | 320 to $2.5*10^5$ cells/mL | 80 cells/mL | (H. R. Zhang et al. 2016) |
| 3D Printed | Closed BPE | Luminol/$H_2O_2$ | Glucose and choline | 100 to 10,000 and 100 to 5,000 | 24 and 10 | (Bhaiyya, Kumar, and Sanket 2022) |
| 3D Printed | Open BPE | Luminol/$H_2O_2$ | $H_2O_2$, $O_2$, and $CO_2$ | 0.08 to 5,000, 0.3 mg/L to 9 mg/L, and 0.6 mg/L to 9 mg/L | 0.069, 0.15 mg/L, and 0.45 mg/L | (Salve et al. 2021) |
| Laser induced graphene | Open BPE | Luminol/$H_2O_2$ | $H_2O_2$ and glucose | 1 to 100 and 1 to 100 | 5.87 and 0.138 | (Bhaiyya et al. 2021) |
| Laser induced graphene | Closed BPE | Luminol/$H_2O_2$ | vitamin C and vitamin $B_{12}$ | 1 to 1,000 and 0.5 to 1,000 nM | 0.96 and 0.109 | (Bhaiyya et al. 2021c) |
| Paper | Three electrodes | $Ru(bpy)_3^{2+}$/TPrA | Dopamine | 1 to 50 µM | 0.5 µM | (Kwon et al. 2020) |
| Paper | Three electrodes | $Ru(bpy)_3^{2+}$/TPrA | vanillic and coumaric acids | 0.25 to 30 and 1 to 50 | 0.26 and 0.68 | (Rivera et al. 2021) |
| Paper | Three electrodes | $Ru(bpy)_3^{2+}$/TPrA | 2(dibutylamino-Ethanol (DBAE) and nicotinamide adenine dinucleotide | 3 to 5,000 and 200 to 10,000 | 0.9 and 72 | (Delaney et al. 2011) |
| Laser-induced graphene | BPE and single electrode | Luminol/$H_2O_2$ | Vitamin $B_{12}$ | 0.5 to 700 nM for BPE and 0.5 to 1,000 nM for single electrode | 0.107 nM and 0.094 nM | (Bhaiyya et al. 2021a) |

- The integration of a mini black box assembly (made using 3D printing or SLA printing) with a smartphone and power supply not only resolves the problem of using PMT and CCD, but also the issue of portability.

## 10.4   CONCLUSION AND FUTURE PERSPECTIVES

Evidently, interest in paper and cloth-based ECL systems is growing, as they offer numerous benefits. These advantages include high sensitivity and selectivity, quick response, less sample requirements, self-pumping to eliminate external pumping, easy disposal, and the ability to be synchronized with ECL platforms. Because of these characteristics, they are used in a varied range of applications, comprising food monitoring, environmental and health monitoring, and molecular analysis. Despite significant advancements in ECL-based POCT, experimental paper and cloth-based CL systems have yet to be transformed into industrial manufacturing or widely used in practice. These persistent demands will provide a unique and promising window for researchers to fabricate a single-step paper and cloth-based ECL system for use in POCTs. Overall, paper and cloth-based ECL systems are important in the field of POCT and will continue to be effective research in the future.

## REFERENCES

Bhaiyya, Manish, Madhusudan B. Kulkarni, Prasant Kumar Pattnaik, and Sanket Goel. 2021. "IoT Enabled PMT and Smartphone Based Electrochemiluminescence Platform to Detect Choline and Dopamine Using 3D-Printed Closed Bipolar Electrodes." *Luminescence* 37: 0–2. 10.1002/bio.4179

Bhaiyya, Manish, Prasant Kumar Pattnaik, and Sanket Goel. 2021a. "Electrochemiluminescence Sensing of Vitamin B 12 Using Laser - Induced Graphene Based Bipolar and Single Electrodes in a 3D - Printed Portable System." *Microfluidics and Nanofluidics* 331: 112831. 10.1007/s10404-021-02442-x

Bhaiyya, Manish, Prasant Kumar Pattnaik, and Sanket Goel. 2021b. "ScienceDirect Electrochemistry A Brief Review on Miniaturized Electrochemiluminescence Devices: From Fabrication to Applications." *Current Opinion in Electrochemistry* 30: 100800. 10.1016/j.coelec.2021.100800

Bhaiyya, Manish, Prasant Kumar Pattnaik, and Sanket Goel. 2021c. "Simultaneous Detection of Vitamin B12 and Vitamin C from Real Samples Using Miniaturized Laser-Induced Graphene Based Electrochemiluminescence Device with Closed Bipolar Electrode." *Sensors and Actuators A: Physical* 331: 112831. 10.1016/j.sna.2021.112831

Bhaiyya, Manish, Prakash Rewatkar, Mary Salve, Prasant Kumar Pattnaik, and Sanket Goel. 2021. "Miniaturized Electrochemiluminescence Platform with Laser-Induced Graphene Electrodes for Multiple Biosensing." *IEEE Transactions on Nanobioscience* 20 (1): 79–85. 10.1109/TNB.2020.3036642

Bhaiyya, Manish, Prasant Pattnaik, and Sanket Goel. 2021. "Miniaturized Electrochemiluminescence Platform with Laser-Induced Graphene Based Single Electrode for Interference-Free Sensing of Dopamine, Xanthine and Glucose." *IEEE Transactions on Instrumentation and Measurement* 70: 1–8. 10.1109/TIM.2021.3071215

Bhaiyya, Manish, Prasant Kumar, and Pattnaik Sanket. 2022. "Multiplexed and Simultaneous Biosensing in a 3D - Printed Portable Six - Well Smartphone Operated Electrochemiluminescence Standalone Point - of - Care Platform." *Microchimica Acta* 189 (79): 1–9. 10.1007/s00604-022-05200-0

Chen, Lu, Chunsun Zhang, and Da Xing. 2016. "Paper-Based Bipolar Electrode-Electrochemiluminescence (BPE-ECL) Device with Battery Energy Supply and Smartphone Readout: A Handheld ECL System for Biochemical Analysis at the Point-of-Care Level." *Sensors and Actuators, B: Chemical* 237: 308–317. 10.1016/j.snb.2016.06.105

Christodouleas, Dionysios C., Balwinder Kaur, and Parthena Chorti. 2018. "From Point-of-Care Testing to EHealth Diagnostic Devices (EDiagnostics)." *ACS Central Science* 4 (12): 1600–1616. 10.1021/acscentsci.8b00625

Delaney, Jacqui L., Conor F. Hogan, Junfei Tian, and Wei Shen. 2011. "Electrogenerated Chemiluminescence Detection in Paper-Based Microfluidic Sensors." *Analytical Chemistry* 83 (4): 1300–1306. 10.1021/ac102392t

Du, Fangxin, Zhiyong Dong, Yiran Guan, Abdallah M. Zeid, Di Ma, Jiachun Feng, Di Yang, and Guobao Xu. 2022. "Single-Electrode Electrochemical System for the Visual and High-Throughput Electrochemiluminescence Immunoassay." *Analytical Chemistry* 94 (4): 2189–2194. 10.1021/acs.analchem.1c04709

Feng, Qiu Mei, Jian Bin Pan, Huai Rong Zhang, Jing Juan Xu, and Hong Yuan Chen. 2014. "Disposable Paper-Based Bipolar Electrode for Sensitive Electrochemiluminescence Detection of a Cancer Biomarker." *Chemical Communications* 50 (75): 10949–10951. 10.1039/c4cc03102d

Fosdick, Stephen E., Kyle N. Knust, Karen Scida, and Richard M. Crooks. 2013. "Bipolar Electrochemistry." *Angewandte Chemie - International Edition* 52 (40): 10438–10456. 10.1002/anie.201300947

Gao, Wenyue, Kateryna Muzyka, Xiangui Ma, Baohua Lou, and Guobao Xu. 2018. "A Single-Electrode Electrochemical System for Multiplex Electrochemiluminescence Analysis Based on a Resistance Induced Potential Difference." *Chemical Science* 9 (16): 3911–3916. 10.1039/c8sc00410b

Guan, Wenrong, Min Liu, and Chunsun Zhang. 2016. "Electrochemiluminescence Detection in Microfluidic Cloth-Based Analytical Devices." *Biosensors and Bioelectronics* 75: 247–253. 10.1016/j.bios.2015.08.023

He, Yong, Yan Wu, Jian Zhong Fu, and Wen Bin Wu. 2015. "Fabrication of Paper-Based Microfluidic Analysis Devices: A Review." *RSC Advances* 5 (95): 78109–78127. 10.1039/c5ra09188h

Jung, Wooseok, Jungyoup Han, Jin Woo Choi, and Chong H. Ahn. 2015. "Point-of-Care Testing (POCT) Diagnostic Systems Using Microfluidic Lab-on-a-Chip Technologies." *Microelectronic Engineering* 132: 46–57. 10.1016/j.mee.2014.09.024

Kost, Gerald J., Sharon S. Ehrmeyer, Bart Chernow, James W. Winkelman, Gary P. Zaloga, R. Phillip Dellinger, and Terry Shirey. 1999. "The Laboratory-Clinical Interface: Point-of-Care Testing." *Chest* 115 (4): 1140–1154. 10.1378/chest.115.4.1140

Kwon, Hyun J., Elmer Ccopa Rivera, Mabio R.C. Neto, Daniel Marsh, Jonathan J. Swerdlow, Rodney L. Summerscales, and Padma P. Tadi Uppala. 2020. "Development of Smartphone-Based ECL Sensor for Dopamine Detection: Practical Approaches." *Results in Chemistry* 2: 100029. 10.1016/j.rechem.2020.100029

Lai, Wei, Yi Liang, Yan Su, and Chunsun Zhang. 2022. "Shared-Cathode Closed Bipolar Electrochemiluminescence Cloth-Based Chip for Multiplex Detection." *Analytica Chimica Acta* 1206: 339446. 10.1016/j.aca.2022.339446

Liu, Cuiling, Dan Wang, and Chunsun Zhang. 2018. "A Novel Paperfluidic Closed Bipolar Electrode-Electrochemiluminescence Sensing Platform: Potential for Multiplex Detection at Crossing-Channel Closed Bipolar Electrodes." *Sensors and Actuators, B: Chemical* 270: 341–352. 10.1016/j.snb.2018.04.180

Liu, Hongxing, Xiaoming Zhou, Weipeng Liu, Xiaoke Yang, and Da Xing. 2016. "Paper-Based Bipolar Electrode Electrochemiluminescence Switch for Label-Free and Sensitive Genetic Detection of Pathogenic Bacteria." *Analytical Chemistry* 88 (20): 10191–10197. 10.1021/acs.analchem.6b02772

Liu, Min, Rui Liu, Dan Wang, Cuiling Liu, and Chunsun Zhang. 2016. "A Low-Cost, Ultraflexible Cloth-Based Microfluidic Device for Wireless Electrochemiluminescence Application." *Lab on a Chip* 16 (15): 2860–2870. 10.1039/c6lc00289g

Liu, Min, Dan Wang, Cuiling Liu, Rui Liu, Huijie Li, and Chunsun Zhang. 2017. "Battery-Triggered Open Wireless Electrochemiluminescence in a Microfluidic Cloth-Based Bipolar Device." *Sensors and Actuators, B: Chemical* 246: 327–335. 10.1016/j.snb.2 017.02.076

Liu, Rui, Cuiling Liu, Huijie Li, Min Liu, Dan Wang, and Chunsun Zhang. 2017. "Bipolar Electrochemiluminescence on Thread: A New Class of Electroanalytical Sensors." *Biosensors and Bioelectronics* 94 (February): 335–343. 10.1016/j.bios.2017.03.007

Liu, Rui, Chunsun Zhang, and Min Liu. 2015. "Open Bipolar Electrode-Electro-chemiluminescence Imaging Sensing Using Paper-Based Microfluidics." *Sensors and Actuators, B: Chemical* 216: 255–262. 10.1016/j.snb.2015.04.014

Marquette, Christophe A., and Loïc J. Blum. 2008. "Electro-Chemiluminescent Biosensing." *Analytical and Bioanalytical Chemistry* 390 (1): 155–168. 10.1007/s00216-007-1631-2

Miao, Wujian. 2008. "Electrogenerated Chemiluminescence and Its Biorelated Applications." *Chemical Reviews* 108 (7): 2506–2553. 10.1021/cr068083a

Richter, Mark M. 2004. "Electrochemiluminescence (ECL)," *Chemical Reviews* 104 (2004): 3003–3036. 10.1016/B978-044453125-4.50009-7

Rivera, Elmer C., Joseph W. Taylor, Rodney L. Summerscales, and Hyun J. Kwon. 2021. "Quenching Behavior of the Electrochemiluminescence of Ru(Bpy)32+/TPrA System by Phenols on a Smartphone-Based Sensor." *ChemistryOpen* 10 (8): 842–847. 10. 1002/open.202100151

Sachdeva, Shivangi, Ronald W. Davis, and Amit K. Saha. 2021. "Microfluidic Point-of-Care Testing: Commercial Landscape and Future Directions." *Frontiers in Bioengineering and Biotechnology* 8 (January 2021): 1–14. 10.3389/fbioe.2020.602659

Sakura, Sachiko. 1992. "Electrochemiluminescence of Hydrogen Peroxide-Luminol at a Carbon Electrode." *Analytica Chimica Acta* 262 (1): 49–57. 10.1016/0003-2670(92)80007-T

Salve, Mary, Aurnab Mandal, Khairunnisa Amreen, B. V.V.S.N.Prabhakar Rao, Prasant Kumar Pattnaik, and Sanket Goel. 2021. "A Portable 3-D Printed Electrochemilumine-scence Platform with Pencil Graphite Electrodes for Point-of-Care Multiplexed Analysis with Smartphone-Based Read Out." *IEEE Transactions on Instrumentation and Measurement* 70. 10.1109/TIM.2020.3023211

Sher, Mazhar, Rachel Zhuang, Utkan Demirci, and Waseem Asghar. 2017. "Paper-Based Analytical Devices for Clinical Diagnosis: Recent Advances in the Fabrication Techniques and Sensing Mechanisms." *Expert Review of Molecular Diagnostics* 17 (4): 351–366. 10.1080/14737159.2017.1285228

Songjaroen, Temsiri, Wijitar Dungchai, Orawon Chailapakul, and Wanida Laiwattanapaisal. 2011. "Novel, Simple and Low-Cost Alternative Method for Fabrication of Paper-Based Microfluidics by Wax Dipping." *Talanta* 85 (5): 2587–2593. 10.1016/j.talanta.-2011.08.024

Wang, Dan, Cuiling Liu, Yi Liang, Yan Su, Qiuping Shang, and Chunsun Zhang. 2018. "A Simple and Sensitive Paper-Based Bipolar Electrochemiluminescence Biosensor for Detection of Oxidase-Substrate Biomarkers in Serum." *Journal of The Electrochemical Society* 165 (9): B361–B369. 10.1149/2.0551809jes

Willmott, Charles, and J. E. Arrowsmith. 2013. "Point of Care Testing." *Surgery (Oxford)* 28 (4): 159–160.

Wu, Suozhu, Zhenyu Zhou, Linru Xu, Bin Su, and Qun Fang. 2014. "Integrating Bipolar Electrochemistry and Electrochemiluminescence Imaging with Microdroplets for Chemical Analysis." *Biosensors and Bioelectronics* 53: 148–153. 10.1016/j.bios.2013.09.042

Yamada, Kentaro, Hiroyuki Shibata, Koji Suzuki, and Daniel Citterio. 2017. "Toward Practical Application of Paper-Based Microfluidics for Medical Diagnostics: State-of-the-Art and Challenges." *Lab on a Chip* 17 (7): 1206–1249. 10.1039/c6lc01577h

Yetisen, Ali Kemal, Muhammad Safwan Akram, and Christopher R. Lowe. 2013. "Paper-Based Microfluidic Point-of-Care Diagnostic Devices." *Lab on a Chip* 13 (12): 2210–2251. 10.1039/c3lc50169h

Zhang, Huai Rong, Yin Zhu Wang, Wei Zhao, Jing Juan Xu, and Hong Yuan Chen. 2016. "Visual Color-Switch Electrochemiluminescence Biosensing of Cancer Cell Based on Multichannel Bipolar Electrode Chip." *Analytical Chemistry* 88 (5): 2884–2890. 10.1021/acs.analchem.5b04716

Zhang, Ruoyuan, Yi Liang, Yan Su, Wei Lai, and Chunsun Zhang. 2021. "Cloth-Based Closed Bipolar Electrochemiluminescence DNA Sensors (CCBEDSs): A New Class of Electrochemiluminescence Gene Sensors." *Journal of Luminescence* 238 (May): 118209. 10.1016/j.jlumin.2021.118209

Zhang, Xiaowei, Qingfeng Zhai, Huanhuan Xing, Jing Li, and Erkang Wang. 2017. "Bipolar Electrodes with 100% Current Efficiency for Sensors." *ACS Sensors* 2 (3): 320–326. 10.1021/acssensors.7b00031

# 11 MXene Materials for Miniaturized Energy Storage Devices (MESDs)

*Jhansi Chintakindi and Afkham Mir*

Department of Chemical Engineering, BITS Pilani Hyderabad Campus, Jawahar Nagar, Hyderabad, Telangana, India

## CONTENTS

## 11.1 INTRODUCTION

2D-layered materials are crystalline structures stacked together by van der Waals bonds and are nearly one atom or a few atoms thick. They are derived from the delamination of pristine 3D bulk materials that are stacked in layers one over the other (Ambrosi and Pumera 2018). Due to their extraordinary features, 2D-layered materials such as graphene, MXenes, phosphorene, and others have found wide application in a variety of sectors. Among them, MXenes, for example, have exceptional physical and chemical properties and are employed in a variety of applications like catalysis (Shukla 2020), wearable electronics (N. Li et al. 2021), energy storage devices (Garg et al. 2020), optoelectronics (Liu and Alshareef 2021), and chemical sensors (Verger et al. 2019).

MXenes were identified as a 2D material for the first time in 2011 after their successful synthesis utilizing the hydrofluoric acid (HF) etching process. It was later classified as a novel 2D material in 2015, and there has been significant progress in terms of synthesis and property enhancement since then [7]. However, there is still a need for improvement in terms of functionalization control, conventional synthesis methods, surface terminations, and other areas.

MXenes with the common formula of $M_{n+1}X_nT_x$ (n can vary from 1, 2, 3, etc.) or $M_{1.3}XT_x$, (where M is a transition metal (e.g. Ti, Mo, Cr, Nb, V, Sc, Zr, Hf or Ta), the letter X represents carbon and/or nitrogen, and the letter $T_x$ represents various terminations such as fluorine, hydroxyl, and/or oxygen atoms), are one of the latest and a

DOI: 10.1201/b23359-11

**173**

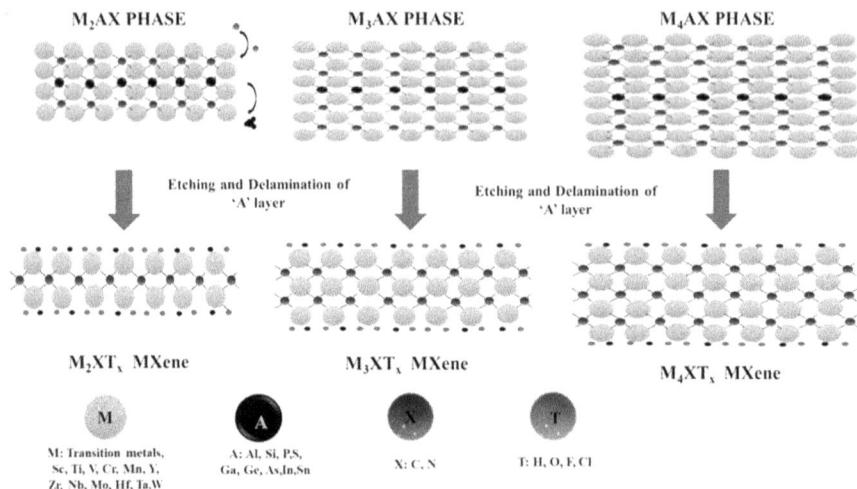

**FIGURE 11.1**  Schematic diagram of etching of MXenes from their pristine MAX phases: $M_2X$, $M_3X_2$, and $M_4X_3$ layers are separated by 'A' layers in different classes of MXenes. Due to the introduction of powerful chemical etchants, selective etching of 'A' layers results in $M_2X$, $M_3X_2$, and $M_4X_3$ delamination, as well as an extra surface termination $T_x$.

large family of 2D transition-metal carbides derived from their parent MAX phase, as shown in Figure 11.1. Because of their hydroxyl- or oxygen-terminated surfaces, MXenes combine the metallic conductivity of transition metal carbides with a hydrophilic nature (Naguib et al. 2011; Naguib et al. 2013). So far, 30 different types of MXenes have been discovered (Gogotsi and Anasori 2019). MXenes have unique properties such as chemical stability at around 700°C (J. Zhang et al. 2018) and tunable surface terminations (Bao et al. 2021), ready-to-bond nature to various species (Magnuson et al. 2017), and high negative Zeta-potential (Tian et al. 2019), which enables the stable colloidal solutions in water have attracted many applications.

In the conventional synthesis routes, concentrated hydrofluoric acid is employed in HF etching to etch out the A layer from the pristine MAX phases over a period of 12–192 h in order to synthesize MXenes (Alhabeb et al. 2018). Because HF is a strong acid that can have a range of negative effects on humans, it is not suitable for laboratory usage, which has led to the development of alternative HF-free procedures such as the minimally intensive layer delamination (MILD) method, molten salts method, alkali method, iodine method, and electrochemical etching methods. The MILD method, which uses extremely concentrated HCl+ LiF as etchants, is an alternative to the HF etching process. For MXene (Wu et al. 2022), the MILD approach was employed to synthesize electrochemical actuators with a remarkable performance of 90% capacity holding after 10,000 cycles by Wang et al. (2021). In the alkali method, concentrated alkaline solution is used as etchants at extreme operating conditions (at temperatures of 270°C for 24 h) to synthesize $Ti_3C_2$ (L. Li et al. 2017). MXene prepared by the alkaline method was used to fabricate a supercapacitor with a capacitance of around 229 F $g^{-1}$, which is much lower than the capacitance exhibited by MXene fabricated through the molten salts method

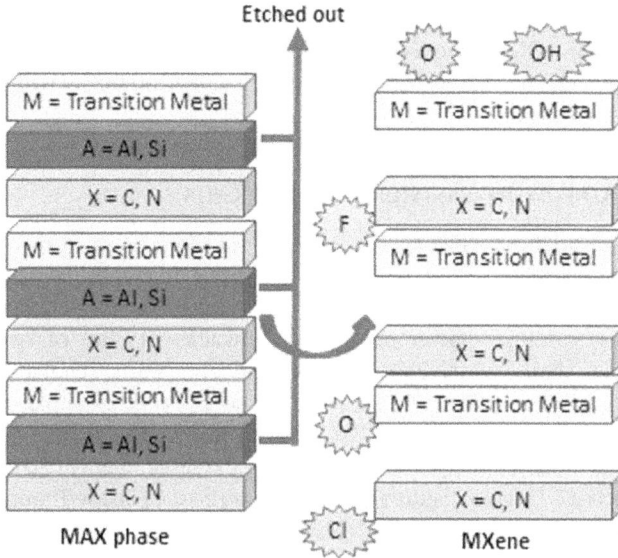

**FIGURE 11.2**   The etching of 'A' layers from parental MAX phases to produce MXene is depicted schematically. The most typical approach for making MXenes is to etch out the A phase from the MAX structure without etching the M sheets at the same time; nevertheless, this is not always a simple operation. Multilayered MXene particles are generated after etching and these can be intercalated and delaminated to form distributed nanomaterials. Surface terminations are an inevitable byproduct of MXene synthesis, which typically includes O, OH, and F groups.

$(785 \text{ C g}^{-1})$ (Y. Li et al. 2020). The molten salt method uses Lewis acids in molten salts at a temperature around 750°C. All of the etching methods mentioned above have harsh operating conditions, whereas etching MXenes with an electrochemical etching process (Sun et al. 2017) is less brusque and is less harsh. However, electrochemical etching may result in the etching of the carbon from the $Ti_3C_2$, resulting in the formation of carbon-derived carbons (CDCs) when operated for a longer period of time. Hence, the method employed to synthesize MXene has an effect on the final exhibited properties of MXenes and the common strategy for delaminate multilayered prepared MXenes is to increase the interlayer spacing of MXene flakes via ion intercalation and the etching of all mentioned processed with surface terminations, as shown in Figure 11.2.

## 11.2   MXENE-BASED MINIATURIZED ENERGY STORAGE DEVICES

Due to the fast evolution of miniaturized and wearable electronics, there is a thriving market for suitable miniature energy storage devices. Miniaturized energy storage devices (MESDs) are an appealing approach for the development of microelectronics due to their infinite lifetime and high power density, but scalable production is typically dependent on electrode materials and design processes. MXenes have recently shown significant promise in advanced MESDs due to their high energy and power densities, strong metallic conductivity, and ease of processing. The fabrication methodologies of MXene micro-solar cells, micro-supercapacitors, micro-batteries, and micro-actuators, as well as their electrochemical performance, will be discussed in the following sections.

### 11.2.1  MXᴇɴᴇ-Bᴀsᴇᴅ Mɪɴɪᴀᴛᴜʀɪᴢᴇᴅ Sᴏʟᴀʀ Cᴇʟʟs

Solar cells are the most efficient and practical way to convert direct sunlight into electricity. For decades, earth-abundant silicon-based solar cells have been the most widely used compounds because they have a long life expectancy of around 25 years [15] and high power conversion efficacies (PCEs) of over 25–26%. However, due to their excessive preliminary fabrication costs, scholars have shifted their attention to more affordable substitute solar cells for example organic solar cells (OSCs), perovskite solar cells (PSCs), dye sensitized solar cells (DSSCs), and the quantum dot solar cells (QDSCs). Ever since the researchers in 2018 discovered the usage of $Ti_3C_2T_x$ as an additive in the photoresist of methyl ammonium lead iodide-based perovskite solar cells (PSCs) (Guo et al. 2018), its applications have expanded to electrodes, hole/electron transport layer (HTL/ETL), additive in HTL/ETL, and as an element of Schottky junction-based solar cells employing silicon (Si) wafers. The role of MXenes in the development of the solar cells has been commendable since then and its functionality in solar cells can be grouped in three broad categories as additives, electrodes, and hole/electron transport layers (HTL/ETL) based on its role in the device manufacturing of solar cells, as shown in Figure 11.3. The role of MXenes in miniature solar cells and their power efficiencies are summarized in Table 11.1.

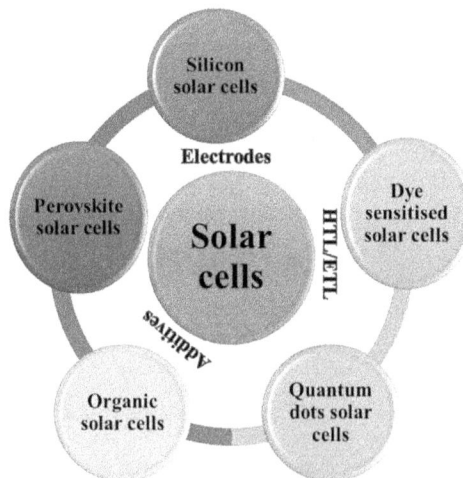

**FIGURE 11.3**  Schematic diagram representing the types of solar cells based on their applications. Organic solar cells (OSCs), perovskite solar cells (PSCs), quantum dot solar cells (QDSCs), and dye sensitized solar cells (DSSCs) are all types of solar cells. MXene's involvement in solar cell device construction can be divided into three categories: additives, electrodes, and hole/electron transport layers (HTL/ETL).

**TABLE 11.1**

**Role of MXenes and Performance of Miniature Solar Cells**

| Fabricated Device Structure | Role of MXene in Miniature Solar Cell | Short-circuit Current Density $(J_{sc})$ (mA cm$^{-2}$) | Open-circuit Voltage $(V_{oc})$ (V) | Power Conversion Efficiency (PCE) (%) |
|---|---|---|---|---|
| $Ti_3C_2T_x$ on ITO-SnO$_2$ (Guo et al. 2018) | Additives | 22.26 | 1.03 | 17.41 |
| $Ti_3C_2T_x$ with PBDB-T (Hou and Yu 2020) | Additives | 17.08 | 0.91 | 11.02 |
| $Ti_3C_2T_x$ with PM$_6$ (Hou and Yu 2020) | Additives | 25.63 | 0.83 | 14.55 |
| $Ti_3C_2Tx$ and PBDB on ITO-SnO$_2$ (Hou and Yu 2020) | Additives | 18.63 | 0.93 | 12.20 |
| $Ti_3C_2T_x$ and PM$_6$ on ITO-ZnO (Hou and Yu 2020) | Additives | 26.38 | 0.83 | 16.51 |
| $Ti_3C_2T_x$ and PTB$_7$ on ITO-ZnO (Hou and Yu 2020) | Additives | 17.53 | 0.77 | 9.36 |
| $Ti_3C_2T_x$ on FTO-TiO$_2$ (Cao et al. 2019) | Electrode | 22. 96 | 0.95 | 13.83 |
| Carbon (CNT) and $Ti_3C_2T_x$ (Mi et al. 2020) | Electrode | 7.16 | 1.35 | 7.09 |
| MXene and AgNW (Tang et al. 2019) | Electrode | 14.62 | 0.79 | 7.16 |
| PSS, MXene with AgNW (Tang et al. 2019) | Electrode | 13.98 | 0.86 | 7.70 |
| $Ti_3C_2T_x$ on FTO-TiO$_2$ (Chen et al. 2019) | HTL-ETL | 8.54 | 1.44 | 9.01 |
| $Ti_3C_2T_x$ on ITO (Yu et al. 2019) | HTL-ETL | 15.98 | 0.89 | 9.02 |
| $Ti_3C_2T_x$ on ITO (HTL) (Hou et al. 2019) | HTL-ETL | 17.85 | 0.88 | 10.53 |
| MXene perovskite (ETL) (Y. Wang et al. 2020) | HTL-ETL | 24.34 | 1.11 | 20.65 |

MXenes serve numerous roles in solar cells: they can be employed as an additive, wherein it speeds up the electron transfer; as an electrode, with high electronic conductivity, good transparency, remarkable flexibility, and tunable work functions to enhance solar cell performance; and as an HTL/ETL, with flexible work functions and carrier conducting properties exploited to achieve higher efficiencies.

In general, the use of MXenes in solar cells has only been reported since the fourth quarter of 2018, and the related research is still in its early stages, focused mostly on determining the feasibility of various solar cells. There is still an opportunity for development in terms of device efficiency and robustness. It's worth noting that there is a need to optimize the physical properties of MXenes with diverse functional groups that is required for more correctly guiding the experiments. The properties of MXenes, on the other hand, such as shape, conductivity, transparency, termination groups, and stability, are all affected by the fabrication technique. It is required to create MXene synthesis processes with precisely regulated characteristics, huge scale, and low cost.

## 11.2.2 MXENE-BASED MICRO-SUPERCAPACITORS (MSC)

The development of miniaturized power units that are congruent with microelectronic devices is necessitated by the introduction of technology in different domains such as ecological monitoring, radio frequency detection, wearable gadgets, and therapeutic applications. Traditional electrochemical supercapacitors are too large to power these micro devices, and traditional supercapacitors' assembly procedures are incompatible with microelectronic fabrication techniques; this reality has sparked a lot of interest in supercapacitor downsizing. The phrase "micro-supercapacitor (MSC)" has been coined to designate supercapacitor designs that have the potential to be integrated with microelectronics, and it is primarily divided into two types: thin-film electrodes with packed structures (thickness less than 10 μm) and clusters of micro-electrodes with sizes in micrometers in two minimum dimensions.

MXene micro-supercapacitor (MSC) fabrication processes are divided as: 1) direct patterning of MXene solids on preferred substrates using laser engraving and reactive ion etching, and 2) depositing MXene ink into varied designs using numerous printing procedures. While fabricating the MSC (which are patterned in millimeter scale), the following aspects must be considered: 1) sufficient pattern resolution, 2) compatibility to allow synergy with other microelectronic elements, and 3) the deposited electrode material configuration on the current collector patterns (Kumar et al. 2021). The various fabrication techniques are schematically represented in Figure 11.4 and include the following techniques:

A. **Laser scribing:** This is a type of laser micromachining in which a laser beam is traced across a substrate to create a pattern of blind cuts or scribe lines. Following that, the scribe lines can be heated to 'break' or 'singulate' wafers or substrates. Brousse et al. (2018) recently used laser-inscription of a bi-layered film to build flexible micro-supercapacitors device based on ruthenium oxide ($RuO_2$) on platinum foil. Because of the electrode columns architecture, the constructed MSC demonstrated high capacitances of 27 mF $cm^{-2}$ in 1 M $H_2SO_4$, as well as strong cycling performance, with 80% capacitance retention after 10,000 cycles. In a similar study, Jiang et al. (2020) used a facile laser inscription technique to create a large-area benzene linked polypyrrole film through polymerization. The fabricated micro-supercapacitors device had a high energy density of more than 50 mW $hcm^{-3}$ and power density 9.6 kW $cm^{-3}$.

Not only does laser scribing eliminate the need for extra templates and complex processes, but it also promotes the production of electrode materials. As a result, it has found a position in the creation of MESDs, particularly as an effective manufacturing method for high-capacitance MSCs.

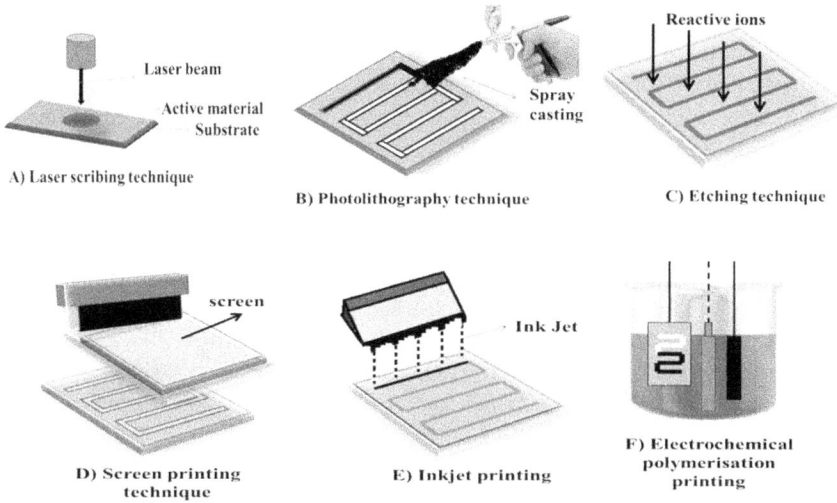

**FIGURE 11.4** Various techniques for fabricating conductive MXene as a micro-supercapacitor. (A) Laser-scribing technique to generate customized designs on several surfaces in a single step. (B) In the photolithography technique, the photoactive material may be shaped to any pattern by exposing it to light through a photomask, rendering it suitable for the manufacturing of MXene-based micro-supercapacitors. (C) Etching technique uses reactive ions to create MSCs. (D) Screen-printing technique supported by a woven mesh above the target substrate to generate electrodes for mass manufacturing. (E) In the inkjet printing technique miniscule volumes of ink are deposited on the substrate to create MSCs. (F) In the electrochemical polymerization technique the deposition of material onto the surface of the substrate electrode material is done via oxidation of the substrate.

It has been extensively employed in the fabrication of many categories of micro-supercapacitor devices, with high expectations placed on the production of devices with good performance.

B. **Photolithography technique:** In this method of printing a master pattern is drawn on top of the surface of a base sheet made of some material (usually a silicon wafer). A thin film of some materials, like silicon dioxide ($SiO_2$), is sprayed to the substrate to generate a pattern of holes. Because only a limited number of electrode substances can be transferred successfully on the current collector while retaining the original properties, causing inadequate energy density, photolithography is not a frequently utilized technique for the synthesis of MXene based micro-supercapacitors. The enhanced spatial capacitance of the MXene-based micro-supercapacitor devices is conceived by generating deep craters/depressions in the device configuration; for example, Xu et al. (2017) built a MXene-based micro-supercapacitors device on chromium and gold current collector and the device's enhanced aspect-ratio framework can incorporate an enhanced mass stacking of the electrode materials. The fabricated device demonstrated a high spatial capacitance of more than 275 mF cm$^{-2}$ and retention in the capacity of 95% after 1,000 cycles after injecting the cured $Ti_3C_2T_x$ solution.

C. **Reactive ion etching technique** is an expression employed in MSC device fabrication to characterize a procedure that eradicates material judiciously from a film deposited on any substrate (whether or not any structures exist on it) to generate a design of the material of interest on the substrate. To make a pattern on a substrate, etching is used. It is a simple and cost-effective approach for creating unique patterns on a variety of substrates in a one-step process. It has a high degree of tractability compared to other substances. To reduce the conundrum of restacking in the MXene sheets, a 3D array of electrodes was devised by constructing a reduced graphene oxide and MXene aerogel. The aerogel is a dense and permeable configuration that can easily and exclusively be designed using a laser engraving technique, and the fabricated final micro-supercapacitor device demonstrated an excellent spatial capacitance of more than 34 mF cm$^{-2}$ and exceptional cycling performance. Wang et al. (2019) initially coated MXene on two faces of the nickel metal sheet (20 m in thickness), and then utilized a cold laser to construct the micro channel in the interdigital arrays. Taking use of the laser's penetration depth, two faces of the current collector may be successfully deployed, which is a novel technique for increasing the spatial energy density of the micro-supercapacitor device in a given space.

D. **The screen print process** is a conventional form of printing that uses ink-blocked template reinforced by an interlaced mesh above the target substrate to generate electrodes for mass manufacturing. The MSC devices prepared by this technique are larger than 100 μm. Li et al. created a thixotropic composite ink for screen printing out of hydrous ruthenium oxide, MXene nanosheets, and conductive Ag nanowires. In another paper, Xu et al. (2017) used a dual-step selective screen-printing approach to create an asymmetric micro-scale hybrid device (MHD) with cobalt aluminum layered hydroxides electrodes on the positive side and Ti$_3$C$_2$T$_x$ on the negative side. The asymmetric MXene-based MSC device has a broader voltage window and a real capacitance than the symmetric MXene MSC. Hence, the asymmetric micro-supercapacitors device fabricated through this process has a substantially greater energy density of 10.8 μWh cm$^{-2}$ while the symmetric MXene MSC exhibited an energy density of around 1.3 μWh cm$^{-2}$ and was manufactured in an exactly similar process. The higher capacity asymmetric MSC device also retained more than 90% of its initial capacity following an extended stability test of 10,000 cycles at a current rate of about 1.25 mA cm$^{-2}$.

E. **The inkjet printing technique:** This is an additive production technique that uses a stream of ink to deposit miniscule droplets on the sheet, comparable to how a 2D printer operates. The ink can be replaced with thermoplastic and wax materials that are smelted during the additive process. The inkjet printing provides exceptional accuracy of printed patterns on a variety of substrates besides having a faster printing speed compared to the conventional screen printing. In this printing technique, a little quantity ink is placed on the preferred substrate before it is changed to a solid form. For fine printing, various polar and non-polar solvents for dispersing MXenes were investigated. Both high concentration (MXene dispersed in NMP with a concentration of 12.5 mg ml$^{-1}$) and small concentration (0.7 mg ml$^{-1}$ of MXene dispersed in ethanol) MXene inks exhibit good printing finesse with no "coffee-ring." Surprisingly, the binding of MXene with the substrate is very sturdy and no amount of MXene could be pulled off even following peeling by a scotch tape. The MXene inkjet designs have minimal resistance and show profound undeviating behavior with printed cycles. For the inkjet-printed micro-supercapacitor a good volumetric capacitance (562 Fcm$^{-3}$) and an energy density of more than 0.3 µW h cm$^{-2}$ was observed, which is among the maximum values obtained for all printed micro-supercapacitor devices.

F. **Electrochemical polymerization technique** is the deposition of the material onto the surface of the substrate electrode material. This happens as a result of the positive radical being created by the oxidation of the base material on the solid substrate, according to a widely established mechanism. Qin et al. (2019) divulged an easy technique for fabricating MXene based micro-supercapacitor devices. He manufactured MXene and a polymer amalgamates by the polymerization of the pre-synthesized MXene dispersion, and the fabricated MSC device demonstrated a spatial capacitance of 47.4 mF cm$^{-2}$.

## 11.2.3  MXene-Based Micro-Batteries (MBs)

Conventional Li-ion micro-batteries are yet another example of miniaturized energy storage technology. Their low energy density limits their application in modern micro-devices with rising electric power consumption. To keep up with the demands of emerging micro- and nano-systems, it is valuable to evaluate new electrochemical energy storage techniques transcending Li-ion micro-batteries (Zhu et al. 2020). A characteristic battery may consists of several cell cases varying between coin cells, cylindrical cells, and prismatic cells or even up to lab scale pouch cells; the latter of three formats are extensively utilized for industrial processes. Conventional Li–S batteries in such formats provide a technical knowledge footing for micro-batteries' development. Miniaturization of electrodes in batteries is aimed for proper utilization in microelectronic devices; however, it ought to be available for integration in miniaturized forms. The electrode architecture's usual varied dimensionalities, such as fibers (one dimensional), stacked/layered (two-dimensional) configurations, and Swiss-roll-like (three-dimensional) shapes, provide uniquely customized solutions for a wide range of applications. Enhanced battery topologies for electrode configurations, electrolytes, and current collectors are required to provide incredible performance features such as long cycle life and higher capacities. As a result, broad ranges of fabricating technologies have been used to create tiny batteries. Miniaturized or micro batteries have been made using a wide range of fabrication techniques. Pyrolysis of polymeric fibers

(Verbrugge et al. 1996; Maitra et al. 2012) or thin-film techniques such as atomic layer (Z. Zhao et al. 2020), chemical vapor (Z. Zhao et al. 2020), electro-deposition (Yufit et al. 2003), or sputtering (Bates et al. 2000) were used to fabricate MXene-based electrodes in the recent past. Lithography (Kinoshita et al. 1999) and laser preparation (Pfleging et al. 2017) become significant for battery miniaturization as fine and highly focused laser rays and the alteration of MXene electrode fabrication equipment became attainable. Apart from all physical and chemical protocols, mechanical electrode synthesis by printing has progressed into an extremely promising avenue for conceptualizing microelectrodes and printing complete batteries for commercialization (Ragones et al. 2020). Various methods so far used and the preparation method, materials used for fabrication, and electrolytes are mentioned in Table 11.2.

Where MXene inks typically reflect solution processed techniques. Laser printing involves the coating of the MXene electrode material from coated substances to the preferred substrate (Piqué et al. 2004). The unpredictability of ink-based printing using MXenes was assessed by Zhang et al. (Y. Z. Zhang et al. 2020). They found that a design that primarily gains from the maximum loading of active materials should be chosen to achieve a micro-battery with the highest performance. One key measure is the quantity of inactive material that needs to be minimized in current collectors, separators, or substrates. MXenes have been initially utilized in positive sulphur electrodes, which have a chemically active electrode structure. To connect the individual sulphur particles, MXene materials provide an almost metallically conducting network. They yield rapid charging carrier transport qualities with minimal effort. Interestingly, the mechanical stability of MXenes demonstrates adequate strength to withstand accumulation due to $Li_2S$ production during sulphur lithiation (C. Zhang et al. 2020). In addition to enhanced electronic conductivity and boosted transport of charge carriers in the MXene electrode, a factor that must be mitigated is the mechanical instability caused by volume fluctuations in the various sulphur phases. In the chemical etching of the $Ti_2SC$ MAX phase, the direct fabrication of an electrode material from a parental MAX phase was done via chemical exfoliation of the carbon/sulfur-nano laminates. After this initial discovery, it was realized that MXenes amalgamates worked in synergy as efficient electrode materials. Apart from $Ti_2SC$, $Ti_2AlC$ was also used to make (M. Zhao et al. 2015) MXenes electrodes for micro-batteries.

Another prominent method for modifying the crystal structure and band architectures of MXenes to improve their characteristics for use in micro-batteries is heteroatom doping. Nitrogen doping is the one of the most commonly used doping technique, and its likelihood and efficiency in improving the charge storage and charge carrier performance of substrate materials has been demonstrated. In numerical simulations, N atoms with adequate electronegativity in a $Ti_3C$-based crystal structure can efficiently improve MXenes charge carrier number and improve the electro-catalytic activity of $Ti_3CN$, ensuring that $Ti_3CN$ anode has a high capacity of at least 98 mA h $g^{-1}$ at 500 mA $g^{-1}$, which is 1.65 times that of $Ti_3C_2$ anode.

**TABLE 11.2**

**MXene-Based MSCs Fabrication with Their Performance**

| Preparation Method | Materials Used | Electrolytes Used | Areal Capacitance ($mF\ cm^{-2}$) | Volumetric Capacitance ($mF\ cm^{-2}$) | Power Density ($mW\ cm^{-2}$) | Energy Density ($\mu W\ cm^{-2}$) |
|---|---|---|---|---|---|---|
| Spray coating + laser cutting (Peng et al. 2016) | $L\text{-}s\text{-}Ti_3C_2T_x$ | $PVA/H_2SO_4$ | 27.3 | 356.8 | 2.34-1.43 | 009-1.95 |
| Vacuum filtration + laser cutting (Paquin et al. 2015) | $Ti_3C_2T_x$ | $PVA/H_2SO_4$ | 71.16 | 151.4 | 3.52-2.51 | 0.33–7.99 |
| Spray casting + shadow mask (Hongyan Li et al. 2017) | $Ti_3C_2T_x$ | $PVA/H_2SO_4$ | 3.26 | 33 | 0.34-0.14 | 0.02–0.16 |
| Vacuum filtration + laser printing (Paquin et al. 2015) | $Ti_3C_2T_x$ | $PVA/H_2SO_4$ | 23 | 57.5 | 1.12-0.92 | 0.09–0.3 |
| Vacuum filtration + oxidative etching (Shen et al. 2016) | $Ti_3C_2T_x$ | $PVA/H_2SO_4$ | 0.07 | 1.44 | 0.01 | N/A |
| Screen printing (Hongpeng Li et al. 2019) | $Ti_3C_2T_x$ | $PVA/H_2SO_4$ | 340 | 183 | 13.64-6.16 | 0.24–4.79 |
| MEMS technique + ink printing (Xu et al. 2017) | $Ti_3C_2T_x$ | $PVA/H_3PO_4$ | 0.43 | 18.9 | 0.01–0.22 | 0.001–0.01 |
| Extrusion printing (C. (John) Zhang et al. 2019) | $Ti_3C_2T_x$ | $PVA/H_2SO_4$ | 43 | – | 0.3-0.11 | 0.1-1.31 |
| Ink jet printing (C. (John) Zhang et al. 2019) | $Ti_3C_2T_x$ | $PVA/H_2SO_4$ | 12 | 562 | – | – |

## 11.3   CONCLUSIONS AND OUTLOOK

Miniaturized energy storage device (MESD) development is a necessary step in the commercialization and development of microelectronics. The development of these tiny devices has been extensive, but scalable production is often dependent on electrode materials and design methods. The MXene group has grown swiftly since it was first discovered in 2011. There's a lot of potential in looking for new MXenes with different functions in order to build sophisticated miniaturized energy storage systems.

We looked at the role of MXene in miniaturized energy storage devices including small solar cells, micro-supercapacitors (MSCs), and micro-batteries in this article. We saw that MXenes play a variety of roles in solar cells: as an additive, it accelerates electron transfer; as an electrode, it improves solar cell performance with high metallic conductivity, high transparency, outstanding flexibility, and adjustable work functions; and as an HTL/ETL, it uses easily tunable work functions and carrier conducting properties to achieve higher efficiencies. We also described the miniaturized device fabrication process, which broadly includes either laser assisted pattern scribing or several printing processes for depositing MXene ink onto a substrate. Finally, we investigated the optimum battery topologies for electrodes, electrolytes, and current collectors that are required to deliver exceptional performance features like long cycle life and larger capacities.

Despite the fact that at least 25 MXenes have been synthesized in the lab, only $Ti_3C_2T_x$ has been thoroughly investigated. There is a great opportunity to look for novel MXenes with varied functions in order to construct sophisticated miniaturised energy storage devices. For instance, in $Ti_2C$ additional Ti atom layers are revealed to the electrolyte; hence, theoretically it has larger gravimetric capacitance than $Ti_3C_2$. Further optimization of these materials, together with computer investigations, may lead to the discovery of new interesting directions. Because of their new physical and chemical features, MXene quantum dots (MQDs) is an interesting family member of MXene materials. MQDs are extensively used in the fields of optical, biological, photo catalytic, and cellular imaging. MQDs have unique qualities such as ultra-small size, good stability, outstanding electronic conductivity, with an abundance of edge dislocations in contrast to MXenes. These exceptional characteristics may allow MQDs to be used in MESDs.

## REFERENCES

Alhabeb, Mohamed, Kathleen Maleski, Tyler S. Mathis, Asia Sarycheva, Christine B. Hatter, Simge Uzun, Ariana Levitt, and Yury Gogotsi. 2018. "Selective Etching of Silicon from $Ti_3SiC_2$ (MAX) To Obtain 2D Titanium Carbide (MXene)." *Angewandte Chemie - International Edition* 57 (19): 5444–5448. doi:10.1002/anie.201802232

Ambrosi, Adriano, and Martin Pumera. 2018. "Exfoliation of Layered Materials Using Electrochemistry." *Chemical Society Reviews* 47 (19): 7213–7224. doi:10.1039/c7cs00811b

Bao, Zhuoheng, Chengjie Lu, Xin Cao, Peigen Zhang, Li Yang, Heng Zhang, Dawei Sha, et al. 2021. "Role of MXene Surface Terminations in Electrochemical Energy Storage: A Review." *Chinese Chemical Letters* 32 (9): 2648–2658. doi:10.1016/j.cclet.2021.02.012

Bates, J.B., N.J. Dudney, B. Neudecker, A. Ueda, and C.D. Evans (2000). "Thin-Film Lithium and Lithium-Ion Batteries." *Solid State Ionics* 135(1–4): 33–45. doi:10.1016/S0167-2738(00)00327-1

Brousse, K., S. Nguyen, A. Gillet, S. Pinaud, R. Tan, A. Meffre, K. Soulantica, et al. 2018. "Laser-Scribed Ru Organometallic Complex for the Preparation of $RuO_2$ Micro-Supercapacitor Electrodes on Flexible Substrate." *Electrochimica Acta* 281: 816–821. doi:10.1016/j.electacta.2018.05.198

Cao, Junmei, Fanning Meng, Liguo Gao, Shuzhang Yang, Yeling Yan, Ning Wang, Anmin Liu, Yanqiang Li, and Tingli Ma. 2019. "Alternative Electrodes for HTMs and Noble-Metal-Free Perovskite Solar Cells: 2D MXenes Electrodes." *RSC Advances* 9 (59): 34152–34157. doi:10.1039/c9ra06091j

Chen, Taotao, Guoqing Tong, Enze Xu, Huan Li, Pengcheng Li, Zhifeng Zhu, Jianxin Tang, Yabing Qi, and Yang Jiang. 2019. "Accelerating Hole Extraction by Inserting 2D Ti3C2-MXene Interlayer to All Inorganic Perovskite Solar Cells with Long-Term Stability." *Journal of Materials Chemistry A* 7 (36): 20597–20603. doi:10.1039/c9ta06035a

Garg, Ruby, Alpana Agarwal, and Mohit Agarwal. 2020. "A Review on MXene for Energy Storage Application: Effect of Interlayer Distance." *Materials Research Express* 7 (2) doi:10.1088/2053-1591/ab750d

Gogotsi, Yury, and Babak Anasori. 2019. "The Rise of MXenes." *ACS Nano* 13 (8): 8491–8494. doi:10.1021/acsnano.9b06394

Guo, Zhanglin, Liguo Gao, Zhenhua Xu, Siowhwa Teo, Chu Zhang, Yusuke Kamata, Shuzi Hayase, and Tingli Ma. 2018. "High Electrical Conductivity 2D MXene Serves as Additive of Perovskite for Efficient Solar Cells." *Small* 14 (47): 1–8. doi:10.1002/smll.201802738

Hou, Chunli, and Huangzhong Yu. 2020. "Modifying the Nanostructures of PEDOT:PSS/$Ti_3C_2T_x$ Composite Hole Transport Layers for Highly Efficient Polymer Solar Cells." *Journal of Materials Chemistry C* 8 (12): 4169–4180. doi:10.1039/d0tc00075b

Hou, Chunli, and Huangzhong Yu. 2020. "ZnO/$Ti_3C_2T_x$ Monolayer Electron Transport Layers with Enhanced Conductivity for Highly Efficient Inverted Polymer Solar Cells." *Chemical Engineering Journal* 407 (July): 127192. doi:10.1016/j.cej.2020.127192

Hou, Chunli, Huangzhong Yu, and Chengwen Huang. 2019. "Solution-Processable $Ti_3C_2T_x$ Nanosheets as an Efficient Hole Transport Layer for High-Performance and Stable Polymer Solar Cells." *Journal of Materials Chemistry C* 7 (37): 11549–11558. doi:10.1039/c9tc03415c

Jiang, Kaiyue, Igor A. Baburin, Peng Han, Chongqing Yang, Xiaobin Fu, Yefeng Yao, Jiantong Li, et al. 2020. "Interfacial Approach toward Benzene-Bridged Polypyrrole Film–Based Micro-Supercapacitors with Ultrahigh Volumetric Power Density." *Advanced Functional Materials* 30 (7): 1–9. doi:10.1002/adfm.201908243

Jiang, Qiu, Yongjiu Lei, Hanfeng Liang, Kai Xi, Chuan Xia, and Husam N. Alshareef. 2020. "Review of MXene Electrochemical Microsupercapacitors." *Energy Storage Materials* 27 (January): 78–95. doi:10.1016/j.ensm.2020.01.018.Elsevier Ltd

Kinoshita, K., X. Song, J. Kim, and M. Inaba. 1999. "Development of a Carbon-Based Lithium Microbattery." *Journal of Power Sources* 81–82: 170–175. doi:10.1016/S0378-7753(99)00189-5

Kumar, Sunil, Malik Abdul Rehman, Sungwon Lee, Minwook Kim, Hyeryeon Hong, Jun Young Park, and Yongho Seo. 2021. "Supercapacitors Based on Ti3C2Tx MXene Extracted from Supernatant and Current Collectors Passivated by CVD-Graphene." *Scientific Reports* 11 (1): 1–9. doi:10.1038/s41598-020-80799-9

Li, Hongpeng, Xiran Li, Jiajie Liang, and Yongsheng Chen. 2019. "Hydrous RuO 2 -Decorated MXene Coordinating with Silver Nanowire Inks Enabling Fully Printed Micro-Supercapacitors with Extraordinary Volumetric Performance." *Advanced Energy Materials* 9 (15): 1–13. doi:10.1002/aenm.201803987

Li, Hongyan, Yang Hou, Faxing Wang, Martin R. Lohe, Xiaodong Zhuang, Li Niu, and Xinliang Feng. 2017. "Flexible All-Solid-State Supercapacitors with High Volumetric Capacitances Boosted by Solution Processable MXene and Electrochemically Exfoliated Graphene." *Advanced Energy Materials* 7 (4): 2–7. doi:10.1002/aenm.201601847

Li, Liang, Gengnan Li, Li Tan, Yumeng Zhang, and Binghan Wu. 2017. "Highly Efficiently Delaminated Single-Layered MXene Nanosheets with Large Lateral Size." *Langmuir* 33 (36): 9000–9006. doi:10.1021/acs.langmuir.7b01339

Li, Neng, Jiahe Peng, Wee Jun Ong, Tingting Ma, Arramel, Peng Zhang, Jizhou Jiang, Xiaofang Yuan, and Chuanfang (John) Zhang. 2021. "MXenes: An Emerging Platform for Wearable Electronics and Looking Beyond." *Matter* 4 (2): 377–407. doi:10.1016/j.matt.2020.10.024

Li, Youbing, Hui Shao, Zifeng Lin, Jun Lu, Liyuan Liu, Benjamin Duployer, Per O.A. Persson, et al. 2020. "A General Lewis Acidic Etching Route for Preparing MXenes with Enhanced Electrochemical Performance in Non-Aqueous Electrolyte." *Nature Materials* 19 (8): 894–899. doi:10.1038/s41563-020-0657-0

Liu, Zhixiong, and Husam N. Alshareef. 2021. "MXenes for Optoelectronic Devices." *Advanced Electronic Materials* 7 (9): 1–28. doi:10.1002/aelm.202100295

Magnuson, Martin, Joseph Halim, and Larsake Naslund. 2017. "Chemical Bonding in Carbide MXene Nanosheets." *Journal of Electron Spectroscopy and Related Phenomena.* doi:10.1016/j.elspec.2017.09.006

Maitra, Tanmoy, Swati Sharma, Alok Srivastava, Yoon Kyoung Cho, Marc Madou, and Ashutosh Sharma. 2012. "Improved Graphitization and Electrical Conductivity of Suspended Carbon Nanofibers Derived from Carbon Nanotube/Polyacrylonitrile Composites by Directed Electrospinning." *Carbon* 50 (5) : 1753–1761. doi:10.1016/j.carbon.2011.12.021

Mi, Longfei, Yan Zhang, Taotao Chen, Enze Xu, and Yang Jiang. 2020. "Carbon Electrode Engineering for High Efficiency All-Inorganic Perovskite Solar Cells." *RSC Advances* 10 (21): 12298–12303. doi:10.1039/d0ra00288g

Naguib, Michael, Murat Kurtoglu, Volker Presser, Jun Lu, Junjie Niu, Min Heon, Lars Hultman, Yury Gogotsi, and Michel W. Barsoum. 2011. "Two-Dimensional Nanocrystals Produced by Exfoliation of Ti 3AlC 2." *Advanced Materials* 23 (37): 4248–4253. doi:10.1002/adma.201102306

Naguib, Michael, Vadym N. Mochalin, Michel W. Barsoum, and Yury Gogotsi. 2013. "25th Anniversary Article: MXenes: A New Family of Two-Dimensional Materials," 26 (7): 992–1005. doi:10.1002/adma.201304138

Paquin, Francis, Jonathan Rivnay, Alberto Salleo, Natalie Stingelin, and Carlos Silva. 2015. "Multi-Phase Semicrystalline Microstructures Drive Exciton Dissociation in Neat Plastic Semiconductors." *J. Mater. Chem. C* 3: 10715–10722. doi:10.1039/b000000x

Peng, You Yu, Bilen Akuzum, Narendra Kurra, Meng Qiang Zhao, Mohamed Alhabeb, Babak Anasori, Emin Caglan Kumbur, Husam N. Alshareef, Ming Der Ger, and Yury Gogotsi. 2016. "All-MXene (2D Titanium Carbide) Solid-State Microsupercapacitors for on-Chip Energy Storage." *Energy and Environmental Science* 9 (9): 2847–2854. doi:10.1039/c6ee01717g

Pfleging, Wilhelm. 2017. "A Review of Laser Electrode Processing for Development and Manufacturing of Lithium-Ion Batteries." *Nanophotonics* 7 (3): 549–573. doi:10.1515/nanoph-2017-0044

Piqué, A., C.B. Arnold, H. Kim, M. Ollinger, and T.E. Sutto. 2004. "Rapid Prototyping of Micropower Sources by Laser Direct-Write." *Applied Physics A: Materials Science and Processing* 79 (4–6): 783–786. doi:10.1007/s00339-004-2586-1

Qin, Leiqiang, Quanzheng Tao, Xianjie Liu, Mats Fahlman, Joseph Halim, Per O.A. Persson, Johanna Rosen, and Fengling Zhang. 2019. "Polymer-MXene Composite Films

Formed by MXene-Facilitated Electrochemical Polymerization for Flexible Solid-State Microsupercapacitors." *Nano Energy* 60: 734–742. doi:10.1016/j.nanoen.2019.04.002

Ragones, Heftsi, Adi Vinegrad, Gilat Ardel, Meital Goor, Yossi Kamir, Moty Marcos Dorfman, Alexander Gladkikh, and Diana Golodnitsky. 2020. "On the Road to a Multi-Coaxial-Cable Battery: Development of a Novel 3D-Printed Composite Solid Electrolyte." *Journal of The Electrochemical Society* 167 (7): 070503. doi:10.1149/2.0032007jes

Shen, Bao Shou, Hao Wang, Li Jun Wu, Rui Sheng Guo, Qing Huang, and Xing Bin Yan. 2016. "All-Solid-State Flexible Microsupercapacitor Based on Two-Dimensional Titanium Carbide." *Chinese Chemical Letters* 27 (10). 1586–1591. doi:10.1016/j.cclet.2016.04.012

Shukla, Vineeta. 2020. "The Tunable Electric and Magnetic Properties of 2D MXenes and Their Potential Applications." *Materials Advances* 1 (9): 3104–3121. doi:10.1039/d0ma00548g

Sun, W., S.A. Shah, Y. Chen, Z. Tan, H. Gao, T. Habib, M. Radovic, and M.J. Green. 2017. "Electrochemical Etching of $Ti_2AlC$ to $Ti_2CT_x$ (MXene) in Low-Concentration Hydrochloric Acid Solution." *Journal of Materials Chemistry A* 5 (41): 21663–21668. doi:10.1039/c7ta05574a

Tang, Honghao, Huanran Feng, Huike Wang, Xiangjian Wan, Jiajie Liang, and Yongsheng Chen. 2019. "Highly Conducting MXene-Silver Nanowire Transparent Electrodes for Flexible Organic Solar Cells." *ACS Applied Materials and Interfaces* 11 (28): 25330–25337. doi:10.1021/acsami.9b04113

Tian, Weiqian, Armin Vahidmohammadi, Michael S. Reid, Zhen Wang, Liangqi Ouyang, Johan Erlandsson, Torbjorn Pettersson, Lars Wagberg, Majid Beidaghi, and Mahiar M. Hamedi. 2019. "Multifunctional Nanocomposites with High Strength and Capacitance Using 2D MXene and 1D Nanocellulose." *Advanced Materials* 1902977. doi:10.1002/adma.201902977

Verbrugge, Mark W., and Brian J. Koch. 1996. "Lithium Intercalation of Carbon-Fiber Microelectrodes." *Journal of The Electrochemical Society* 143 (1): 24–31. doi:10.1149/1.1836382

Verger, Louisiane, Varun Natu, Michael Carey, and Michel W. Barsoum. 2019. "MXenes: An Introduction of Their Synthesis, Select Properties, and Applications." *Trends in Chemistry* 1 (7): 656–669. doi:10.1016/j.trechm.2019.04.006

Wang, Na, Jinzhang Liu, Yi Zhao, Mingjun Hu, Ruzhan Qin, and Guangcun Shan. 2019. "Laser-Cutting Fabrication of Mxene-Based Flexible Micro-Supercapacitors with High Areal Capacitance." *ChemNanoMat* 5 (5): 658–665. doi:10.1002/cnma.201800674

Wang, Tong, Tianjiao Wang, Chuanxin Weng, Luqi Liu, Jun Zhao, and Zhong Zhang. 2021. "Engineering Electrochemical Actuators with Large Bending Strain Based on 3D-Structure Titanium Carbide MXene Composites." *Nano Research* 14 (7): 2277–2284. doi:10.1007/s12274-020-3222-x

Wang, Yunfan, Pan Xiang, Aobo Ren, Huagui Lai, Zhuoqiong Zhang, Zhipeng Xuan, Zhenxi Wan, et al. 2020. "MXene-Modulated Electrode/SnO2Interface Boosting Charge Transport in Perovskite Solar Cells." *ACS Applied Materials and Interfaces* 12 (48): 53973–53983. doi:10.1021/acsami.0c17338

Wu, Jiabin, Qun Li, Christopher E. Shuck, Kathleen Maleski, Husam N. Alshareef, Jun Zhou, Yury Gogotsi, and Liang Huang. 2022. "An Aqueous 2.1 V Pseudocapacitor with MXene and V-MnO2 Electrodes." *Nano Research* 15 (1): 535–541. doi:10.1007/s12274-021-3513-x

Xu, Sixing, Wei Liu, Xia Liu, Xuanlin Kuang, and Xiaohong Wang. 2017. "A MXene Based All-Solid-State Microsupercapacitor With 3D Interdigital Electrode." *2017 19th International Conference on Solid-State Sensors, Actuators and Microsystems (TRANSDUCERS)*, Kaohsiung, Taiwan, 706–709. doi:10.1109/TRANSDUCERS.2017.7994146

Yu, Zhimeng, Wei Feng, Wanheng Lu, Bichen Li, Hongyan Yao, Kaiyang Zeng, and Jianyong Ouyang. 2019. "MXenes with Tunable Work Functions and Their Application as Electron- and Hole-Transport Materials in Non-Fullerene Organic Solar Cells." *Journal of Materials Chemistry A* 7 (18): 11160–11169. doi:10.1039/c9ta01195a

Yufit, V., M. Nathan, D. Golodnitsky, and E. Peled. 2003. "Thin-Film Lithium and Lithium-Ion Batteries with Electrochemically Deposited Molybdenum Oxysulfide Cathodes." *Journal of Power Sources* 122 (2): 169–173. doi:10.1016/S0378-7753(03)00401-4

Zhang, Chuanfang (John), Lorcan McKeon, Matthias P. Kremer, Sang Hoon Park, Oskar Ronan, Andres Seral-Ascaso, Sebastian Barwich, et al. 2019. "Additive-Free MXene Inks and Direct Printing of Micro-Supercapacitors." *Nature Communications* 10 (1): 1–9. doi:10.1038/s41467-019-09398-1

Zhang, Chuanfang, Yonglu Ma, Xuetao Zhang, Sina Abdolhosseinzadeh, Hongwei Sheng, Wei Lan, Amir Pakdel, Jakob Heier, and Frank Nüesch. 2020. "Two-Dimensional Transition Metal Carbides and Nitrides (MXenes): Synthesis, Properties, and Electrochemical Energy Storage Applications." *Energy and Environmental Materials* 3 (1): 29–55. doi:10.1002/eem2.12058

Zhang, Jing, Shibo Li, Shujun Hu, and Yang Zhou. 2018. "Chemical Stability of $Ti_3C_2$ MXene with Al in the Temperature Range 500–700°C." *Materials* 11 (10): 1–9. doi:10.3390/ma11101979

Zhang, Yi Zhou, Yang Wang, Qiu Jiang, Jehad K. El-Demellawi, Hyunho Kim, and Husam N. Alshareef. 2020. "MXene Printing and Patterned Coating for Device Applications." *Advanced Materials* 32 (21): 1–26. doi:10.1002/adma.201908486

Zhao, Meng-qiang, Morgane Sedran, Zheng Ling, Maria R. Lukatskaya, Olha Mashtalir, Michael Ghidiu, Boris Dyatkin, et al. 2015. "Synthesis of Carbon / Sulfur Nanolaminates by Electrochemical Extraction of Titanium from $Ti_2SC$ **." *Angewandte Chemie* 54 (6): 1–6. doi:10.1002/anie.201500110

Zhao, Zhe, Ye Kong, Zhiwei Zhang, Gaoshan Huang, and Yongfeng Mei. 2020. "Atomic Layer-Deposited Nanostructures and Their Applications in Energy Storage and Sensing." *Journal of Materials Research* 35 (7): 701–719. doi:10.1557/jmr.2019.329

Zhu, Zhe, Ruyu Kan, Song Hu, Liang He, Xufeng Hong, Hui Tang, and Wen Luo. 2020. "Recent Advances in High-Performance Microbatteries: Construction, Application, and Perspective." *Small* 16 (39): 1–28. doi:10.1002/smll.202003251

# 12 Microsupercapacitors for Miniaturized Electronic Device Applications

*Asmita Dileep Gaonkar, Shraddha Paniya, and Kiran Vankayala*

## CONTENTS

## 12.1  INTRODUCTION

With growing technological development, the use of portable electronic devices has increased tremendously. Miniaturized electronic devices such as intelligent robots, foldable mobile phones, wearable electronics, etc. are emerging as advanced devices to enhance the living standards of people. Since the size of portable electronic devices is shrinking due to advancements in nanotechnology, low-power integrated circuits in various devices like microprocessors, wireless communication chips, etc. will lead to the growing usage of miniature embedded micro-electromechanical systems (MEMS) which work in controlled/uncontrolled environments to collect, process, save, and transfer the data.[1] In order to achieve advancements in next-generation electronics, it is imperative to explore miniaturized energy storage devices which possess appreciable flexibility as well as portability. The effective energy storage strategies to supplement the energy needs of microdevices are still in the

DOI: 10.1201/b23359-12

developing phase. Also, there is a lack of energy devices that supply power in a continuous manner to complete the charging process. These two factors are known to limit the future development of microdevices. To overcome these issues, substituting the long-established electric supply mode with contactless charging is essential, which can improve the feasibility of energy storage by eradicating the inconvenient external circuit connections in microdevices. To supplement the energy needs of these miniaturized devices, electrochemical energy systems (EES) such as micro-batteries (microBs), micro-fuel cells (microFCs), micro-supercapacitors (microSCs), etc. play vital role for the development of portable electronic devices, which are compatible to miniaturized electronic devices. Among micro-EES devices, microBs have been integrated with miniaturized devices.[2] In spite of the fact that microBs are commercially available and their demand in the market is rising, their low power density and limited cycle life impede their widespread use in miniaturized devices. In order to mitigate these issues, microSCs are projected as a promising alternative to microBs in micro-electronic systems as miniaturized power sources owing to superior cycle life, faster charge/discharge rates, and robust operating limits. Figure 12.1A shows a schematic illustration depicting brief timeline of the development of microSCs. MicroSCs have been considered to complement devices with varying power consumption requirements (Figure 12.1A,B).[3] For instance, microSCs have been used as an energy source for low power consumption devices such as sensing systems (humidity, temperature, etc.), and also for substantially more power demanding communication units such as Wi-Fi, Bluetooth, and 4G-network.[4]

This chapter focuses on various aspects of microSCs such as topology, performance metrics, various electrode materials, and applications of microSCs in miniaturized devices. The chapter also discusses various ways to improve the performance of microSCs.

## 12.2 FUNDAMENTALS

Supercapacitors (SCs) are known to successfully fill the existing void between conventional capacitors and batteries. Conventional capacitors offer low energy density and low power density as compared to that of SCs but the latter has relatively higher power density and lower energy density than batteries.[5] Thus, SCs hold the properties of both conventional capacitors and batteries. Figure 12.1B shows the energy and power density of different energy storage devices.

SCs are electrochemical devices which can store and supply high-power electricity very quickly, for a large number of cycles (up to millions of cycles) without any notable decrease in performance. *The simplest SC comprises mainly of a pair of electrodes (anode & cathode) and an electrolyte that separates the electrodes.* In order to enhance the applicability of SCs in various electronic applications, continuous progress in the direction of flexible SCs having a small size which are compatible with portable electronic devices is essential. *In this line, microSCs (Figure 12.1A & B) that are typically at the scale of $cm^2$ or $mm^2$ footprint have gained considerable attention as miniaturized energy storage systems which possess adequate power density and maintain a fast frequency response.* Since

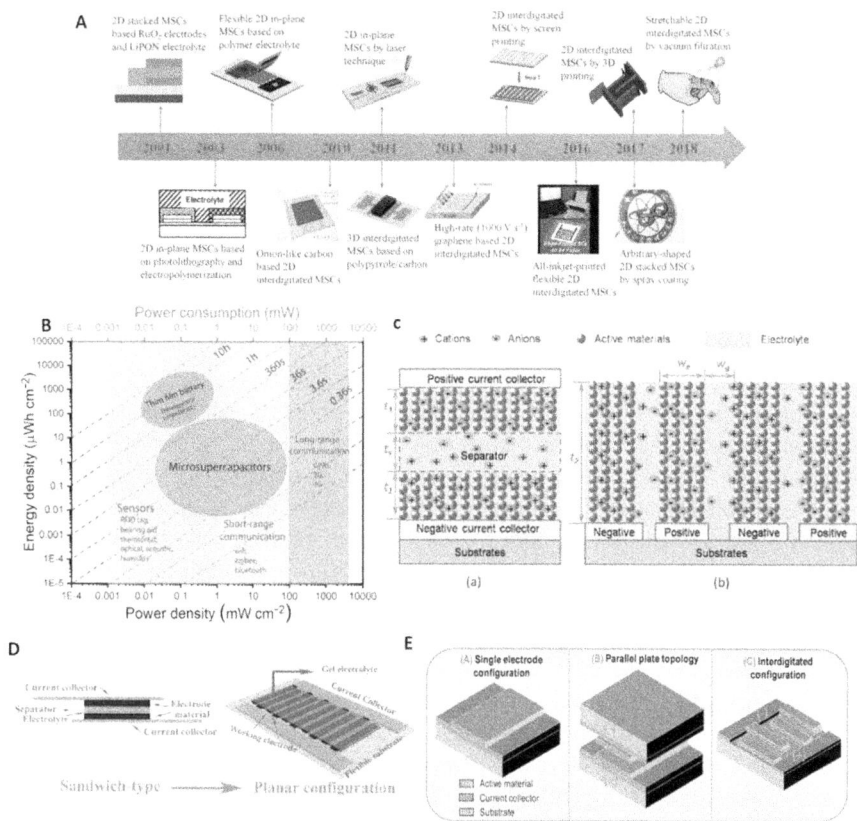

**FIGURE 12.1** A. Schematic illustration depicting a brief timeline of the historical development of microSCs. Adapted with permission from ref.[3] Copyright (2019), Wiley-VCH; B. power density and energy density of various energy-storage technologies. Adapted with permission from ref.[5] Copyright (2020), Elsevier; C. structural differences between conventional SC (sandwich configuration) (left) and microSCs (an interdigitated finger structure) (right). Adapted with permission from ref.[1] Copyright (2013), Wiley-VCH; D. schematic illustration of sandwich type (left) and in-planar type configuration (right). Adapted with permission from ref.[6] Copyright (2014), Royal Society of Chemistry; E. various topologies of microSCs. Adapted with permission from ref.[7] Copyright (2019), Royal Society of Chemistry.

microSCs withstand repeated bending, they can be used in wearables and Internet of Things (IoT) applications.

Based on charge storage mechanism, microSCs can be categorized into various types, namely electrochemical double layer capacitors (EDLCs) and pseudocapacitors (PCs).[4]

In EDLCs, an electrical double-layer, develops at the interface between electrode and electrolyte results in the capacitance. During the charging process, the surfaces of two electrodes will either be positively or negatively charged under the application of an external electric field, which will lead to the formation of compact electric double layer at the electrode surface due to the accumulation of oppositely charged ions present in the electrolyte via electrostatic interactions and results in storage of energy. During the discharging, the flow of electrons is from the negative to the positive electrode through the load, and current is produced in the external circuit and the energy is released.[4] Since the physical process of accumulation (adsorption of ions on electrode surface) of charges at electrode surface is limited, it warrants high surface area electrode materials to accumulate reasonable electric charges in order to exhibit appreciable capacitance. In simple words, EDLC is a double-layer capacitor with high capacitance but with low voltage limits, especially with porous carbon-based materials in aqueous electrolytes. Further, leakage currents in SCs are reported to be low and are appropriate for various applications that work in the potential range of 1.8 V–2.5 V.

In PCs, the rapid adsorption/desorption or surface redox reactions are responsible for the charge storage to produce capacitance. In PCs, during charging, under the applied electric field, the accumulation of charges on the electrode/electrolyte interface followed by rapid reversible redox reactions takes place at the electrode surface, which leads to the storage of energy and opposite processes occur at electrode/electrolyte interface during discharging that releases the stored energy. It is to be noted that the capacitance offered by PCs is reported to be 10–100 times higher compared to the capacitance offered by EDLCs.[8,9] This is due to the fact that in case of PCs, ions are capable to enter into the electrodes and get involved in the redox reactions, whereas in EDLCs charges get stored at the surface of electrodes. EDLCs employ high surface area porous carbon-based materials as electrodes while PCs usually employ metals, metal oxides, conducting polymers, etc. that can offer rich Faradaic reactions as electrode materials.

The low energy density offered by SCs warrants further development of SCs despite the efforts dedicated in developing PCs to achieve improved energy density. Asymmetric capacitors (ACs) have been projected as promising candidates to achieve improved energy density. In asymmetric microSCs, properties of EDLC type and those of PC type microSCs are combined together to give asymmetric microSCs configuration which has the advantage of possessing properties from both types of microSCs.[4]

## 12.3 TOPOLOGIES AND PERFORMANCE METRICS

The operation and charge storage mechanism of SCs and microSCs are similar. As shown in Figure 12.1C, similar to SCs, *microSCs also consist of two electrodes separated with a liquid electrolyte but with miniaturized dimensions, possessing footprint areas around cm$^2$ or mm$^2$*. In order to develop microSCs, rational designing of electrodes and device architecture is highly needed, which requires proper performance metrics to be used for comparing the performance of various reported microSCs.[7]

In order to express the performance metrics of SC-based EES devices, a clear distinction should be made between conventional SCs at the macroscale and microSCs. A traditional way to report the performance of SC is by normalizing the parameters, namely energy, power, and capacitance by the weight or volume of the device.[10] However, as reported by Gogotsi and Simon, the gravimetric performance metric (normalized with weight) is inappropriate when compared to areal performance metric (normalized with area), especially for microSCs since the amount of active material (typically μm thin films) used in microSCs is minute.[11]

Further, the volumetric performance metric (normalized with volume) is considered as a suitable metric compared to gravimetric performance. The volumetric property involves the comparison of performance of different electrode materials without considering their thickness, which can lead to misunderstanding if there exists significant difference in the thickness of the electrodes. *Therefore in case of miniaturized devices, it is more significant to consider the performance of the microSCs normalized to its footprint area on the chip.*[10] Lethien et al. have suggested that surface metrics which is normalized to the thickness of the active layer could be compared to get a reliable comparison in the performance of microSCs similar to the metrics used in microBs.[7] Balducci et al. reported that *the most important performance metrics for the microSCs are area-normalized capacitance (F cm$^{-2}$), power and energy density based on the footprint area of device.* In the case of *redox materials such as battery type materials that exhibit non-electrostatic behaviour, expressing charge storage capacity in units of mA h g$^{-1}$ or C g$^{-1}$ is appropriate metric instead of capacitance.*[12]

The performance metric depends on both, the intrinsic properties of microSCs like electrode material, electrolyte, design, etc. as well as on the topology of the microSC device. The reported topologies used in microSCs are i) single electrode configuration, ii) parallel plate configuration, and iii) interdigitated configuration, as shown in Figure 12.1E. Among these, the parallel plate configuration is considered as a proper topology when the area is constrained, however not a promising configuration as it involves the use of two substrates that may increase the thickness of microSCs, and thus it consumes a large volume, and makes it non-compatible to electronic devices. To precisely maintain the thickness of the microSCs, it is suggested to place both the electrodes on a single substrate and this type of configuration is termed an interdigitated configuration. This configuration is reported to be the most efficient configuration for microSCs' applications. Further, devices with this topology exhibit cell surface capacitance, which is one-fourth of the areal capacitance of the single electrode. Most of the reported publications on microSCs use a classical pattern of interdigitated configuration.[7]

As mentioned earlier, microSCs are different from microBs not only by their characteristic high power but also by the long cycle life. This necessitates cycling stability as an important metric when assessing new materials and/or new device configurations for microSCs. The GCD method is the most commonly used method to determine cyclic stability. *The potential ranges and applied current densities used during GCD tests must be clearly reported, while reporting cyclic stability data. It has been reported that for EDLC based SCs, cyclic stability for at least 10000 cycles, whereas for PCs, 5000 cycles should be reported.*[12] Further, similar to SC, rectangular-shape CVs and time constant ($\tau = RC_{microSC}$) are the characteristics of microSCs, where R and $C_{microSC}$ represent cell resistance contribution and capacitance of microSC, respectively. As per the definition of $\tau$, adjustment should be taken into consideration between the power and capacitance as higher surface capacitance leads to higher time constant.[7] The *float test* is another way to determine cyclic stability in which the device is held at constant voltage for 50–100 hours while regularly sampling for every 20 hours with impedance and GCD or CV measurements.[12] Electrolytes also play a critical role in dictating the performance of microSCs. The rate at which the charge is transferred between the pair electrodes in SC determines how fast the energy can be stored or released in SC devices. The suitable electrolyte should have the factors that include (i) a large potential window, (ii) high electrochemical and chemical stability, (iii) good ionic conductivity, (iv) a huge operative temperature range, (v) should be inert with current collectors, electrode materials, etc., (vi) should be eco-friendly, and (vii) less volatile. The electrolytes are mainly divided into various types namely liquid, solid-state, or quasi-solid-state electrolytes and redox active electrolyte. Zhong et al. have comprehensively reviewed the role of electrolytes in microSCs.[13]

## 12.4   ELECTRODE MATERIALS FOR MICROSCS

Various electrode materials such as carbon-based, metal-based materials and conducting polymers have been used in microSCs. Porous carbon-based and graphene-based materials are some of the largely studied electrode materials for SC and microSC applications. As a result of their exceptional mechanical, electronic, and physicochemical properties, two-dimensional (2D) materials like graphene, transition metal dichalcogenides (TMDs), boron nitride (h-BN), black phosphorus (BP), MXenes, etc. have been used as electrode materials in microSCs.[14] There have been several comprehensive reviews reported on the use of 2D materials such as graphene, TMDs, etc.[15] The present section discusses some of the relatively recent and less explored 2D materials that have been used as electrode materials in microSCs.

### 12.4.1   MXENES

MXenes belong to the category of 2D inorganic compounds consisting of layers that are a few atoms thick transition metal carbides, nitrides, or carbonitrides. The general formula of MXenes is $M_{n+1}X_nT_x$, where M is a transition metal, X is carbon and/or nitrogen, and T is the surface termination groups (like –F, –OH etc.). MXenes like $Ti_3C_2T_x$ possess good metallic conductivity with hydrophilic nature due to the presence of hydroxyl- or oxygen-terminated surfaces. MXenes have attracted much attention in recent years owing to their distinctive properties that include good volumetric capacitance, high electronic conductivity ($\sim$6,000–8,000 S cm$^{-1}$), good chemical stability, etc.[16] In 2011, Naguib et al. discovered MXenes and since then, MXenes have been proposed as a preferred choice for microSCs as micro-electrodes for developing on-chip electronic devices.[17] The multi-layered MXene flakes were obtained through wet-chemical etching of corresponding MAX phases. In 2013, single-layer MXene flakes were obtained by the intercalation of large organic molecules and delaminating the sheets from each other, and opened a new way to explore the truly 2D nature of MXenes.[18]

Recently, a scalable production of MXene/carbon nanotube (CNT) ($Ti_3C_2$/CNT) based materials as electrodes in microSCs with the interdigitated configuration was reported by Kim et al.[19] High value of areal capacitance of 317.3 mF cm$^{-2}$ was achieved at a 50 mV s$^{-1}$ scan rate using S-DWCNT/MXene in PVA-H$_2$SO$_4$ gel electrolyte. Improved ion transfer rate was observed upon decreasing the gap between the electrodes from 10 µm to 500 nm, which leads to improved areal capacitance and thus energy density. Chen et al. reported free-standing microSCs fabricated using MXene-MoS$_2$-based materials by easy and cost-effective vacuum filtration method, and then carving of interdigitated patterns with a laser source was carried out.[20] The introduction of MoS$_2$ into MXenes increases the electrochemical performance by nearly 60% in comparison to pristine MXene. The as-prepared microSC device exhibits a high volumetric specific capacitance of 173.6 F cm$^{-3}$ at the scan rate of 1 mV s$^{-1}$ with capacitance retention of around 98% and columbic efficiency of 89% even after 6,000 cycles along with the bending angle of device up to 150°.[20] This study demonstrated the potentiality of TMDs which can be integrated with MXenes to realize high-performance microSC devices. In another report, Gogotsi et al. have reported a semi-transparent Ti$_3$C$_2$T$_x$/poly-(3,4-ethylene dioxythiophene) (PEDOT) heterostructure-based electrochromic microSCs with interdigitated topology (Figure 12.2A). A device with a thickness of PEDOT-MXene film as 100 nm showed a high capacitance of 2.4 mF cm$^{-2}$ at 10 mV s$^{-1}$ scan rate. A 58% capacitance retention was observed at a scan rate of 1,000 mV s$^{-1}$. The PEDOT/ Ti$_3$C$_2$T$_x$ also showed a good electrochromic behaviour as color changes were observed in the voltage range of 0.6-0 V and –0.6-0 V[21] (Figure 12.2B). Recently, Zhu et. al. reported microSCs using 2D Ti$_3$C$_2$T$_x$-based MXene electrodes employing a water-in-LiCl (WIL) salt gel electrolyte and achieved a large operating voltage window.[22] It was observed that at high anodic potential, oxidation of Ti$_3$C$_2$T$_x$ was

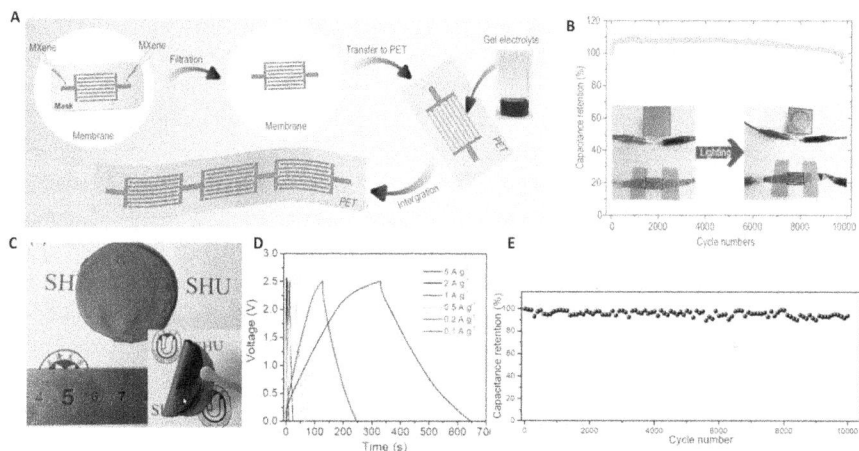

**FIGURE 12.2** A. Schematic illustration of the fabrication of microSCs based on Ti$_3$C$_2$T$_x$ electrodes; B. long-term cyclic stability of MXene microSCs. (Inset of B shows a logo which is powered by as-fabricated MXene based microSCs.) Adapted from ref.[22] Copyright (2022), National Science Review; C. digital images of the flexible and freestanding COF@rGO-2 film; D. GCD curves of COF@rGO-microSC recorded at 0.1 A g$^{-1}$ to 5 A g$^{-1}$; E. long-term cyclic stability of the as-prepared COF@rGO-microSC. Adapted from ref.[23] Copyright (2022), American Chemical Society.

suppressed in WIL electrolyte due to high content of WIL that regulated the anion intercalation in MXenes electrodes, and thereby broadens the potential window. A high volumetric energy density of 31.7 mWh cm$^{-3}$ was obtained along with an operating voltage of 1.6 V when the symmetric planar MXene-microSCs were used. Long-term operation was observed even at $-40°C$, indicating the potentiality of WIL for microSCs that could be useful in extreme environments.[22]

### 12.4.2 2D-ORGANIC FRAMEWORK MATERIALS

2D-organic framework materials comprising 2D metal organic frameworks (MOFs) and 2D covalent organic frameworks (COFs) are an upcoming category of porous crystalline materials possessing regular porous structure and controllable functionality, are projected as electrode materials in SCs. It has been reported that the pseudocapacitance may originate from the redox activity of heteroatoms (B, N, O, and S) located at specific positions of framework whereas the electrical double layer capacitance originates due to the existence of highly porous organic molecular assembly in 2D organic frameworks. Thus, 2D MOFs and 2D COFs are known to have the potential for utilization as electrode material for EES applications. However, the use of organic frameworks is limited in microSC applications due to the inadequacy of suitable microfabrication techniques.

He et al. reported an asymmetric in-plane microSC constructed by using MOF [Cu$_3$(BTC)$_2$]$^-$ films doped with electron acceptors (7,7,8,8-tetracyanoquinododimethane (TCNQ) as electrodes to tune the electrical conductivity. Asymmetric in-plane microSCs consisting of TCNQ@Cu$_3$(BTC)$_2$ as a cathode and activated carbon as an anode were fabricated. An areal capacitance of 95.1 mF cm$^{-2}$ was obtained at a scan rate of 5 mV s$^{-1}$. In addition, long-term stability at 10 mA cm$^{-2}$ with 94.1% capacitance retention was observed up to 5,000 charge-discharge cycles.[24] In another report, Yao et al. reported hybrid films of anthraquinone-containing COFs and reduced graphene oxide (rGO)–based electrodes for microSCs. Efficient electrolyte ion transportation was maintained as COFs can eliminate accumulation of rGO nanosheets in hybrids. A specific capacitance of 451.96 F g$^{-1}$ and energy density of 44.22 W h kg$^{-1}$ was achieved[23] (Figure 12.2C–E).

Transition metal silicides (TMSi) are another class of less explored non-2D materials as electrodes for microSC applications. Insertion of silicon atoms into a crystal lattice of transition metal yields intermetallic TMSi with diversified electronic and geometric structure. In 2019, Lee et al. demonstrated free-standing single-crystalline Co$_2$Si nanowires (NW) on silicon (Si) substrate for on-chip microSC applications. A high areal capacitance of 983 μF cm$^{-2}$ was obtained at 2μA cm$^{-2}$, along with a high energy density of 629 μF cm$^{-2}$ at 2μA cm$^{-2}$. The system also showed a high cyclability with ~94% retention in capacitance even after 4,000 cycles. The group claims that this study was the first demonstration of TMSi NWs-based on-chip SCs.[25]

## 12.5 FABRICATION METHODS

The microSCs consist of two types of design configurations, i) sandwich configuration and ii) planar configuration (Figure 12.1D).[6] In the primitive stages, microSCs have adopted a "sandwich configuration" in which a solid electrolyte sets apart two stacked electrodes (positive and negative electrodes). In 2001, Lim et. al. reported the first sandwich configuration–type microSC where a lithium phosphorousoxynitride (LiPON) solid electrolyte was sandwiched between two $RuO_2$ electrodes. The configuration showed volumetric capacitance of ~380 μF cm$^{-3}$.[26] Though this approach is cost-effective, but there are drawbacks such as inappropriate position of the electrode film and the higher possibility of short circuit as the electrodes lie one on each other in sandwich configuration. Further, due to lack of precise control on the distance between thick solid electrolytes and electrodes, this causes power loss and also increases resistance for ion movement.[27] In light of these issues, the planar configuration (Figure 12.1D) gained more attention than the sandwich configuration due to its ease of scalability and availability of different methods of fabrication. The planar design was demonstrated by Sung et al. in 2003 with polypyrrole and poly-(3-phenylthiophene) as electrode materials with liquid electrolytes. The configuration exhibited a capacitance of ~5.2 mF.[28]

The following section briefly discusses the typical methods reported to fabricate microSCs.

The fabrication method needs to be properly selected as per the type of electrode material without impacting the ion and electron transport ability. For instance, slight variations in the fabrication method would lead to differences in microstructures of a given electrode material, which will eventually affect the performance of the device. It should be noted that the fabrication method defines the gap between the two electrodes, dimensions of the electrode, accuracy of placement of electrodes, etc. and thus they have an impact on the electrochemical performance of microSCs.[29] There have been reviews which comprehensively discuss various methods used for fabrication of microSCs and the suitable method of fabrication depending on the type of material.[4,30–38] Figure 12.3A summarizes various methods reported for the fabrication of microSCs. Briefly, in some fabrication methods, active materials will be deposited/coated onto desired pre-patterned substrates such as interdigitated configuration, while in some methods, the patterns are made on pre-deposited active material on substrates. The deposition methods include electrolytic, electrophoretic and chemical vapour deposition techniques which have been used to deposit active materials on desired substrates. Etching methods, namely, plasma etching and laser etching are some of the methods used to create interdigitated patterns. Various printing techniques such as screen printing, laser printing, inkjet printing and Gravure printing have been reported for the fabrication of microSCs. Photolithography is another method used to pattern the substrates with desired designs. More specific details about various fabrication methods and the parameters that control the performance of microSCs can be found in recent reviews.[4,30]

It should be noted that there exists no fabrication strategy that is currently dominant over the other strategies in constructing microelectrodes for microSCs. Therefore, enhancing the existing assembly strategies and selecting appropriate fabrication methods for practical design with the overall factors, such as active materials, electrolytes, and the interface between microelectrodes and electrolytes, should be underlying aspects in order to obtain high-performance microSCs, which can be integrated in advanced miniaturized devices.[29]

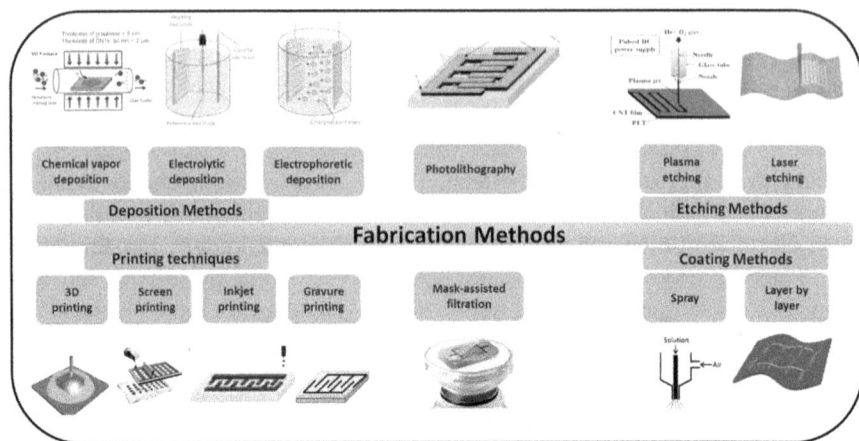

**FIGURE 12.3**  Schematic illustration of various types of fabrication methods. Inset images: "Electrolytic deposition, Electrophoretic deposition, Inkjet printing" Adapted with permission from ref.[10] Copyright (2016), Macmillan publishers, 3D printing, Laser etching, Photolithography, Mask assisted filtration". Adapted with permission from ref.[2] Copyright (2020), IOP science publishers. "Chemical vapour deposition, Plasma etching, layer by layer, Screen printing". Adapted with permission from ref.[29] Copyright (2022), MDPI publishers. "Gravure printing" Adapted with permission from ref.[31] Copyright (2015), AIP Publishing. "Spray coating" Adapted with permission from ref.[32] Copyright (2015), Elsevier.

## 12.6  VARIOUS WAYS TO IMPROVE THE PERFORMANCE OF MICROSCS

To realize high-performance microSCs, it is essential to develop strategies to improve energy density of the device without affecting other parameters like power density, cyclic stability, area of the microSCs, etc. The performance of microSCs depends on the distance between the pair of electrodes width (gap width, $W_g$) and thickness of electrodes. If $W_g$ is more then there will be a reduction in the mobility of ions which will reduce power density. Similarly, if the width of the electrode ($W_e$) is large, then it will cause a long diffusion length for ions. In case of devices with electrodes of smaller values of $W_e$ will have less active material on the electrodes. Thus, the optimization of $W_e$ is necessary in fabricating microSCs. *MicroSCs comprising of electrodes with suitable $W_e$ and narrow $W_g$ is known to exhibit high performance.* As reported, enhancing the ratio of electrode width to gap width ($W_e/W_g$) can reduce equivalent series resistance (ESR) which leads to improved power density and energy density.[4] The other way to increase the surface energy density is by fabricating microSCs with thicker electrodes as areal density increases with increasing the thickness of active material which in turn increases the capacitance since areal energy density is directly proportional to the surface capacitance and square of cell voltage. However, this approach cannot be generalized, especially when the active layer is of less conducting material.[7] Use of hierarchical electrodes is another way to improve the performance of microSCs. The active electrode material is typically deposited on the hierarchical template that forms the three-dimensional (3D) microstructure scaffold.[39] The two main advantages of 3D microSCs are i) to increase the energy density in the limited footprint area of the device and ii) to decrease the transport path of ions.[40] The 3D microSCs have been fabricated by etching a deep hole on the substrate in which a large amount of active electrode material was placed to produce high mass loading so the performance of microSCs will be enhanced.[7]

The electrolyte is another component that affects the performance of microSCs as the cell voltage is related to the electrolyte. To increase the capacitive performance in the case of graphene electrodes, redox electrolytes ($Cu/Cu^{2+}$, $Fe^{2+}/Fe^{3+}$) have been used. However, it should be noted that these redox electrolytes can provide capacitance for only one electrode; either negative or positive. Recently Wu et al. demonstrated graphene-based microSCs using water in salt (WIS) ambipolar redox electrolyte ($ZnI_2$) that was composed of a pair of redox species, namely $I^-/I_2$ and $Zn/Zn^{2+}$. The device exhibited appreciable pseudocapacitance of both positive and negative electrodes simultaneously. It should be noted that one ambipolar redox electrolyte can provide two pair of redox species which will allow two electrons for the oxidation at the positive electrode and reduction at the negative electrode simultaneously and separately, thus providing a huge pseudocapacitance. The authors obtained a high volumetric capacity and energy density of 106 mAh $cm^{-3}$, and 111 mWh $cm^{-3}$ along with long-lasting cycle life with a 92.1% retention of capacity even after 5,300 cycles.[41]

## 12.7    APPLICATIONS OF MICROSCS

As mentioned earlier, microSCs possess added advantages than microBs and can be integrated with various electronic devices such as smart watches and phones, ultra-thin laptops/notebooks, smart healthcare bands, biocompatible energy sources in pacemakers, etc.[42]

In this section, the integration of microSCs with some of the microdevices will be discussed.

A photodetector is a device that involves the conversion of light energy into electrical energy. The integration of microSC with a photodetector enables the realization of self-powered microelectronic systems.[43] Photoactive materials such as titanium dioxide ($TiO_2$) is coated on the microelectrodes of microSCs. It functions as micro-electrodes of microSCs as well as light-absorbing active component of photodetectors. Upon illumination of UV light, electrons and holes will be generated in $TiO_2$ layer due to absorption of UV-light by $TiO_2$. These photogenerated charge carriers will move towards positive and negative micro-electrodes of microSCs under an applied electric field, leading to an increase in leakage current.

The integration of microSCs with nanogenerator allows the direct conversion of mechanical energy into electrochemical energy which can be stored. In these devices, nanogenerators act as an energy harvesters and microSCs serve as energy-storage units. Luo et al reported piezoelectric generator integrated with microSCs that are fabricated using the laser engraving method. It was observed that the mechanical energy generated due to the application of external stress on the device, is directly converted into electrochemical energy and is stored in microSCs. It has shown a peak power density of 0.8 $Wm^{-2}$ at 20 MΩ with the high capacitance of 10.29 mF $cm^{-2}$ at 0.01 mA $cm^{-2}$.[44]

Further, microSCs are projected as a potential substituent to conventional aluminium electrolytic capacitors (AECs) in alternative current (AC) line filters. The fast frequency response, small size, high power density, and easy-to-integrate features of microSCs make them suitable candidates to replace AECs. It functions as AC line filters to correct pulse energy or as a current ripple filter in the kHz range.[45] Xu et al. have demonstrated high-frequency microSCs based on 2D pseudocapacitive MXene and MWCNT electrodes that showed areal capacitance of 6 mF $cm^{-2}$ at 120 Hz. The MWCNT provides fast ion transport path and MXene offers high capacitance and power density for high-frequency response compared to the commercially available tantalum capacitors at 120 Hz.[46] In short, microSCs with their unique characteristics provide large potential for IoTs and AC filters applications.

## 12.8  CONCLUSIONS

Nanotechnology has allowed the ever-growing development in small wearable electronic devices. This calls for flexible and micro-energy storage devices. As discussed in the beginning of the chapter, microSCs have more merits over microBs owing to their excellent areal power density, long cycle life, and rapid charge-discharge cycles. This chapter discusses the fundamental aspects of microSCs, various 2D-electrode materials used for microSCs, a brief section on fabrication methods, and some recent applications of microSCs. For the assessment and comparison of various microSCs, the appropriate performance metrics are discussed. The important metrics are suggested to be area normalized capacitance, foot print area of normalized energy and power of the device. Further, fabrication methods are briefly discussed. The performance of microSCs on device configuration (width of electrode, electrode separation etc.) is also discussed. At last, some applications of microSCs which are integrated with microdevices are discussed.

## 12.9  CHALLENGES AND FUTURE OUTLOOK

In order to enhance the wide range applicability of microSCs in miniaturized devices, it is essential to address a few challenges, as mentioned below.

Owing to the growing interest in use of microSCs in flexible electronics, a simple and reliable approach is required for measuring the mechanical flexibility of the device during its mechanical deformation in order to gain better insights into the changes in the microstructure of active materials and the electrolytes. The upcoming research should focus on developing simple, cost-effective, reliable, and rapid methods for the fabrication of microSCs that can be upscaled. The stability of the pseudocapacitive materials is another aspect that requires attention. Further, it is important to design and synthesize eco-friendly electrolytes with a wide potential window that are compatible to the electrodes. By employing ionic liquid electrolytes, the potential window of the electrolyte can be significantly increased. Recently, WIS-type electrolytes ($NaNO_3$, $NaClO_4$, and LiCl) received considerable attention and are projected as promising systems; however, this requires future research due to limited ion transport, and their poor performance at low temperatures. Similarly, there is also a need for the development of gel electrolytes that offer stability at high temperatures with outstanding ionic conductivity. The high interfacial resistance commonly observed in solid-state microSCs is another challenge to be given high attention in future studies. Also, an easy and deployable fabrication method that yields high resolution of interspacing of electrodes in microSCs is highly essential.

## ACKNOWLEDGMENTS

ADG indebted to SERB-DST for financial assistance. SP acknowledges BITS Pilani, Goa for financial assistance. KV acknowledges BITS Pilani, Goa for ACG grant and SERB-DST (SRG/2020/000719), New Delhi, India for funding.

## REFERENCES

[1] Xiong, Guoping, Chuizhou Meng, Ronald G. Reifenberger, Pedro P. Irazoqui, and Timothy S. Fisher. "A Review of Graphene-Based Electrochemical Microsupercapacitors." *Electroanalysis*, 2014, 26, 30–51. doi:10.1002/elan.201300238

[2] Liu, Huaizhi, Guanhua Zhang, Xin Zheng, Fengjun Chen, and Huigao Duan. "Emerging Miniaturized Energy Storage Devices for Microsystem Applications: From Design to Integration." *International Journal of Extreme Manufacturing*, 2020, 2, 042001 (28 pp) doi:10.1088/2631-7990/abba12

[3] Zheng, Shuanghao, Xiaoyu Shi, Pratteek Das, Zhong-shuai Wu, and Xinhe Bao. "The Road Towards Planar Microbatteries and Micro- Supercapacitors: From 2D to 3D Device Geometries." *Advanced Materials*, 2019, 31. 1900583. doi:10.1002/adma.201900583

[4] Bu, Fan, Weiwei Zhou, Yihan Xu, Yu Du, Cao Guan, and Wei Huang. "Recent Developments of Advanced Micro-Supercapacitors: Design, Fabrication and Applications." *Npj Flexible Electronics*, 2020, 4, 31 (1–16) doi:10.1038/s41528-020-00093-6

[5] Jiang, Qiu, Yongjiu Lei, Hanfeng Liang, Kai Xi, Chuan Xia, and Husam N. Alshareef. "Review of MXene Electrochemical Microsupercapacitors." *Energy Storage Materials*, 2020, 27, 78–95. doi:10.1016/j.ensm.2020.01.018

[6] Peng, Xu, Lele Peng, and Yi Xie. "Two Dimensional Nanomaterials for Flexible Supercapacitors." *Chemical Society Reviews*, 2014, 43, 3303–3323. doi:10.1039/c3cs60407a

[7] Lethien, Christophe, and Le Bideau. "Challenges and prospects of 3D Micro-Supercapacitors for Powering the Internet of Things." *Energy and Environmental Science*, 2019, 12, 96–115. doi:10.1039/c8ee02029a

[8] Zhao, Xiaoyu, Yingbing Zhang, Yanfei Wang, and Huige Wei. "Battery-Type Electrode Materials for Sodium-Ion Capacitors." *Batteries and Supercaps*, 2019, 2, 899–917. doi:10.1002/batt.201900082

[9] Da, Yumin, Jinxin Liu, Lu Zhou, Xiaohui Zhu, Xiaodong Chen, and Lei Fu. "Engineering 2D Architectures toward High-Performance Micro-Supercapacitors." *Advanced Materials*, 2019, 31, 1802793. doi:10.1002/adma.201802793

[10] Kyeremateng, Nana Amponsah, Thierry Brousse, and David Pech. "Microsupercapacitors as Miniaturized Energy-Storage Components for on-Chip Electronics." *Nature Nanotechnology*, 2017, 12, 7–15. doi:10.1038/nnano.2016.196

[11] Gogotsi, Y., and P. Simon. "True Performance Metrics in Electrochemical Energy Storage." *Science*, 2011, 334, 917–918. doi:10.1126/science.1213003

[12] Balducci, A., D. Belanger, T. Brousse, J.W. Long, and W. Sugimoto. "Perspective—A Guideline for Reporting Performance Metrics with Electrochemical Capacitors: From Electrode Materials to Full Devices." *Journal of The Electrochemical Society*, 2017, 164, 1487–1489. doi:10.1149/2.0851707jes

[13] Zhong, Cheng, Yida Deng, Wenbin Hu, Jinli Qiao, Lei Zhang, and Jiujun Zhang. "A Review of Electrolyte Materials and Compositions for Electrochemical Supercapacitors." *Chemical Society Reviews*, 2015, 44, 7484–7539. doi:10.1039/c5cs00303b

[14] Zhang, Panpan, and Faxing Wang. "Two-Dimensional Materials for Miniaturized Energy Storage Devices: From Individual Devices to Smart Integrated Systems." *Chemical Society Reviews*, 2018, 47, 7426–7451. doi:10.1039/c8cs00561c

[15] Qin, Jieqiong, Pratteek Das, Shuanghao Zheng, and Zhong Shuai Wu. "A Perspective on Two-Dimensional Materials for Planar Micro-Supercapacitors." *APL Materials*, 2019, 7, 090902(1–17). doi:10.1063/1.5113940

[16] Xu, Xueqin, Li Yang, Wei Zheng, Heng Zhang, Fushuo Wu, Zhihua Tian, Peigen Zhang, and Zheng Ming Sun. "MXenes with Applications in Supercapacitors and Secondary Batteries: A Comprehensive Review." *Materials Reports: Energy*, 2022, 2, 100080. doi:10.1016/j.matre.2022.100080

[17] Naguib, Michael, Murat Kurtoglu, Volker Presser, Jun Lu, Junjie Niu, Min Heon, Lars Hultman, Yury Gogotsi, and Michel W. Barsoum. "Two-Dimensional Nanocrystals Produced by Exfoliation of $Ti_3AlC_2$." *Advanced Materials*, 2011, 23, 4248–4253. doi:10.1002/adma.201102306

[18] Mashtalir, Olha, Michael Naguib, Vadym N. Mochalin, Yohan Dall'Agnese, Min Heon, Michel W. Barsoum, and Yury Gogotsi. "Intercalation and Delamination of Layered Carbides and Carbonitrides." *Nature Communications*, 2013, 4, 1716 (1–7). doi:10.1038/ncomms2664

[19] Kim, Eunji, Byeong-joo Lee, Kathleen Maleski, Yoonjeong Chae, Yonghee Lee, Yury Gogotsi, and Chi Won. "Microsupercapacitor with a 500 nm gap between MXene/CNT Electrodes." *Nano Energy*, 2021, 81, 105616 (1–11). doi:10.1016/j.nanoen.2020.105616

[20] Chen, Xing, Siliang Wang, Junjie Shi, Xiaoyu Du, Qinghua Cheng, Rui Xue, Qiang Wang, Min Wang, Limin Ruan, and Wei Zeng. 2019. "Direct Laser Etching Free-Standing MXene-$MoS_2$ Film for Highly Flexible Micro-Supercapacitor". *Advanced Materials Interfaces*, 2019, 6, 1901160. doi:10.1002/admi.201901160

[21] Li, Jianmin, Ariana Levitt, Narendra Kurra, Kevin Juan, Natalia Noriega, Xu Xiao, Xuehang Wang, Hongzhi Wang, Husam N. Alshareef, and Yury Gogotsi. "MXene-Conducting Polymer Electrochromic Microsupercapacitors." *Energy Storage Materials*, 2019, 20, 455–461. doi: 10.1016/j.ensm.2019.04.028

[22] Zhu, Yuanyuan, Shuanghao Zheng, Pengfei Lu, Jiaxin Ma, Pratteek Das, Feng Su, Hui-Ming Cheng, and Zhong-Shuai Wu. "Kinetic Regulation of MXene with Water-in-LiCl Electrolyte for High-Voltage Micro-Supercapacitors." *National Science Review*, 2022, 9, nwac024 (1–9). doi:10.1093/nsr/nwac024

[23] Yao, Mengyao, Chaofei Guo, Qianhao Geng, Yifan Zhang, Xin Zhao, Xin Zhao, and Yong Wang. "Construction of Anthraquinone-Containing Covalent Organic Frameworks/Graphene Hybrid Films for a Flexible High-Performance Micro-supercapacitor." *Industrial and Engineering Chemistry Research*, 2021, 61, 7480–7488. doi:10.1021/acs.iecr.1c04638

[24] He, Yafei, Sheng Yang, Yubin Fu, Faxing Wang, Ji Ma, Gang Wang, Guangbo Chen, et al. "Electronic Doping of Metal-Organic Frameworks for High-Performance

Flexible Micro-Supercapacitors." *Small Structures*, 2021, 2, 2000095, doi:10.1002/sstr.202000095

[25] Lee, Jiyoung, Chung Yul Yoo, Yeong A. Lee, Sang Hyun Park, Younghyun Cho, Jae Hyun Jun, Woo Youn Kim, Bongsoo Kim, and Hana Yoon. "Single-Crystalline Co2Si Nanowires Directly Synthesized on Silicon Substrate for High-Performance Micro-Supercapacitor." *Chemical Engineering Journal*, 2019, 370, 973–979. doi:10.1016/j.cej.2019.03.269

[26] Lim, Jae Hong, Jin Choi, Han-ki Kim, Il Cho, and Young Soo. "Thin Film Supercapacitors Using a Sputtered $RuO_2$ Electrode," *Journal of Electrochemical Society*, 2001, 148, 275–278. doi:10.1149/1.1350666

[27] Wang, Jinhui, Fei Li, Feng Zhu, and Oliver G. Schmidt. "Recent Progress in Micro-Supercapacitor Design, Integration, and Functionalization." *Small Methods*, 2019, 3, 1800367. doi:10.1002/smtd.201800367

[28] Sung, Joo-hwan, Se-joon Kim, and Kun-hong Lee. "Fabrication of Microcapacitors Using Conducting Polymer Microelectrodes." *Journal of Power Sources*, 2003, 124, 343–350. doi:10.1016/S0378-7753(03)00669-4

[29] Vandeginste, Veerle. "A Review of Fabrication Technologies for Carbon Electrode-Based Micro-Supercapacitors." *Applied Sciences*, 2022, 12, 862. doi:10.3390/app12020862

[30] Zhang, Jihai, Gaixia Zhang, Tao Zhou, and Shuhui Sun. "Recent Developments of Planar Micro-Supercapacitors: Fabrication, Properties, and Applications." *Advanced Functional Materials*, 2020, 30, 1910000. doi:10.1002/adfm.201910000

[31] Xiao, Yuxiu, Lei Huang, Qi Zhang, Shuhua Xu, Qi Chen, and Wangzhou Shi. "Gravure Printing of Hybrid $MoS_2$@S-RGO Interdigitated Electrodes for Flexible Microsupercapacitors." *Applied Physics Letters*, 107, 013906. doi:10.1063/1.4926570

[32] Aziz, F., and A.F. Ismail. "Spray Coating Methods for Polymer Solar Cells Fabrication: A Review." *Materials Science in Semiconductor Processing*, 2015, 39, 416–425. doi:10.1016/j.mssp.2015.05.019

[33] Li, Yang, Joel Henzie, Teahoon Park, Jie Wang, Christine Young, Huaqing Xie, Jin Woo Yi, et al. "Fabrication of Flexible Microsupercapacitors with Binder-Free ZIF-8 Derived Carbon Films via Electrophoretic Deposition." *Bulletin of the Chemical Society of Japan*, 2020, 93, 176–181. doi:10.1246/BCSJ.20190298

[34] Wang, Yang, Yi Zhou Zhang, David Dubbink, and Johan E. ten Elshof. "Inkjet Printing of δ-$MnO_2$ Nanosheets for Flexible Solid-State Micro-Supercapacitor." *Nano Energy*, 2018, 49, 481–488. doi:10.1016/j.nanoen.2018.05.002

[35] Tran, Tuan Sang, Naba Kumar Dutta, and Namita Roy Choudhury. "Graphene-Based Inks for Printing of Planar Micro-Supercapacitors: A Review." *Materials*, 2019, 12, 978 (1–21). doi:10.3390/ma12060978

[36] Hu, Guohua, Joohoon Kang, Leonard W.T. Ng, Xiaoxi Zhu, Richard C.T. Howe, Christopher G. Jones, Mark C. Hersam, and Tawfique Hasan. "Functional Inks and Printing of Two-Dimensional Materials." *Chemical Society Reviews*, 2018, 47, 3265–3300. doi:10.1039/c8cs00084k

[37] Zhang, Panpan, Faxing Wang, Sheng Yang, Gang Wang, Minghao Yu, and Xinliang Feng. "Flexible In-Plane Micro-Supercapacitors: Progresses and Challenges in Fabrication and Applications." *Energy Storage Materials*, 2020, 28, 160–187. doi:10.1016/j.ensm.2020.02.029

[38] Jia, Rui, Guozhen Shen, Fengyu Qu, and Di Chen. "Flexible On-Chip Micro-Supercapacitors: EfFicient Power Units for Wearable Electronics." *Energy Storage Materials*, 2020, 27, 169–186. doi:10.1016/j.ensm.2020.01.030

[39] Kim, Cheolho, and Jun Hyuk Moon. "Hierarchical Pore-Patterned Carbon Electrodes for High-Volumetric Energy Density Micro-Supercapacitors." *ACS Applied Materials Interfaces*, 2018, 10, 19682–19688. doi:10.1021/acsami.8b03958

[40] Park, Seong Hyeon, Geordie Goodall, and Woo Soo Kim. "Perspective on 3D-Designed Micro-Supercapacitors." *Materials and Design*, 2020, 193, 108797. doi: 10.1016/j.matdes.2020.108797

[41] Meng, Caixia, Feng Zhou, Hanqing Liu, Yuanyuan Zhu, Qiang Fu, and Zhong-shuai Wu. "Water-in-Salt Ambipolar Redox Electrolyte Extraordinarily Boosting High Pseudocapacitive Performance of Micro- Supercapacitors." *ACS Energy Letters*, 2022, 7, 1706–1711. doi: 10.1021/acsenergylett.2c00329

[42] Rani, Shalu, Nagesh Kumar, and Yogesh Sharma. "Recent Progress and Future Perspectives for the Development of Micro-Supercapacitors for Portable / Wearable Electronics Applications." *Journal of Physics*, 2021, 3, 032017. doi: 10./1088/2515-7655/ac01c0

[43] Zhong, Yan, Guanggui Cheng, Chen Chen, Zirong Tang, Shuang Xi, and Jianning Ding. 2021. "In-Plane Flexible Microsystems Integrated with High-Performance Microsupercapacitors and Photodetectors." *Journal of Electronic Materials*, 2021, 50, 3517–3526. doi: 10.1007/s11664-021-08871-2

[44] Luo, Jianjun, Feng Ru Fan, Tao Jiang, Zhiwei Wang, Wei Tang, Cuiping Zhang, and Mengmeng Liu. "Integration of Micro-Supercapacitors with Triboelectric Nanogenerators for a Flexible Self-Charging Power Unit". *Nano Research*, 2015, 8, 3934–3943. doi: 10.1007/s12274-015-0894-8

[45] Feng, Xin, Xiaoyu Shi, Jing Ning, Dong Wang, Jincheng Zhang, Yue Hao, and Zhong-Shuai Wu. "Recent Advances in Micro-Supercapacitors for AC Line-Filtering Performance: From Fundamental Models to Emerging Applications." *EScience*, 2021, 1, 124–140. doi: 10.1016/j.esci.2021.11.005

[46] Xu, Sixing, Wei Liu, Bingmeng Hu, and Xiaohong Wang. "Circuit-Integratable High-Frequency Micro Supercapacitors with Filter/Oscillator Demonstrations." *Nano Energy*, 2019, 58, 803–810. doi: 10.1016/j.nanoen.2019.01.079

# 13 Tetracyanoquinodimethane Based Small Organic Molecules as Potential Memristor and Solar Cell Devices

*Subbalakshmi Jayanty*

## CONTENTS

## 13.1 INTRODUCTION

Tetracyanoquinodimethane (TCNQ), an organic $\pi$-electron acceptor, gained attention in organic electronics due to its ability to form charge-transfer salts with high

DOI: 10.1201/b23359-13

electrical conductivity [1,2]. The reaction of TCNQ with various amines in certain stoichiometric quantities results in zwitterionic bis-substituted TCNQ derivatives, generally called diaminodicyanoquinodimethanes (DADQs) with a D-π-A framework. Incidentally, the choice of amine plays a vital role to fine-tune precise structures along with the captivating and unforeseeable properties that inspire to design and develop novel TCNQ derivatives with remarkable characteristics in the solids and solutions; this holds DADQs to be still in focus with disparate applications, namely as molecular organic semiconductors [3], hopper crystals [4], second harmonic generation [5,6] and as phosphorescent materials [7]. Initially, TCNQ derivatives were explored notably for second-order non-linear optical (NLO) features, due to large hyperpolarizabilities (β) [8]. Only in the recent past, significant investigations on optical property of TCNQ derivatives was surveyed. Especially, mono/di-substituted TCNQs attained with aliphatic [9,10], heterocyclic[11–14] amines and limited (~5) aromatic amines [15,16] have evolved in contrasting phenomenon like solid-state emission, phase change accredits [17] and second harmonic generation (SHG) [18]. While accountability for the equitable fluorescence in DADQ solutions was owing to the weak interactions, strong hydrogen bonding and molecular self-assembly; aggregation [19,20] robust intermolecular desirability, excited states geometry relaxation and lesser π-π/CH-π, interactions [21] prevailed in solids. Interestingly, BPZDQ and its *p*-toulenesulphonate salt [22,23] were utilized in imaging epidermal of decotyledon plant leaves and selective bacterial endospores.

Recently, exploration on the effect of ethylene spacer group on TCNQ with amine functionality on one end and (i) heterocyclic moiety on the other end, predominantly led to the SHG activity [14]; while (ii) simple aliphatic group (*N,N*- dimethyl) on the other end primarily resulted in the DADQ with the highest solid-state quantum yield [24], a demonstration on the structure-property correlation. Interestingly, resistive switching (RS) memory could be attained on the reaction of TCNQ with hydroxyethyl piperazine (HEP) (ethylene spacer with OH functionality) resulting in bis (Hydroxyethyl piperizino) dicyanoquinodimethane (BHEPDQ). Solar cell property noticed with a varied design aspect wherein TCNQ was reacted with cyclopropyl carbonyl piperazine (CCP) (a derivative of heterocyclic amine) resulting in di-cyclopropylcarbonyl piperazine substituted dicyanoquinodimethane (BCCPDQ) (Scheme 13.1); in contrast to bis (*N*-benzoylpiperazino) dicyanoquinodimethane (BBPDQ) [17,25] with amine being a benzene derivative, manifested phosphorescence property.

The stored information is conserved even on withdraw of external power in non-volatile memory device, criteria for many electronic applications. Thereby RS memory devices are known to have the potential to replace the enduring compatible semiconductors based on metal oxides as non-volatile memory devices surmounting practical limitations, for farther progression and evolution in the memory section. Molecular organic materials could replace Si-based electronic devices in certain applications [26] related to organic electronics. Furthermore, organic molecules encompass simple architects, high-density related to consolidation, affordable creation, and good instinctive pliability, thereby eligible for flexible gadget applications [27]. Though organic polymers have been known for RS memory devices; small molecules regarding the same are little utilized and progress is subtlely slow.

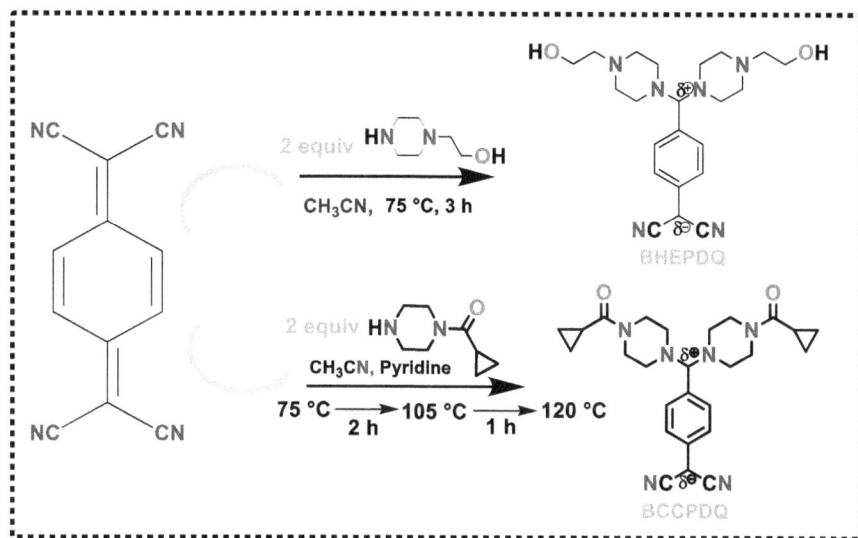

**SCHEME 13.1** Synthesis of BHEPDQ and BCCPDQ.

Henceforth, efforts were committed to the investigation of a small organic molecular materials based RS memory device. Lately, pivot is transforming technology from rigid based into a soft low-power flexible electronic applications along with the low-cost, high scalability and process compatibility.

Surprisingly, so far, among any DADQs, the solar cell property is unveiled. Of late, solar cell–based organic molecules, also known as organic solar cells (OSCs), accomplished power conversion efficiency (PCE) ~ 18% besides employing interfacial, bulk heterojunctions (BHJ) [28,29]. Moreover, small-molecule OSCs are gaining attentiveness due to their wide spectral ranges emerging from conjugated frameworks, and swift functionalization credibility. In several optoelectronic devices [30,31]. TCNQ-based compounds were particularly applied as dopants to ease, inject and transport charge carriers. In this regard, a TCNQ doped active layer was developed by Guillain et al. for polymer-based solar cells, to mitigate high charge extraction capability [31]. Of late, flexible solar cells are gaining more recognition than the rigid due to their strength and prospective in material and congenial electronics [32,33].

Owing to sturdy crystalline nature, TCNQ thin film, exhibited high absorption in the visible region; however, it displayed more recombination access on the account of currents leaked in the effective domain. $TiO_2$ NPs, which act as an electron transport layer, were assimilated to suppress recombination of generated hole-electron pair at the juncture. Also, $TiO_2$ NPs possess considerable mobility of their electrons (0.1–10 $cm^2$/V/s) besides profound conduction assembly (coincide along BCCPDQ), thereby initiating the chemical stability of the PV device.

Primarily, to date, except for optical/non-linear optical property, resistive switching (RS) memory device or solar cell attribute of either TCNQ-based derivatives or DADQs has not been investigated and remain as a gap with respect to

TCNQ-based molecular materials. Therefore, this chapter's emphasis is on the comprehensive exploration on the design, photophysical property, electrochemical characteristics, fabrication and flexible organic memristor and solar cell employment of BHEPDQ and BCCPDQ each, respectively. Since recent developments on memristor and solar cell devices were on rigid substrates like ITO which are inappropriate for flexible electronic applicability; and its accomplishment; therefore demands a pathway for a systematized analysis on flexible substrates. PET is nontoxic and substrates based on PET showed affinity with inscription procedures, furnished optimistic results, and therefore emerged as the potential candidate in the development of practical electronic devices. Hence, in this chapter we describe the efforts devoted to explore RS switching and PV properties of BHEPDQ and BCCPDQ each, respectively.

## 13.2  PHOTOPHYSICAL, ELECTROCHEMICAL PROPERTY BESIDES NON-VOLATILE RS MEMORY DEVICE EXECUTION OF BHEPDQ

### 13.2.1  PHOTOPHYSICAL PROPERTY

BHEPDQ was synthesized according to the modification of the reported procedure [10]. To a 0.2 g TCNQ solution at 75°C, 0.267 g of HEP was added beneath nitrogen atmosphere (by-product HCN is collected and taken off by scrubber which is dipped in the $FeSO_4$ solution). After 3 hr of stirring, a yellow-green microcrystalline precipitate was obtained. Attained solid filtered, followed by washing with cold acetonitrile and then dried under vacuum. BHEPDQ of 82% yield was achieved; further crystallized as thick blocks, at room temperature by slow evaporation from DMF. M.P. 300–305°C. BHEPDQ was characterized rigorously by varied spectroscopic and microscopy techniques [34].

Intermolecular H-bonding between N2 of piperazine and H1 of OH (2.930 Å), short contacts led to supramolecular self-assemblies with molecular dipoles. Dihedral angle (average) C2–C1–C5–N1 was ~ −34.06°. Generally, DADQs exhibit zwitterionic structure because they possess large dipole moment in their ground state, in solution as well as solids.

$\lambda_{max}$ absorption was at ~419 nm, indicative of intramolecular charge transfer in zwitterionic DADQs. In polar solvents, hypsochromic shift noted; indicative of the ground state polar nature; agreeing with earlier reports [12,19]. A stokes shift observed within ~82 to 110 nm. Low quantum yields were ($\Phi_f = 0.07$ to $\Phi_f = 0.20\%$) observed in solutions when compared to the solid BHEPDQ ($\Phi_f = 0.43\%$), probably because zwitterionic DADQs are prone to aggregation and herein HEP facilitate particular and directed intermolecular interactions. Enhancement of fluorescence in solids is ascribed through restricted relaxation of the excited state, and no radiation decay aspect within solution [25]. Fluorescence lifetime decay in solids was ~1.26 ns. Acid sensing of BHEPDQ with HCl manifested quenching of fluorescence with red shift monitored by a shift in $CH_2$ protons signal of piperazine; which supports the tertiary nitrogen protonation of piperazine moieties.

## 13.2.2 SCANNING ELECTRON MICROSCOPY (SEM) AND THERMAL STABILITY

Acetonitrile solution of BHEPDQ (1 mmol) spin coated for 60 sec at about 1,400 rpm on the silicon wafer substrate revealed micro-needles and rod-type features, rare among DADQs [12]. Similar study in 1:1 volume ratio of toluene-acetonitrile exhibited microcrystalline flowers. Furthermore, morphology in DMF showed uniformly distributed aggregates. Variation in the morphology is accounted due to solvent polarity and dispersal capability. Crystals of BHEPDQ exhibited ~298.16°C as the melting temperature (studied by the differential scanning calorimetry) under $N_2$ atmosphere at 10°C/min. Heat capacity was calculated as ~−328.44 mJ. Thermogravimetric analysis under similar conditions revealed single phase decomposition ($T_d$ ~303°C). Persisting compound notably goes off fully and the corresponding 68.84% weight loss of BHEPDQ (~3.16 mg) is correlated to 282.2 molecular weight, perhaps due to the loss of one HEP moiety.

## 13.2.3 ELECTROCHEMICAL STUDY

Cyclic voltammetry (CV) study was carried out at room temperature in the dry acetonitrile. Three reduction waves (reversible) owing to three successive reductions (of dicyano fragment) toward disparaging radical anions [35] was observed in BHEPDQ. Further, three irreversible oxidation peaks (due to HEP) were observed. Considering utmost oxidation and the reduction outsets correspondingly each from the CV, the LUMO and HOMO energy levels, relative band gap was determined as −5.69 eV and −2.93 eV; from empirical equations [36,37] given below:

$$E_{LUMO} = -[E_{onset}^{Red} - 0.49 + 4.8]eV \text{ and } E_{HOMO} = -[E_{onset}^{ox} - 0.49 + 4.8]eV$$

$$Band\ Gap = E_{LUMO} - E_{HOMO}$$

Therefore, the obtained bandgaps from the electrochemical method (~2.76 eV, within the range of semiconducting materials) matched well with the optical band gap (2.75 eV) determined by the diffuse reflectance analysis (DRS). Cyclic voltammetry was executed by scan rate dependent study to gain an insight on the mechanistic pathway. Herein, the anodic peak potential was 1.06 V; fitted ($R^2$ being 0.998) primarily with the square root of the scan rate, this reveals that BHEPDQ molecular system is distinctly diffusion-controlled.

## 13.2.4 FLEXIBLE AL/BHEPDQ/ITO DEVICE WITH NON-VOLATILE RESISTIVE SWITCHING MEMORY PROPERTY

A voltage variation among −3 V to +3 V showed the RS conduct in the BHEPDQ device. Figure 13.1 presents the fabricated device. An indigent hysteresis was noted at positive and negative boundaries when the applied voltage varied from 0 V → +3 V → 0 V → −3 V → 0 V; this discloses the device as bipolar. When the voltage was applied from 0 → 3 V, at first the device maintained a high resistance state (HRS).

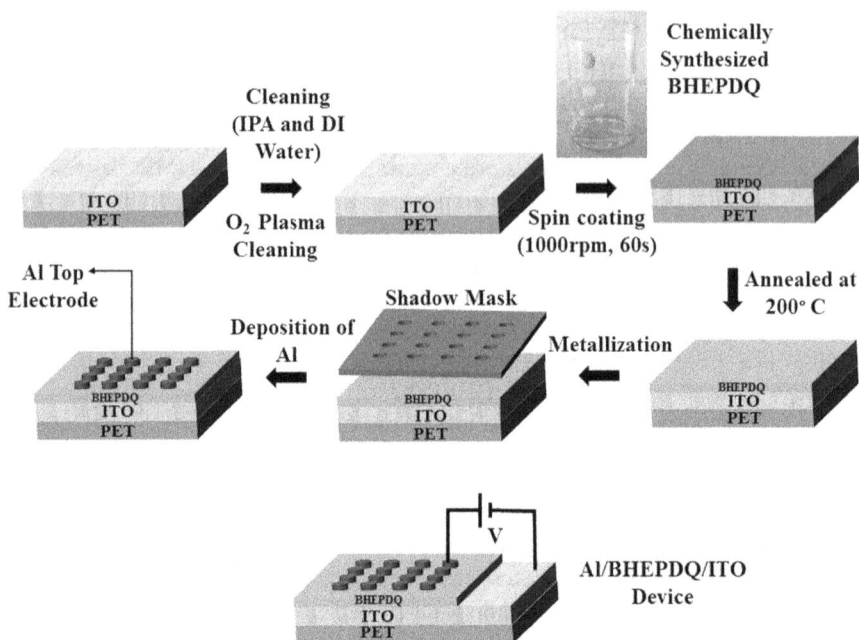

**FIGURE 13.1** Construction of non-volatile resistive switching memory BHEPDQ device. BHEPDQ solution spin-casted on the ITO-PET (plasma cleaned) substrate followed by annealing at 150°C. Top electrodes (Al) were placed by E-Beam evaporator, besides shadow mask was used to attain relevant pattern. *"Adapted with permission from* [34]. *Copyright (2022) RSC Publishers.*

SET process was indicated by an increase in the current at ~ 2.26 V and RESET process was noticed when concerned voltage sweep was from 0 to ~–2.88 V; during this, the device switched midst LRS to HRS. Accordingly, +2.26 V/– 2.88 V, were confirmed as the SET/RESET voltages; distinctly illustrate the RS attributed for BHEPDQ. An insignificant variation in the RS property (especially at 10 and 30 cycles) produced with undeviating swap in the SET and RESET values each, respectively, contrast through original gadgets, recommend outrageous reliability of the constructed devices. Furthermore, attained voltages were indeed low at the SET as well as RESET processess, collated to further organic molecular materials, utilized for RS [38,39]. Consequently, the described device from BHEPDQ is appropriate for low-power organic electronics.

To verify the potential application of the fabricated BHEPDQ device having Al has the top electrode and ITO as the bottom electrode; for resilient and apparel electronics; bending tests were performed. RS behavior of the above said device was explored [40] by studying the outcome of convex and concave i.e. upward and downward bends. Figures 13.2a,b manifest the bend under the influence of an upward applied force, i.e. convex bend. Figure 13.2c portrays the I-V aspects of the same beneath convex bend. A part from this, impact of downward bending force

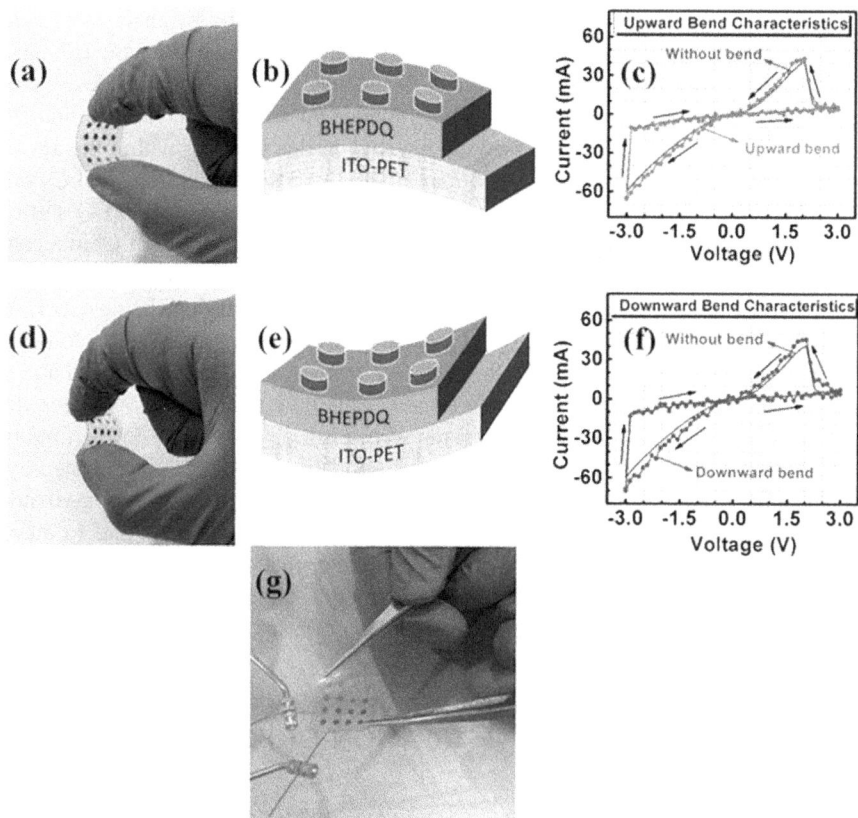

**FIGURE 13.2** (a) Upward bending; (b) schematic presentation of the upward bend; (c) I-V properties of upward bend; (d) demonstration of downward bend; (e) representation of downward bending; (f) I-V plot of downward bend. *Adapted with permission from* [34]. *Copyright (2022) RSC Publishers.*

escorted concave bending (Figure 13.2d and e). However, devices preserved nearly similar I-V characteristics even after the corresponding bends. In addition, the SET, RESET voltages were established as ~2.63V and ~−2.88V each respectively for the convex and concave bends, that are the same for devices with no stress. Due to good electrical execution in spite of varied mechanical stress; the devices outspread their resilience for low power flexible devices. Figure 13.2f displays the device examined on top as well as bottom electrodes succeeding bending. A bending diameter of 5 mm was observed as a captious flexed state. LRS along with the HRS was considered in the current-time retention characteristic features beneath the sequel of mechanical stress under convex and concave bends, the retention noted up to $10^4$ s; found to be stable across the evaluation task. To be noted, constant switching ratio maintained in the course of the study divulged its pertinence as a non-volatile switching memory device. Also, mechanical stress did not have any considerable effect on the electrical performance.

In contemplation of exploring the switching as well as the mechanism in conduction process of the device; achieved I–V properties were shaped for the HRS and LRS sections in log-log scale under + ve and – ve bias. Originally the RS device was found to be in HRS mode when the voltage applied across was increased from 0 V. Thus, a higher electric field increased the injected carrier numbers within the device layer. During this first part of the curve in HRS (+ve realm) stipulated SCLC conduction action with slope $\alpha$ as $\geq 2$. At the SET voltage +2.26 V, BHEPDQ shifted from HRS to LRS. A decrease in voltage from 3 V to 0 V, manifested Ohmic conduction and on further reducing the voltage to −2.90 V from 0 V, sustained LRS, yet the charge transport was directed by the Ohmic conduction with slope being $\alpha$ as 1. At RESET voltage −2.88 V, BHEPDQ device transposed its state through LRS to HRS, during swept back of the applied voltage to 0 V from – 3 V. The SCLC mechanism was accountable considering charge transport. This investigation validates that the Ohmic conduction and SCLC mechanisms were primarily dominant in the time of HRS and LRS states each respectively. A transport model showcasing the step-by-step detailed band diagram was suggested for the fabricated (BHEPDQ) non-volatile RS memory device in order to inspect the performance at interface and to attain greater apprehension on acquired resistive switching mechanism besides electrical properties [34]. Further, within the proposed model, aluminium ions and oxygen vacancies were initiated for the resistive switching effects over the devices [40].

## 13.3 PHOTOPHYSICAL PROPERTY, ELECTROCHEMICAL STUDY AND SOLAR CELL INVESTIGATION OF BCCPDQ

### 13.3.1 PHOTOPHYSICAL PROPERTY

BCCPDQ was synthesized under similar lines of BHEPDQ [34]. 0.316 g of cyclopropyl carbonyl piperazine added to a hot solution of TCNQ (0.2 g) dissolved in acetonitrile; under $N_2$ atmosphere at 75°C. Pyridine (Py) was added in a catalytic amount at 75°C, after nearly 10 min, simultaneously temperature was allowed to raise from ~100–105°C for nearly 2 h then temperature was raised ~120°C for 1 h. Reaction mixture permitted towards cooling ~30°C; kept standing for 2 days in a $N_2$ atmosphere. Crude compound attained was filtered and washed with acetonitrile (cold); further solvated in DMF and taken out with diethyl ether. Finally, BCCPDQ could be crystallized from the DMF. M.P. 245–250°C.

BCCPDQ pertained to a triclinic system (*P-1* space group). Asymmetric unit had three entities i.e. a complete BCCPDQ, one DMF and four $H_2O$ molecules [41]. Molecular structure of BCCPDQ indicated (C1–C6) benzenoid character of the central ring (average bond C-C bond distance ~1.3895 Å). Average torsional angle was 45.46°; rather less than BBPDQ (~48°, with phenyl carbonyl as substituent) [17]; plausible owing to the pliable cyclopropyl carbonyl substituent. Substantial short contacts/H-bonding network was observed since water and DMF molecules were sandwiched among the BCCPDQ dipoles. Remarkable non-covalent interactions (Å) and H-bonds extend from weak, modest into strong; exemplified orientations among the BCCPDQs. Therefore, the progressing reverberations of diligent intermolecular interactions followed by DMF besides water molecules together with carbonyl functionality has presumably navigated intensification of fluorescence, besides demonstration of solar cell attribute, not observed yet among any TCNQ derivatives.

Though solvatochromism was not significantly observed among various solutions; nevertheless, a little blue shift was observed in MeCN, MeOH, EtOH, DMF and DMSO with reference to *i*-propyl alcohol (IPA) ($\lambda_{max,abs}$; 442 nm). A wavelength of maximum emission in solvents considered in this study was ~ 529 nm. A larger dipole moment of the ground state could be noticed due to the comparative absorption spectral shifts, than emission spectra. Fluorescence emission inflated ~$14 \times 10^4$ in IPA having a viscosity ($\eta$ = 2.04 cP); then in DMSO (~$13 \times 10^4$, $\eta$ = 1.99) followed by EtOH (~$9 \times 10^4$, $\eta$ = 1.07); DMF (~$7 \times 10^4$, $\eta$ = 0.79); MeOH (~$5 \times 10^4$, $\eta$ = 0.54); MeCN (~$2.3 \times 10^4$, $\eta$ = 0.37). This study perhaps suggests the contribution of viscosity; confirmed by a controlled experiment done in IPA at different temperatures ranging from 0°C to 15°C with 5°C raise. Fluorescence emission reduced simultaneously as the viscosity decreased. Moreover, accountability of viscosity also established performing experiments in a mixture of glycerol-water; herein enhancement in the fluorescence emission was observed when glycerol quantity was increased. A subsided emission intensity in acetonitrile, methanol and DMF contrast to IPA, DMSO and EtOH is assigned due to $S_1/T_1$ vibronic modes overlap with $S_0$ [42]. Absorption in solid BCCPDQ was at a higher wavelength and quite broad characteristic. However, an explicit ~fourfold enhanced fluorescence emission in solids was observed when compared to solutions; probably ascribed due to the subduing relaxation of vertical excited state to a non-emitting one. Further increased solid state emission is attributed to H-bonds guiding supramolecular assemblies, besides aggregation phenomenon. Also, aggregation was supported by adding 0 to 90% water fraction ($f_w$) of Millipore quality having 18 MΩ resistance; to BCCPDQ DMSO solution (10 ppm). Fluorescence emission strikingly lessened amid 20% $f_w$, evocative of quenching caused due to aggregation. Notably, in DMSO and IPA quantum yield was ~ 0.31% and for BCCPDQ solid quantum yield was ~ 10%. BCCPDQ manifested yellow in ambient light and greenish emission beneath 365 nm (along UV light). This study conveys that BCCPDQ is one among a few of the small organic molecules as solid-state emitters, while design of such small organic molecules still continues to be a challenge.

## 13.3.2  CYCLIC VOLTAMMETRY

Cyclic voltammetry (CV) study was performed on 10 μM BCCPDQ solution at the scan rate of 25 mVs$^{-1}$. Investigation regulated at various scan rates commencing from 10 mV/s to 100 mV/s. Owing to consecutive reduction processes (three) of dicyano methylene moiety to their relevant radical anions, three quasi-reversible reduction potentials were perceived. Apart from reduction, irreversible oxidation peaks (three) were disclosed $E^{ox}$ at 0.67 V, 1.06 V and 1.377 V, attributed to the appearance of CCP (donor) in the BCCPDQ zwitterion. The oxidation process was ascribed to one nitrogen linked to the CO group and two nitrogen of piperazine near to the $\delta^+$ carbon. Further, these oxidation peaks are feasibly assigned due to the formation of corresponding radical cations i.e. neutral BCCPDQ to BCCPDQ radical cation and further to BCCPDQ$^{\cdot 2+}$ in the nitrogen centers. Third, peak due to oxidation noticed at the higher potential is possibly ascribed for neutral BCCPDQ to BCCPDQ radical cation oxidation at the nitrogen centers attached to carbonyl functionality, because of lesser electron density, contrast to further nitrogen centers. Selecting respective extreme data points of the oxidation and reduction onsets, HOMO and LUMO energy of the BCCPDQ were acquired to get a greater instinct on energy levels besides band formation.

HOMO and LUMO energy levels were intended, utilizing the equation discussed sec. 13.2.3 in BHEPDQ [34], correspondingly HOMO and LUMO energy levels were obtained as −5.687 eV and −3.017 eV for each respectively. The electrochemical bandgap (∼2.67 eV, range of semiconducting materials) matched well with the optical bandgap (∼2.68). The scan rate dependent cyclic voltammetry study was done to gain mechanistic insight regarding BCCPDQ's electrochemical responses. To verify the mechanistic pathway of the redox phenomenon, anodic peak current densities were raised followed by gradual increase of the scan rate; further the anodic peak potential vs. scan rate and the anodic peak potential versus square root of scan rate were plotted, linearly fitted individually. Accordingly, the anodic peak potential at 1.06 V fitted best with the scan rate ($R^2$ = 0.9962) rather than square root of scan rate ($R^2$ = 0.9870); this specifies the BCCPDQ redox system as electron transferred-controlled.

### 13.3.3 SCANNING ELECTRON MICROSCOPY AND THERMOGRAVIMETRIC ANALYSIS

BCCPDQ was stable to ∼250°C. Decomposition *via* single step emerged in weight loss of 67.46% with a corresponding mass of 308.9 gm from overall mass ∼458 gm. At first, two 1-(cyclopropyl carbonyl) piperazine moieties were found to undergo decomposition. After the entire decomposition process, the final mass remained was 32.54% (149 gm) and identified to be TCNQ possessing the dicyanomethylene moiety, without amine substituents.

One mmol drop casted films obtained from acetonitrile and methanol solutions of BCCPDQ distinctly showed close packed, uniformly dispersed spherical characteristics with 1.71 µm and 1.43 µm average particle diameter, respectively. Spherical features with dense packing were perceived in acetonitrile than in methanol. On the other hand, drop casted films through ethanol and IPA exhibited both isolated/spherical and non-uniform irregular feature with 0.68 µm and 0.75 µm diameter of the particles in each discretely [41]. Contrast morphological attributes are assigned perhaps owing to different rates of the solvents evaporation such as acetonitrile, methanol and ethanol, emanating in diversified growth kinetics [12]. Spherical features in thin films coated from acetonitrile were observed among amorphous nanoparticles of certain DADQs possessing aromatic amine substituents having conformationally versatile bonds [43]; nevertheless, herein, BCCPDQ solutions, with BCCPDQ possessing simple aliphatic amine as substituent showed spherical features.

### 13.3.4  SOLAR CELL PROPERTY OF **BCCPDQ**

#### 13.3.4.1  UV-Vis Absorption and Fluorescence Emission in BCCPDQ Film

The wavelength of maximum absorption corresponds to 390 nm in pure BCCPDQ films. In the tandem film a broader absorption feature was observed and the peak maximum was noticed to be red shifted (415 nm), probably due to the enhanced intermolecular π-π stacking. Fluorescence spectroscopy was done in order to delineate the entrapment of the carriers, besides transport phenomena of the photoluminescence. Emission intensity of $TiO_2$ doped BCCPDQ was lower than basic BCCPDQ, divulges existence of either sub-bands or surface irregularities and advantageous in the separation of the charge carriers to stimulate the carry away of excitons.

#### 13.3.4.2  Photovoltaic Characteristics

For testing the photovoltaic property, in the first instance a BCCPDQ solar cell was constructed by a typical structure coated with Al as the top and ITO as the bottom electrode. I-V measurements accomplished over the device under standard conditions i.e. 1 Sun and 1.5 G air mass irradiation. A BCCPDQ device showed a short circuit current density ($J_{SC}$) and an open circuit voltage ($V_{OC}$) of 2.56 mA cm$^{-2}$ and 0.50 V, respectively, relative to −1 to 1 V sweep voltage; 0.25 V was the maximum voltage ($V_M$) and 1.445 mA/cm$^2$, maximum current density ($J_M$). The fill factor (FF) value was 29.4%. The ratio of total output power to the input irradiated (power conversion efficiency) power of the designed cell is identified by expression (13.1):

$$\text{Efficiency } \eta = \frac{V_{OC} * J_{sc} * FF}{P_{in}} \qquad (13.1)$$

Efficiency of the ordinary self-sustained BCCPDQ device (1 cm$^2$/unit cell dimension) was attained as 0.40%. The little η value accredited was due to the cumulated electric field that led to the recombination of light induced charge carriers generated in active layer. Hence, to thwart rigorous recombination rate within the device, $TiO_2$ nanoparticles were incorporated, which act as an electron transport layer (ETL) across BCCPDQ and ITO-coated PET layer (Figure 13.3).

An ETL not only takes part in the extraction and transit of light induced charge carriers from the operative BCCPDQ layer to anode/cathode [44] but also transforms the confluence effects among the photogenerated layer besides the electrode, therefore minimize the rate of recombination. Moreover, $TiO_2$ NPs boost the junction possessions and intensify diffusion of charges by clearing the way for directive transit of the charges [44,45].

The $TiO_2$ NPs implanted PV device cell manifested $J_{SC}$, $V_{OC}$ as 6.72 mA cm$^{-2}$ and 1 V; 0.54 V and 3.68 mA/cm$^2$ as $V_M$ and $J_M$ values individually. Interestingly the FF value was 33.6% and efficiency of 2.26% could be achieved. This is ~tenfold elevated in comparison to the self-sustained BCCPDQ device. The enhancement in η value is perhaps attributed to the modification of the juncture among BCCPDQ/$TiO_2$ NPs, resulting in the adjustment of bandgap connecting both the layers. Aforementioned attenuates sequel in instantaneous migration of charge over

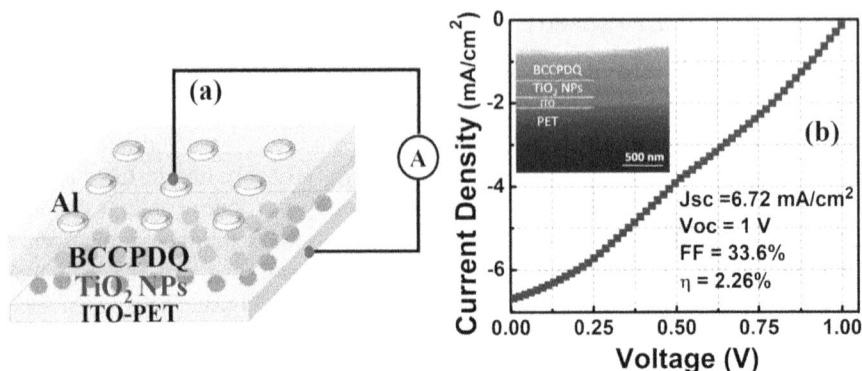

**FIGURE 13.3** (a) $TiO_2$ NPs implemented on the fabricated Al/BCCPDQ/$TiO_2$ NPs/ITO photovoltaic device. (b) J-V attributes attained under 1 Sun and 1.5 G air mass; inset depicts micro-image of cross-sectional BCCPDQ device with ~430 nm thickness and the coated $TiO_2$ NPs ~ 215 nm. *Adopted from Ref.* [41].

juncture. Reduction of series resistance ($R_s$) leads to the increase in the FF value. Consequently, BCCPDQ may be easily considered in the OSC employment.

In the self-sustained BCCPDQ device, a BCCPDQ layer is sandwiched among the Al and bottom electrode, a feeble electric field is originated and low $V_{OC}$ and $J_{SC}$ are produced due to the internal photo-induced excitons, since these experience prompt recombination as they jump towards the electrodes. Contrastingly, the energy levels among valence band and the conduction band of the BCCPDQ and $TiO_2$ NPs eased the segregation of charges at the interface. Considering coherent collection of charges, bottom electrode positioned dynamically along with junction in the formation of ohmic contact. In the process of illumination, excitons generated in photoactive BCCPDQ surface and experienced diffusion at the interface under a reverse bias, so as to undergo dissociation by the internal electric field. Consequently, peripheries of bands capitulated and their bending facilitated the carrier separation activity. Eventually, $TiO_2$ NPs offered a well-ordered channel for charge extraction besides electron transportation towards the electrode regulating recombination estimate of electrons at the interface of the fabricated BCCPDQ/$TiO_2$ NPs device. Furthermore, the inflated Fermi level offset at the top electrode/BCCPDQ junction gave rise to a Schottky barrier hastening hole extraction through the photoactive layer by simultaneously limiting and combination around the juncture. As a consequence, higher photovoltaic performance was noticed in the fabricated device compared to the standalone BCCPDQ device [41].

### 13.3.4.3 Stability Study

The impact of tensile and compressive bending over photovoltaic accomplishment of Al/BCCPDQ/$TiO_2$ NPs/ITO-PET device has been examined, since the effectiveness of the pliable PV device was established to be dependent on the robustness of the electrical effects in the process of varied bending states. This study establishes the device as sturdy and definitive; the two play a crucial factor for rotation processing in the low-power apparel organic electronics. During the maximum bending, little increase in the current density was noted, possibly owing to the rapid elevation in the resistance of the substrate; nevertheless, the J-V attributes were noticed as same when the device was set to the usual position.

Steadiness of the fabricated OSC was investigated for $\sim$100 hr. Tests were conducted beneath 100 mW/cm$^2$ (1 Sun) illumination and 1.5 G air mass. Amidst the starting phase i.e. < 20 hr amicable PCE was displayed by the cell, additionally with impressive stability and no photo degradation. Additionally, scarcely variation in the PCE was noted due to the outside climate ($\sim$25 to 30°C). Declining of the heterojunction resulted soon, compared to the fundamental BCCPDQ cell, possibly due to continued exposure of the device regarding light and also owing to the transient effects produced by heat. Nonetheless, the degradation percentage was serene less and interestingly, in spite of studying for 100 hr, the designed/fabricated cell manifested magnificent 94% of the primary efficiency.

## 13.4 OVERVIEW

Employment of ethylene spacer and polar group possessing heterocyclic aliphatic amine led to the development of BHEPDQ with enhanced fluorescence in solids ($\Phi_f = \sim$10%) than solutions ($\Phi_f = \sim$0.30%). Contrasting morphology was noticed among different drop casted films, indicative of varied solvent interaction as well as different growth kinetics. However, film obtained from a DMF solution exhibited uniform morphological characteristics, thus examined for the device pertinence. Remarkably, the devices underwent 5 mm bending diameter with unaltered electrical accomplishment. The acquired results establish BHEPDQ's effectiveness towards practical flexible electronic applications. The current work shows a direction towards the development of small organic molecule based advanced flexible memory devices for data retentions and synaptic execution for flexible neuromorphic/neural network solicitation.

On the other hand, usage of 1-(cyclopropylcarbonyl)piperazine gave thermally stable ($\sim$250°C) BCCPDQ fluorophore. BCCPDQ manifested supramolecular assemblies steered by several weak intermolecular interactions between BCCPDQ dipoles, water and DMF. The OSC BCCPDQ device, with TiO$_2$ NPs, showed a magnificent stable property under tensile and compressive stress, exemplifying 2.26% PCE; $\sim$ tenfold enhancement than self-supported gadget. Electrical findings were demonstrated correspondingly; for with and without bending, it showed minimum disparity and revealed device desirability for flexible electro-optic entreaty. Cyclic voltammetry disclosed diffusion phenomenon in BHEPDQ and an electron transfer mechanistic path in BCCPDQ. Contemporary work demonstrates inception regarding design and progress of solution prepared flexible organic memristors and solar cell appliances, specifically based on TCNQ derivatives to fulfill the demand and gap.

## ACKNOWLEDGMENTS

SJ thanks all the contributors and collaborators in this work. SJ also thanks the Department of Science and Technology (DST), Govt. India under the project

(DST-EMR/2016/002209), Central Analytical Laboratory (CAL) facility by the BITS-Pilani, Hyderabad campus and DST-FIST grant facility sanctioned to the Department of Chemistry, BITS-Pilani Hyderabad Campus.

## REFERENCES

[1] Alves, H.; Molinari, A.S.; Xie, H.; Morpurgo, A.F. Metallic Conduction at Organic Charge-Transfer Interfaces. *Nat. Mater.* 2008, *7* (7), 574–580.

[2] Goetz, K.P.; Vermeulen, D.; Payne, M.E.; Kloc, C.; McNeil, L.E.; Jurchescu, O.D. Charge-Transfer Complexes: New Perspectives on an Old Class of Compounds. *J. Mater. Chem. C* 2014, *2* (17), 3065–3076.

[3] Qu, X.; Lu, J.; Zhao, C.; Boas, J.F.; Moubaraki, B.; Murray, K.S.; Siriwardana, A.; Bond, A.M.; Martin, L.L. (Pro2H+)2(TCNQ.–)2·TCNQ: An Amino Acid Derived Semiconductor. *Angew. Chemie Int. Ed.* 2011, *50* (7), 1589–1592.

[4] Senthilnathan, N.; Radhakrishnan, T.P. Molecular Hopper Crystals and Electron Beam-Triggered Reversible Actuation. *Chem. Mater.* 2020, *32* (19), 8567–8575.

[5] Szablewski, M.; Thomas, P.R.; Thornton, A.; Bloor, D.; Cross, G.H.; Cole, J.M.; Howard, J.A.K.; Malagoli, M.; Meyers, F.; Brédas, J.-L.; Wenseleers, W.; Goovaerts, E. Highly Dipolar, Optically Nonlinear Adducts of Tetracyano-p-Quinodimethane: Synthesis, Physical Characterization, and Theoretical Aspects. *J. Am. Chem. Soc.* 1997, *119*, 3144–3154.

[6] Patra, A.; Venkatram, N.; Rao, D.N.; Radhakrishnan, T.P. Optical Limiting in Organic Molecular Nano/Microcrystals: Nonlinear Optical Effects Dependent on Size Distribution. *J. Phys. Chem. C* 2008, *112* (42), 16269–16274.

[7] Sudhakar, P.; Radhakrishnan, T.P. A Strongly Fluorescent Molecular Material Responsive to Physical/Chemical Stimuli and Their Coupled Impact. *Chem.: Asian. J.* 2019, *14*, 4754–4759.

[8] Ravi, M.; Cohen, S.; Agranat, I.; Radhakrishnan, T.P. Molecular and Crystal Structures of a Class of Push-Pull Quinonoid Compounds with Potential Nonlinear Optical Applications. *Struct. Chem.* 1996, *7*, 225–232.

[9] Gangopadhyay, P.; Sharma, S.; Rao, A.J.; Rao, D.N.; Cohen, S.; Agranat, I.; Radhakrishnan, T.P. Optical Second Harmonic Generation in Achiral Bis(n-Alkylamino)Dicyanoquinodimethanes: Alkyl Chain Length as the Design Element. *Chem. Mater.* 1999, *11* (2), 466–472.

[10] Hertler, W.R.; Hartzler, H.D.; Acker, D.S.; Benson, R.E. Substituted Quinodimethans. III. Displacement Reactions of 7,7,8,8-Tetracyanoquinodimethan. *J. Am. Chem. Soc.* 1962, *84* (17), 3387–3393.

[11] Boyineni, A.; Jayanty, S. Supramolecular Helical Self-Assemblies and Large Stokes Shift in 1-(2-Cyanophenyl)Piperazine and 4-Piperidinopiperidine Bis-Substituted Tetracyanoquinodimethane Fluorophores. *Dyes Pigmt.* 2014, *101*, 303–311.

[12] Szablewski, M.; Bloor, D.; Kagawa, Y.; Mosurkal, R.; Cole, J.M.; Clark, S.J.; Cross, G.H.; Pålsson, L.O. Matrix Dependence of Blue Light Emission from a Novel NH$_2$-Functionalized Dicyanoquinodimethane Derivative. *J. Phys. Org. Chem.* 2006, *19*, 206–213.

[13] Srujana, P.; Gera, T.; Radhakrishnan, T.P. Fluorescence Enhancement in Crystals Tuned by a Molecular Torsion Angle: A Model to Analyze Structural Impact. *J. Mater. Chem. C* 2016, *4*, 6510–6515.

[14] Raghavaiah, P.; Kuladeep, R.; Narayana Rao, D.; Jyothi Lakshmi, A.; Srujana, P.; Subbalakshmi, J. Bis -(1-(2-Aminoethyl)Piperidino), (2-(2-Aminoethyl)Pyridino) and (1-(2-Aminoethyl)Pyrrolidino)-Substituted Dicyanoquinodimethanes: Consequences of Flexible Ethylene Spacers with Heterocyclic Moieties and Amine Functionalities. *Acta Crystallogr. Sect. B Struct. Sci. Cryst. Eng. Mater.* 2016, *72*, 709–715.

[15] Patra, A.; Radhakrishnan, T.P. Molecular Materials with Contrasting Optical Responses from a Single-Pot Reaction and Fluorescence Switching in a Carbon Acid. *Chem. Eur. J.* 2009, *15*, 2792–2800.

[16] Srujana, P.; Sudhakar, P.; Radhakrishnan, T.P. Enhancement of Fluorescence Efficiency from Molecules to Materials and the Critical Role of Molecular Assembly. *J. Mater. Chem. C* 2018, *6*, 9314–9329.

[17] Sudhakar, P.; Radhakrishnan, T.P. Stimuli Responsive and Reversible Crystalline–Amorphous Transformation in a Molecular Solid: Fluorescence Switching and Enhanced Phosphorescence in the Amorphous State. *J. Mater. Chem. C* 2019, *7*, 7083–7089.

[18] Radhakrishnan, T.P. Molecular Structure, Symmetry, and Shape as Design Elements in the Fabrication of Molecular Crystals for Second Harmonic Generation and the Role of Molecules-in-Materials. *Acc. Chem. Res.* 2008, *41* (3), 367–376.

[19] Srujana, P.; Radhakrishnan, T.P. Establishing the Critical Role of Oriented Aggregation in Molecular Solid State Fluorescence Enhancement. *Chem. Eur. J.* 2018, *24*, 1784–1788.

[20] Jayanty, S.; Radhakrishnan, T.P. Enhanced Fluorescence of Diaminodicyano quinodimethanes in the solid state and Fluorescence Switching in a Doped Polymer by Solvent Vapors. *Chem. Eur. J.* 2004, *10*, 791–797.

[21] Scott, J.L.; Yamada, T.; Tanaka, K. Guest Specific Solid-State Fluorescence Rationalised by Reference to Solid-State Structures and Specific Intermolecular Interactions. *New J. Chem.* 2004, *28*, 447–450.

[22] Senthilnathan, N.; Chandaluri, C.G.; Radhakrishnan, T.P. Efficient Bioimaging with Diaminodicyanoquinodimethanes: Selective Imaging of Epidermal and Stomatal Cells and Insight into the Molecular Level Interactions. *Sci. Rep.* 2017, *7* (1), 10583–10594.

[23] Senthilnathan, N.; Gaurav, K.; Venkata Ramana, C.; Radhakrishnan, T.P. Zwitterionic Small Molecule Based Fluorophores for Efficient and Selective Imaging of Bacterial Endospores. *J. Mater. Chem. B* 2020, *8* (21), 4601–4608.

[24] Anwarhussaini, S.D.; Mishra, S.; Subbalakshmi J.N. N-(diethylethylenediamino) 8,8-dicyanoquinodimethane: Effect of ethyl moiety on photophysical property besides thermal stability *J. Fluoresc.* 2022, *32* (1), 115–124.

[25] Chandaluri, C.G.; Radhakrishnan, T.P. Hierarchical Assembly of a Molecular Material through the Amorphous Phase and the Evolution of Its Fluorescence Emission. *J. Mater. Chem. C* 2013, *1*, 4464–4471.

[26] Prime, D.; Paul, S.; Josephs-Franks, P.W. Gold Nanoparticle Charge Trapping and Relation to Organic Polymer Memory Devices. *Phil. Trans. R. Soc. A*, 2009, *367*, 4215–4225.

[27] Meena, J.S.; Sze, S.M.; Chand, U.; Tseng, T.Y. Overview of Emerging Nonvolatile Memory Technologies. *Nanoscale Res. Lett.* 2014, *9*, 526–559.

[28] Nian, L.; Gao, K.; Jiang, Y.; Rong, Q.; Hu, X.; Yuan, D.; Liu, F.; Peng, X.; Russell, T.P.; Zhou, G. Small-Molecule Solar Cells with Simultaneously Enhanced Short-Circuit Current and Fill Factor to Achieve 11% Efficiency. *Adv. Mater.* 2017, *29* (29), 1700616–1700622.

[29] Lin, Y.; Firdaus, Y.; Isikgor, F.H.; Nugraha, M.I.; Yengel, E.; Harrison, G.T.; Hallani, R.; El-Labban, A.; Faber, H.; Ma, C.; Zheng, X.; Subbiah, A.; Howells, C.T.; Bakr, O.M.; McCulloch, I.; Wolf, S. De; Tsetseris, L.; Anthopoulos, T.D. Self-Assembled Monolayer Enables Hole Transport Layer-Free Organic Solar Cells with 18% Efficiency and Improved Operational Stability. *ACS Energy Lett.* 2020, *5* (9), 2935–2944.

[30] Yu, R.; Yao, H.; Hong, L.; Gao, M.; Ye, L.; Hou, J. TCNQ as a Volatilizable Morphology Modulator Enables Enhanced Performance in Non-Fullerene Organic Solar Cells. *J. Mater. Chem. C* 2020, *8* (1), 44–49.

[31] Guillain, F.; Endres, J.; Bourgeois, L.; Kahn, A.; Vignau, L.; Wantz, G. Solution-Processed p-Dopant as Interlayer in Polymer Solar Cells. *ACS Appl. Mater. Interfaces* 2016, *8* (14), 9262–9267.

[32] Fukuda, K.; Yu, K.; Someya, T. The Future of Flexible Organic Solar Cells. *Adv. Energy Mater.* 2020, *10* (25), 2000765–2000774.

[33] Lei, T.; Peng, R.; Song, W.; Hong, L.; Huang, J.; Fei, N.; Ge, Z. Bendable and Flexible Organic Solar Cells Based on Ag Nanowire Films with 10.30% Efficiency. *J. Mater. Chem. A*, 2019, *7*, 3737–3744.

[34] Anwarhussaini, S.D.; Himabindu, B.; Pavan Kumar, B.; Souvik, K.; Chakraborty, C.; Subbalakshmi, J. Photophysical, Electrochemical and Flexible Organic Resistive Switching Memory Device Application of a Small Molecule: 7,7-Bis(Hydroxyethyl-piperazino) Dicyanoquinodimethane. *Org. Electron.* 2020, *76*, 105457–105468.

[35] Perepichka, D.F.; Bryce, M.R.; Pearson, C.; Petty, M.C.; McInnes, E.J.L.; Zhao, J.P. A Covalent Tetrathiafulvalene–Tetracyanoquinodimethane Diad: Extremely Low HOMO–LUMO Gap, Thermoexcited Electron Transfer, and High-Quality Langmuir–Blodgett Films. *Angew. Chem. Int.* 2003, *42*, 4636–4639.

[36] Keshtov, M.L.; Kuklin, S.A.; Radychev, N.A.; Nikolaev, A.Y.; Ostapov, I.E.; Krayushkin, M.M.; Konstantinov, I.O.; Koukaras, E.N.; Sharma, A.; Sharma, G.D. New Low Bandgap Near-IR Conjugated D–A Copolymers for BHJ Polymer Solar Cell Applications. *Phys. Chem. Chem. Phys.* 2016, *18*, 8389–8400.

[37] Chakraborty, C.; Layek, A.; Ray, P.P.; Malik, S. Star-Shaped Polyfluorene: Design, Synthesis, Characterization and Application towards Solar Cells. *Eur. Polym. J.* 2014, *52*, 181–192.

[38] Sun, B.; Zhang, X.; Zhou, G.; Li, P.; Zhang, Y.; Wang, H.; Xia, Y.; Zhao, Y. An Organic Nonvolatile Resistive Switching Memory Device Fabricated with Natural Pectin from Fruit Peel. *Org. Electron.* 2017, *42*,181–186.

[39] Islam, S.M.; Banerji, P.; Banerjee, S.; Electrical Bi-stability, Negative Differential Resistance and Carrier Transport in Flexible Organic Memory Device Based on Polymer Bilayer Structure. *Org. Electron.* 2014, *15*,144–149.

[40] Won, S.; Lee, S.Y.; Park, J.; Seo, H. Forming-less and non-Volatile resistive switching in WOX by oxygen vacancy control at interfaces, *Sci. Rep.* 2017, *7*, 10186–10193.

[41] Anwarhussaini, S.D.; Renuka, H.; Anuradha, M.; Raghavaiah, P.; Mahadev Sai Karthik, C.; Sanket, G.; Subbalakshmi, J. Photophysical, electrochemical properties and flexible organic solar cell application of 7,7-bis(1-cyclopropyl carbonyl piperazino)-8,8 dicyanoquinodimethane, *Materials Advances* 2022, *3*, 3151–3164.

[42] J.R. Lakowicz, *Principles of Fluorescence Spectroscopy*, Springer US, Boston, 2006.

[43] Chandaluri, C.G.; Radhakrishnan, T.P. Amorphous-to-Crystalline Transformation with Fluorescence Enhancement and Switching of Molecular Nanoparticles Fixed in a Polymer Thin Film. *Angew. Chemie Int. Ed.* 2012, *51*, 11849–11852.

[44] Shahvaranfard, F.; Altomare, M.; Hou, Y.; Hejazi, S.; Meng, W.; Osuagwu, B.; Li, N.; Brabec, C.J.; Schmuki, P. Engineering of the Electron Transport Layer/Perovskite Interface in Solar Cells Designed on TiO 2 Rutile Nanorods. *Adv. Funct. Mater.* 2020, *30* (10), 1909738–1909746.

[45] Zheng, S.; Wang, G.; Liu, T.; Lou, L.; Xiao, S.; Yang, S. Materials and Structures for the Electron Transport Layer of Efficient and Stable Perovskite Solar Cells. *Sci. China Chem.* 2019, *62* (7), 800–809.

# 14 Hydrodynamic Performance of a Submerged Piezoelectric Wave Energy Converter Device in Real Sea Conditions

*Kshma Trivedi and Santanu Koley*

## CONTENTS

## 14.1 INTRODUCTION

In the recent decades, wave energy is an appealing source for scientists and engineers in view of fact that the ocean spans about 70% of earth's surface. Further, ocean wave energy is among the most effective and significant renewables in view of its enormous potential for capturing energy from the waves. Numerous wave energy extracting technologies have been developed and classified as: (i) oscillating water column, (ii) two-body heaving systems, and (iii) overtopping converters. Regardless, those WECs have been powered by large, heavy generators such as hydraulic systems, spinning turbines, or linear turbines. To overcome these issues, piezoelectric devices are proposed that have unique features such as small size and high energy density. Further, piezoelectric materials are remarkable because of their capacity to produce electricity when distorted by the sensor effect (see [1]). [2] proposed the wave energy converters (WECs) consist of the number of piezoelectric linked cantilevers that are fastened to a floating buoy framework. This research led to the following results: (i) it is noticed that a buoy harvester with a lengthier,

DOI: 10.1201/b23359-14

narrower floater and a higher sinker is beneficial to extract more power by the PWEC device, and (ii) the productiveness of the energy harvester remarkably increases with an increase in the buoy harvester's sinker's cross-section area for certain range, whereas no significant effect is observed in the efficiency beyond the certain range of buoy harvester's sinker's cross-section area. [3] examined the combined hydro-electromechanical characteristics of a PWEC device. It was depicted that the wave period was shown to have a remarkable impact on the wave power extraction and dissipation caused by the piezoelectric effects. [4] performed the experimental setup to investigate the performance of the PWEC device employing sway motion. The proposed research is useful in the configuration of multipurpose PWEC device for low frequency vibration, as well as in "sea-based" applications utilizing buoys and boats. [5] investigated the working mechanism of a painting type flexible PWEC device experimentally. It was found that the power extraction by the PWEC device is linearly proportional to the current velocity and wave steepness. Further, it was noticed that flexibility of PWEC device improves the serviceability of the device. [6] examined the power extraction by the PWEC device integrated with the coastal structures. It was concluded that wave pressure and voltage generated by the PWEC is significantly affected by the higher wave with longer period. The effect of structural configuration and wave parameters on the power extraction by the PWEC device is illustrated theoretically and experimentally by [7]. This article led to the following conclusions: (i) it was seen that the power extraction by the device significantly influenced by the vibrated amplitude for all of the painted flexible PWEC device, wave period, and wave steepness, and (ii) for the strong stiffness case, the wave power extraction by the PWEC device increases with an increase in deformation rate. [8] studied the working mechanism of PWEC device with the various stopper configuration utilizing the Galerkin approximation. It was found that the performance of the PWEC device is intrinsically enhanced by the stopper configuration. Recently, [9] employed a semi-analytical approach to examine the productivity of breakwater coupled PWEC device in the context of linear wave theory. In this article, following results are achieved: (i) it was reported that the device's performance is not significantly affected by breakwater width and submergence depth, and (ii) nature of the edge conditions of the plate is essential to improve the device's hydrodynamics. [10] carried an experiment to examine the working mechanism of a PWEC device includes a buoy, and a frequency up-conversion mechanism. It was shown that the frequency up-conversion mechanism plays a crucial role to transform low-frequency wave movement into mechanical movement with a sixfold higher frequency.

In all the aforementioned research, the efficiency and power extraction by a PWEC device are investigated in the influence of normal incident waves. Nonetheless, in reality, ocean waves are rather erratic. As a result, it is vital to examine the PWEC device's performance and power extraction in the involvement of irregular incoming waves.

The framework of this research article is listed as: Section 14.2 includes the mathematical formulation. The "boundary element method (BEM)"–based solution is explained in Section 14.3. Section 14.4 incorporates the results and outcomes. Eventually, in Section 14.5, the research findings are discussed.

## 14.2   MATHEMATICAL FORMULATION

The mathematical modeling of a PWEC device positioned over the uniform seabed in the involvement of irregular incident waves is described in this section. The illustration of the physical problem as well as the orientation of the coordinate axes are shown in Figure 14.1. A 2D coordinate system is utilized for modeling purposes. Here, $L$ represent the plate length that is partially submerged to the ocean with the submergence depth $d$. To solve the physical problem, the BEM is employed. In BEM, the computational domain should be closed. Here, the computational domain is categorized into two subdomains $R_1$ and $R_2$, respectively. To close the sub-domain $R_1$, two fictitious boundaries are located far away from the structure at $x = -l$ and $x = r$, respectively. These fictious boundaries are represented by $\Gamma_l$ and $\Gamma_r$, namely. Further, $\Gamma_b$ represents the bottom boundary. Moreover, free surfaces are denoted by $\Gamma_{f1}$ and $\Gamma_{f3}$, respectively. Further, $R_2 = \Gamma_{i2} \cup \Gamma_p \cup \Gamma_{i1} \cup \Gamma_{f2}$. Here, $\Gamma_{i1}$ and $\Gamma_{i2}$ are the fictitious boundaries, respectively. In addition, $\Gamma_p$ and $\Gamma_{f2}$ are termed plate boundary and free surface boundary, respectively. Furthermore, the fluid flow is "time harmonic" with $\omega$ "angular frequency". In this regard, the "velocity potential" $\Psi_j(x, z, t), j = 1, 2$ exists and will take the form $\Psi_j(x, z, t) = Re\{\varphi_j(x, z)e^{-i\omega t}\}$ (see [11–13]). The governing equation in this context is given as (see [14–16]):

$$\nabla^2 \varphi_j(x, z) = 0, j = 1, 2. \tag{14.1}$$

The boundary conditions (bcs) at $z = 0$ are given by

$$\frac{\partial \varphi_j}{\partial z} - K\varphi_j = 0, \text{ on } \Gamma_{f1} \cup \Gamma_{f2} \cup \Gamma_{f3}, \tag{14.2}$$

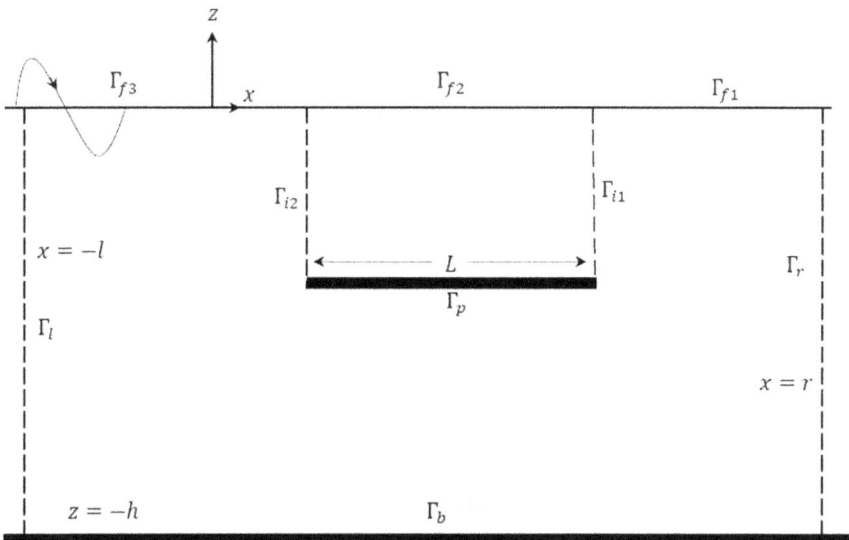

**FIGURE 14.1**   Illustration of the physical problem.

where $K = \omega^2/g$. Now, the bottom bcs at $z = -h$ is given by

$$\frac{\partial \varphi_1}{\partial z} = 0, \text{ on } \Gamma_b, \tag{14.3}$$

Further, the linearized kinematic bcs on at $z = -d$ is given as

$$\frac{\partial \varphi_1}{\partial z} = \frac{\partial \varphi_2}{\partial z} = -i\omega\zeta, \text{ on } \Gamma_p, \tag{14.4}$$

where $\zeta$ represents the plate deflection. Furthermore, the dynamic bcs on the plate is written as

$$g\eta\left(1 + \frac{\alpha^2 \xi \omega}{i + \xi \omega}\right)\frac{\partial^4 \zeta}{\partial x^4} - \omega^2 \gamma \zeta = i\omega(\varphi_1 - \varphi_2), \text{ on } \Gamma_p. \tag{14.5}$$

Here, $\eta = \frac{\grave{C}}{\rho g}$, $\alpha = \frac{\grave{\vartheta}}{\sqrt{\grave{C}B}}$, $\xi = \frac{\grave{B}}{\grave{L}}$, $\gamma = \frac{I'_b}{\rho}$, (see [9] for details). Now, the fixed edge bcs on the plate are given by

$$\zeta = 0, \frac{\partial \zeta}{\partial x} = 0 \text{ on the edges of } \Gamma_p. \tag{14.6}$$

Further, the pressure and normal velocities are continuous with the two fictitious boundaries $\Gamma_{i1}$ and $\Gamma_{i2}$ at $x = b$ and $x = L$, respectively, are written as

$$\varphi_1 = \varphi_2, \frac{\partial \varphi_1}{\partial x} = \frac{\partial \varphi_2}{\partial x} \text{ on } \Gamma_{i1} \cup \Gamma_{i2}. \tag{14.7}$$

Finally, the far-field bcs yields (see [17–20])

$$\begin{cases} \dfrac{\partial(\varphi_1 - \varphi^I)}{\partial x} - ik_0(\varphi_1 - \varphi^I) = 0, & \text{on } \Gamma_l, \\ \dfrac{\partial \varphi_1}{\partial x} - ik_0 \varphi_1 = 0, & \text{on } \Gamma_r. \end{cases} \tag{14.8}$$

Here, $\varphi^I = e^{ik_0 x} f_0(k_0, z)$ with $f_0(k_0, z) = \left(-\frac{igA}{\omega}\right)\frac{\cosh k_0(z + h)}{\cosh(k_0 h)}$. Moreover, $k_0$ satisfies the dispersion relation $\omega^2 = gk_0 \tanh(k_0 h)$ (see [21–24]).

## 14.3  SOLUTION METHODOLOGY

The solution procedure focused on the BEM to handle the BVP discussed in the previous section is provided here, utilizing "Green's second identity" on $\varphi_j(x, z)$ along with the "fundamental solution $G(x, z; \xi, \chi)$,"

$$\frac{1}{2}\varphi_j(x,z) = \int_\Gamma \left( \varphi_j(x,z)\frac{\partial G(x,z;\xi,\chi)}{\partial n} - G(x,z;\xi,\chi)\frac{\partial \varphi_j(x,z)}{\partial n} \right) d\Gamma, \quad (\xi,\eta) \in \Gamma,$$

(14.9)

where $(x,z)$ and $(\xi,\chi)$ are the "field" and "source points", namely. The "fundamental solution" appears in the form (see [25–27])

$$G(x,z;\xi,\chi) = \frac{1}{2\pi}\ln \tilde{r}; \quad \tilde{r} = \sqrt{(x-\xi)^2 + (z-\chi)^2}.$$

(14.10)

Utilizing the bcs (14.2–14.8) into Equation (14.9), we get

$$-\frac{1}{2}\varphi_1 + \int_{\Gamma_l}\left(\frac{\partial G}{\partial n} - ik_0 G\right)\varphi_1 d\Gamma + \int_{\Gamma_b}\frac{\partial G}{\partial n}\varphi_1 d\Gamma + \int_{\Gamma_r}\left(\frac{\partial G}{\partial n} - ik_0 G\right)\varphi_1 d\Gamma$$

$$+ \int_{\Gamma_{f1}}\left(\frac{\partial G}{\partial n} - KG\right)\varphi_1 d\Gamma + \int_{\Gamma_{i1}}\left(\varphi_1\frac{\partial G}{\partial n} - G\frac{\partial \varphi_1}{\partial n}\right)d\Gamma + \int_{\Gamma_p}\left(\varphi_1\frac{\partial G}{\partial n} - i\omega G\zeta\right)d\Gamma$$

$$+ \int_{\Gamma_{i2}}\left(\varphi_1\frac{\partial G}{\partial n} - G\frac{\partial \varphi_1}{\partial n}\right)d\Gamma + \int_{\Gamma_{f3}}\left(\frac{\partial G}{\partial n} - KG\right)\varphi_1 d\Gamma$$

$$= \int_{\Gamma_r}\left(\frac{\partial \varphi^I}{\partial n} - ik_0\varphi^I\right)G d\Gamma.$$

(14.11)

$$-\frac{1}{2}\varphi_2 + \int_{\Gamma_{i2}}\left(\varphi_1\frac{\partial G}{\partial n} + G\frac{\partial \varphi_1}{\partial n}\right)d\Gamma + \int_{\Gamma_p}\left(\varphi_2\frac{\partial G}{\partial n} - i\omega G\zeta\right)d\Gamma + \int_{\Gamma_{i1}}\left(\varphi_1\frac{\partial G}{\partial n} + G\frac{\partial \varphi_1}{\partial n}\right)d\Gamma$$

$$+ \int_{\Gamma_{f2}}\left(\frac{\partial G}{\partial n} - KG\right)\varphi_2 d\Gamma = 0.$$

(14.12)

Now, utilizing BEM, Equations (14.11) and (14.12) are adapted into a system of equations (see [28–31] for details):

$$\sum_{j=1}^{nbl}\varphi_{1j}(H_{ij} - ik_0 G_{ij})\Big|_{\Gamma_l} + \sum_{j=1}^{nbb}H_{ij}\varphi_{1j}\Big|_{\Gamma_b} + \sum_{j=1}^{nbf1}\varphi_{1j}(H_{ij} - KG_{ij})\Big|_{\Gamma_{f1}}$$

$$+ \sum_{j=1}^{nbi1}\left(H_{ij}\varphi_{1j} - G_{ij}\frac{\partial \varphi_{1j}}{\partial n}\right)\Big|_{\Gamma_{i1}} + \sum_{j=1}^{nbp}(H_{ij}\varphi_{1j} - i\omega G_{ij}\zeta_j)\Big|_{\Gamma_p}$$

$$+ \sum_{j=1}^{nbi2}\left(H_{ij}\varphi_{1j} - G_{ij}\frac{\partial \varphi_{1j}}{\partial n}\right)\Big|_{\Gamma_{i2}} + \sum_{j=1}^{nbf3}\varphi_{1j}(H_{ij} - KG_{ij})\Big|_{\Gamma_{f3}}$$

$$= \sum_{j=1}^{nbl}G_{ij}\left(\frac{\partial \varphi_j^I}{\partial n} - ik_0\varphi^I\right)\Big|_{\Gamma_l}.$$

(14.13)

$$\sum_{j=1}^{nbi2} \left( H_{ij}\varphi_{1j} + G_{ij}\frac{\partial\varphi_{1j}}{\partial n} \right)\Bigg|_{\Gamma_{i2}} + \sum_{j=1}^{nbp} (H_{ij}\varphi_{2j} - i\omega G_{ij}\zeta_j)\Bigg|_{\Gamma_p} + \sum_{j=1}^{nbi1} \left( H_{ij}\varphi_{1j} + G_{ij}\frac{\partial\varphi_{1j}}{\partial n} \right)\Bigg|_{\Gamma_{i1}}$$

$$+ \sum_{j=1}^{nbf2} \varphi_{2j}(H_{ij} - KG_{ij})\Bigg|_{\Gamma_{f2}} = 0. \tag{14.14}$$

It should be mentioned that Equations (14.13) and (14.14) include the unknown plate deflection $\zeta_j$. To solve the unknown in Equations (14.13) and (14.14), coupling is required with Equation (14.5). Now, utilizing the "central difference formulae" to discretize the Equation (14.5), we get (see [32]):

$$\left( \frac{\zeta_{j+2} - 2\zeta_{j+1} + 6\zeta_j - 2\zeta_{j-1} + \zeta_{j-2}}{\Delta_j^4} \right) + S\left( \frac{\zeta_{j+1} - 2\zeta_j + \zeta_{j-1}}{\Delta_j^2} \right) = T\left( \varphi_{1j} - \varphi_{2j} \right), \tag{14.15}$$

where $S = -\dfrac{\omega^2\gamma}{g\eta\left(1 + \frac{\alpha^2\xi\omega}{i+\xi\omega}\right)}$, and $T = -\dfrac{i\omega}{g\eta\left(1 + \frac{\alpha^2\xi\omega}{i+\xi\omega}\right)}$. Similarly, discretized in Equation (14.6), we get

$$\zeta_j = 0, \quad \zeta_{j+1} = \zeta_{j-1}. \tag{14.16}$$

## 14.4   RESULTS AND DISCUSSIONS

Wave power extraction by the PWEC device is written as

$$P_{ext} = \frac{\rho g A^2 [\sinh kh \cosh kh + kh}{8\omega \cosh^2(kh)}(1 - |\tilde{R}|^2 - |\tilde{T}|^2), \tag{14.17}$$

Here $|\tilde{R}|$ and $|\tilde{T}|$ are signified as "reflection and transmission coefficients", namely. Thus, the incident wave energy flux for random waves is as (see [33–36])

$$P_{inc} + \rho g \int_0^\infty S_I = (\omega)C_g d\omega. \tag{14.18}$$

Here, $C_g$ is termed as group velocity. Further, the expression for the incident wave spectrum is given as (see [37–39])

$$S_I(\omega) = \frac{1.25}{4}\frac{\omega_m^4}{\omega^5}H_{1/3}^2 exp\left(-1.25\left(\frac{\omega_m}{\omega}\right)^4\right). \tag{14.19}$$

Here, $\omega_m$ and $H_{1/3}$ are model frequencies and significant wave heights, respectively (Table 14.1).

**TABLE 14.1**
**Significant Wave Heights and Modal Frequencies for Various Sea States**

| Sea state | $H_{1/3}$(m) | $\omega_m$ |
|---|---|---|
| 3 | 0.875 | 0.837 |
| 4 | 0.714 | 1.875 |

Finally, the PWEC's wave power extraction efficiency is expressed as the following:

$$\eta = 1 - |\tilde{R}|^2 - |\tilde{T}|^2. \tag{14.20}$$

Parameters related with the irregular incoming waves and shape parameters of the PWEC device are as follows: $h = 10m$, $L/h = 2.0$, $l/h = 2.0$, $b/h = 0.5$, $r = b + L + 2.0h$, $\rho = 1025kg/m^3$, $g = 9.8m/s^2$.

Figures 14.2 and 14.3 illustrate the change in power extraction, $P_{ext}$, by the PWEC device as a function of incident wave frequency, $\omega_{av}$, for various plate length, $L/h$, and submergence depth, $d/h$, respectively for various sea climates. In Figures 14.2(a) and 14.3(a), it is noticed that more resonances occur as $L/h$ rises. This occurs because as the plate length increases, the natural vibration frequencies of the plate increase. As a result, the number of resonances increases. In addition, the amplitude of the resonance is higher for a plate length with smaller values. Moreover, Figures 14.2(b) and 14.3(b) depict the effect of change in submergence depth on the wave power extraction by the PWEC device. It is reported that the PWEC device's power extraction is higher for the moderate incident wave frequencies. It is illustrated that the PWEC device's power extraction is higher for the smaller $d/h$. The same is justified by the fact that wave energy is more concentrated close to $z = 0$. It is also

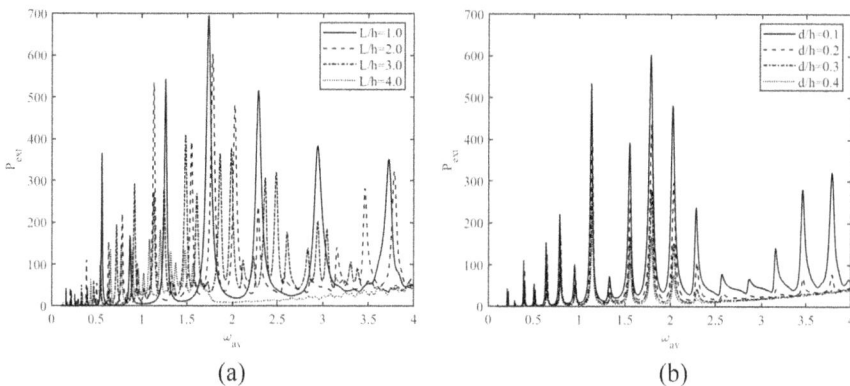

(a)                                    (b)

**FIGURE 14.2**   Change in $P_{ext}$ vs. $\omega_{av}$ for numerous (a) $L/h$, and (b) $d/h$ for sea state 3.

revealed that the resonance appears at the same incoming wave frequencies inattentive of the change in $d/h$. Further, a comparison between the Figures 14.2 and 14.3 shows that the wave power extraction by PWEC device is higher for sea state 4 than sea state 3.

In Figures 14.4 and 14.5, the change in wave power extraction efficiency, $\eta$, by the PWEC device vs $\omega_{av}$ is plotted for various $L/h$, and $d/h$, respectively for various sea climates. In Figure 14.4(a), it is found the wave power extraction efficiency is higher for smaller $L/h$, whereas the number of resonances increases as $L/h$ increases. Additionally, in Figure 14.4(b), it is illustrated that the wave power extraction efficiency by the PWEC device increases with the decrease in $d/h$. The reason for the same is provided in the discussions of Figures 14.2(b) and 14.3(b). The overarching pattern of Figures 14.5 (a) and (b) is same as that of Figures 14.4 (a) and (b).

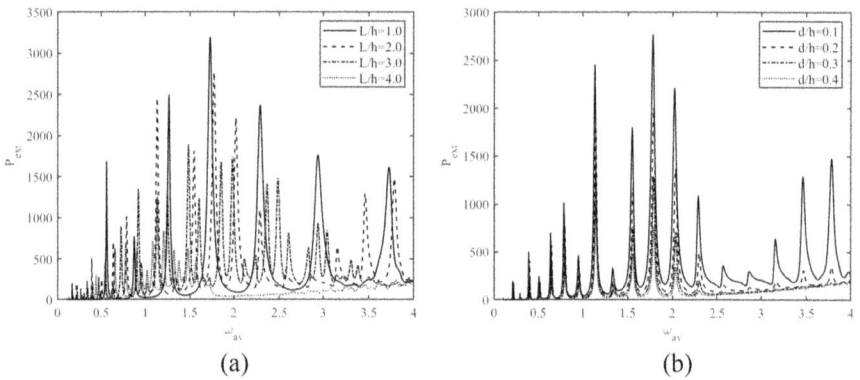

(a)                                                                  (b)

**FIGURE14.3**  Change in $P_{ext}$ vs. $\omega_{av}$ for numerous (a) $L/h$, and (b) $d/h$ for sea state 4.

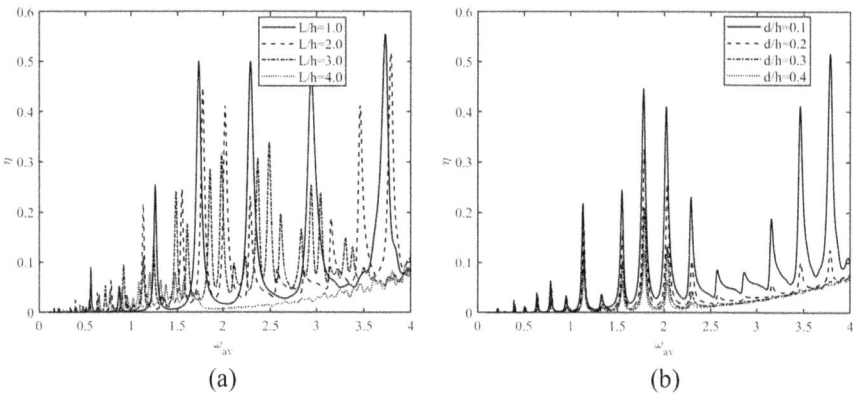

(a)                                                                  (b)

**FIGURE 14.4**  Change in $\eta$ vs. $\omega_{av}$ for several (a) $L/h$, and (b) $d/h$ for sea state 3.

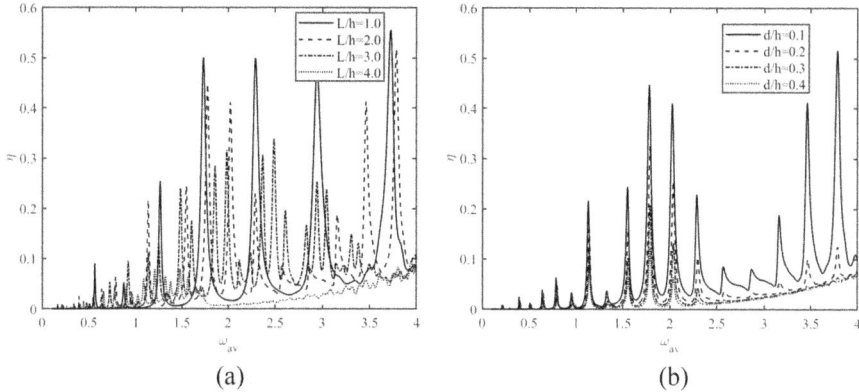

(a)                                                          (b)

**FIGURE 14.5**   Change in $\eta$ vs. $\omega_{av}$ for numerous (a) $L/h$, and (b) $d/h$ for sea state 4.

## 14.5   CONCLUSIONS

In this research, the hydrodynamics of the PWEC device is analyzed in the involvement of irregular incident waves. The BEM is used to handle the corresponding BVP. To construct the erratic wave climate around the PWEC plant, the Bretschneider spectrum along with the sea states are taken as the incident wave spectrum. It is noticed that the wave power extraction and wave power extraction efficiency by the PWEC device is higher for the smaller plate length and the number of resonances increases as the plate length increases. Further, it is observed that the wave power extraction and wave power extraction efficiency by the PWEC device is higher for the smaller submergence depth. Moreover, the amplitude of the resonance is higher for sea state 4 than sea state 3.

## ACKNOWLEDGMENT

"DST Project: DST/INSPIRE/04/2017/002460" and "SERB Project: CRG/2021/001550".

## REFERENCES

[1] Buriani, F. *Mathematical modelling of wave–structure interactions with application to wave energy conversion*. Diss. Loughborough University, 2019.
[2] Wu, N., Q. Wang, and X.D. Xie. Ocean wave energy harvesting with a piezoelectric coupled buoy structure. *Applied Ocean Research*, 2015, 50, 110–118.
[3] Renzi, E. Hydroelectromechanical modelling of a piezoelectric wave energy converter. *Proceedings of the Royal Society A: Mathematical, Physical and Engineering Sciences*, 2016, 472.2195, 20160715.
[4] Hwang, W.S., J.H. Ahn, S.Y. Jeong, H.J. Jung, S.K. Hong, J.Y. Choi, J.Y. Cho, J.H. Kim, and T.H. Sung. Design of piezoelectric ocean-wave energy harvester using sway movement. *Sensors and Actuators A: Physical*, 2017, 260, 191–197.

[5] Mutsuda, H., Y. Tanaka, R. Patel, Y. Doi, Y. Moriyama, and Y. Umino. A painting type of flexible piezoelectric device for ocean energy harvesting. *Applied Ocean Research*, 2017, 68, 182–193.

[6] Kim, K.-H., S.-B. Cho, H.-D. Kim, and K.-T. Shim. Wave power generation by piezoelectric sensor attached to a coastal structure. *Journal of Sensors*, 2018, 2018, 7986438.

[7] Mutsuda, H., Y. Tanaka, Y. Doi, and Y. Moriyama. Application of a flexible device coating with piezoelectric paint for harvesting wave energy. *Ocean Engineering*, 2019, 172, 170–182.

[8] Zhou, K., H.L. Dai, A. Abdelkefi, and Q. Ni. Theoretical modeling and nonlinear analysis of piezoelectric energy harvesters with different stoppers. *International Journal of Mechanical Sciences*, 2020, 166, 105233.

[9] Zheng, S., M. Meylan, X. Zhang, G. Iglesias, and D. Greaves. Performance of a plate-wave energy converter integrated in a floating breakwater. *IET Renewable Power Generation*, 2021, 15, 3206–3219.

[10] Chen, S.-E., R.-Y. Yang, G.-K. Wu, and C.-C. Wu. A piezoelectric wave-energy converter equipped with a geared-linkage-based frequency up-conversion mechanism. *Sensors*, 2020, 21(1), 204.

[11] Trivedi, K., and S. Koley. Hydrodynamic performance of the dual-chamber oscillating water column device placed over the undulated sea bed. *Energy Reports*, 2022, 8, 480–486.

[12] Koley, S., H. Behera, and T. Sahoo. Oblique wave trapping by porous structures near a wall. *Journal of Engineering Mechanics*, 2015, 141(3), 04014122.

[13] Koley, S., R.B. Kaligatla, and T. Sahoo Oblique wave scattering by a vertical flexible porous plate. *Studies in Applied Mathematics*, 2015, 135(1), 1–34.

[14] Koley, S., A. Sarkar and T. Sahoo. Interaction of gravity waves with bottom-standing submerged structures having perforated outer-layer placed on a sloping bed. *Applied Ocean Research*, 2015, 52, 245–260.

[15] Kaligatla, R.B., S. Koley, and T. Sahoo. Trapping of surface gravity waves by a vertical flexible porous plate near a wall. *Zeitschrift für angewandte Mathematik und Physik*, 2015, 66(5), 2677–2702.

[16] Koley, S., and T. Sahoo. Oblique wave scattering by horizontal floating flexible porous membrane. *Meccanica*, 2017, 52(1), 125–138.

[17] Koley, S., and T. Sahoo. Scattering of oblique waves by permeable vertical flexible membrane wave barriers. *Applied Ocean Research*, 2017, 62, 56–168.

[18] Koley, S., and T. Sahoo. Oblique wave trapping by vertical permeable membrane barriers located near a wall. *Journal of Marine Science and Application*, 2017, 16(4), 490–501.

[19] Koley, S., and T. Sahoo. Wave interaction with a submerged semicircular porous breakwater placed on a porous seabed. *Engineering Analysis with Boundary Elements*, 2017, 80, 18–37.

[20] Koley, S., R. Mondal, and T. Sahoo. Fredholm integral equation technique for hydroelastic analysis of a floating flexible porous plate. *European Journal of Mechanics-B/Fluids*, 2018, 67, 291–305.

[21] Koley, S., and T. Sahoo. An integro-differential equation approach to study the scattering of water waves by a floating flexible porous plate. *Geophysical & Astrophysical Fluid Dynamics*, 2018, 112(5), 345–356.

[22] Koley, S. Hypersingular Integral Equation Approach for Hydroelastic Analysis of a Submerged Elastic Plate. In *International Conference on Applied and Computational Mathematics*, pp. 321–330. Springer, Singapore, 2018.

[23] Koley, S. Wave Trapping by Trapezoidal Porous Breakwater. In *Advances in Fluid Mechanics and Solid Mechanics*, pp. 83–90. Springer, Singapore, 2020.

[24] Trivedi, K., and S. Koley. Effect of Varying Bottom Topography on the Radiation of Water Waves by a Floating Rectangular Buoy. *Fluids*, 2021, 6(2), 59.

[25] Vipin, V., K. Trivedi, and S. Koley. Performance of a submerged piezoelectric wave energy converter device floating over an undulated seabed. *Energy Reports*, 2022, 8, 182–188.

[26] Koley, S., and K. Trivedi. Mathematical modeling of oscillating water column wave energy converter devices over the undulated sea bed. *Engineering Analysis with Boundary Elements*, 2020, 117, 26–40.

[27] Trivedi, K., S. Koley and K. Panduranga. Performance of an U-shaped oscillating water column wave energy converter device under oblique incident waves. *Fluids*, 2021, 6(4), 137.

[28] Behera, H., S. Koley, and T. Sahoo. Wave transmission by partial porous structures in two-layer fluid. *Engineering Analysis with Boundary Elements*, 2015, 58, 58–78.

[29] Koley, S., and T. Sahoo. Integral equation technique for water wave interaction by an array of vertical flexible porous wave barriers. *ZAMM-Journal of Applied Mathematics and Mechanics/Zeitschrift für Angewandte Mathematik und Mechanik*, 2021, 101(5), e201900274.

[30] Koley, S., and K. Panduranga. Energy balance relations for flow through thick porous structures. *International Journal of Computational Methods and Experimental Measurements*, 2021, 9(1), 28–37.

[31] Panduranga, K., and S. Koley. Water Wave Interaction with Very Large Floating Structures. In *Advances in Industrial Machines and Mechanisms*, pp. 531–540. Springer, Singapore, 2021.

[32] Koley, S. Water wave scattering by floating flexible porous plate over variable bathymetry regions. *Ocean Engineering*, 2020, 214, 107686.

[33] Kar, P., S. Koley, K. Trivedi and T. Sahoo. Bragg Scattering of Surface Gravity Waves Due to Multiple Bottom Undulations and a Semi-Infinite Floating Flexible Structure. *Water*, 2021, 13(17), 2349.

[34] Koley, S. Wave transmission through multilayered porous breakwater under regular and irregular incident waves. *Engineering Analysis with Boundary Elements*, 2019, 108, 393–401.

[35] Trivedi, K., and S. Koley. Mathematical modeling of breakwater-integrated oscillating water column wave energy converter devices under irregular incident waves. *Renewable Energy*, 2021, 178, 403–419.

[36] Trivedi, K., and S. Koley. Time-domain analysis of quarter-circle shaped oscillating water column device. *Energy Reports*, 2022, 8, 431–437.

[37] Trivedi, K., and S. Koley. Annual mean efficiency of the duct type OWC in regional ocean environments. *Energy Reports*, 2022, 8, 346–351.

[38] Trivedi, K., and S. Koley. Performance of an L-shaped Duct Oscillating Water Column Wave Energy Converter Device Under Irregular Incident Waves. In *Advances in Energy Technology*, pp. 719–728. Springer, Singapore, 2022.

[39] Trivedi, K., and S. Koley. Irregular water wave interaction with oscillating water column wave energy converter devices placed over undulated seabed. In *AIP Conference Proceedings*, vol. 2357, no. 1, p. 100007. AIP Publishing LLC, 2022.

# 15 The Emerging Nanostructured Field Effect Transistors for Dielectrically Modulated Biosensing Applications
## *Review of the Present Development*

*Aditya Tiwari*
Department of Electrical & Electronics Engineering, Birla Institute of Technology and Science-Pilani, Hyderabad, India

*Sangeeta Jana Mukhopadhyay*
Department of Electrical and Electronics Engineering, Dr. Sudhir Chandra Sur Institute of Technology and Sports Complex, JIS Group, Surermath, Kolkata, India

*Sayan Kanungo*
Department of Electrical & Electronics Engineering and Materials Center for Sustainable Energy & Environment, Birla Institute of Technology and Science-Pilani, Hyderabad, India

## CONTENTS

DOI: 10.1201/b23359-15

## 15.1   INTRODUCTION: BACKGROUND

Over the last two decades, the application of nanotechnology in biosensing led to rapid progress in highly sensitive, reliable, and faster biosensor design with inherent compatibility with integrated circuit (IC) technology. Owing to this progress, the successful emergence of so-called lab-on-chip (LOC) diagnostic systems has created exciting opportunities in the fields of medical biology spanning from gene sequencing to early detection of lethal diseases. In this context, the recent developments in the metal-oxide-semiconductor field-effect transistors (MOSFET) based electrochemical biosensor has exhibited substantial promise for label-free biomolecule detection on the basis of intrinsic charge-states. The compatibility with the existing MOSFET process flow, as well as the scope of integration for MOSFET biosensors with integrated circuit technology, is considered highly promising for LOC-based emerging application areas such as point-of-care (POC) diagnosis. Furthermore, the recently reported class of MOSFET-based biosensors which operates on the principle of dielectric modulation is drawing more and more attention from the global research community due to their capability of label-free detection for both charged as well as charge-neutral biological species. Structurally, the dielectrically modulated Biological FET (DM-BioFET) biosensors closely resemble the conventional MOSFET, which incorporates a relatively thicker gate insulator region, where a nano-gap cavity region is introduced during fabrication. The cavity region is functionalized with an appropriate bio-receptor having a specific affinity towards the target biomolecules, i.e. analyte. Subsequently, after analyte incubation, the receptor/analyte binding in the cavity region alters the effective dielectric constant as well as the surface charge density of this region. The biomolecular interaction in the cavity modulates the gate electrostatic coupling over the channel, which in turn alters the carrier injection component into the channel and thereby the terminal current of the biosensor device. The detection and quantifications of the target biomolecule in DM-BioFET biosensors have been accomplished from the change in terminal current and threshold voltage of the device. However, despite their inherent advantages, the incorporation of a thick nano-gap cavity region compromises the overall electrostatic integrity and leads to higher drain-induced barrier lowering effects in the DM-BioFET device. Consequently, the sensing performance of conventional DM-BioFET gets substantially degraded even with moderate channel length.

On the other side, the ever-increasing demand for miniaturization and expanding functionalities in IC technology leads to rapid downscaling of conventional MOSFET devices, which imposed a significant technological challenge in device design by simultaneously sustaining the device downscaling and limiting the detrimental short channel effects (SCE) in the scaled-down MOSFET. Historically, this challenge has been most effectively addressed by incorporating different emerging nanostructures in MOSFET design and subsequently exploiting their inherently superior electrostatic integrity to subdue the SCE. Driven by this paradigm, the integration of different emerging nanostructures for conventional charge modulated biological FET (CM-BioFET) design leads to significant improvement in the limit of detection (LOD) and sensitivity of such biosensors.

However, to date, no exclusive review is available on nanostructure-based DMFET biosensors. Subsequently, this book chapter presents a comprehensive summary of emerging nanostructure-based dielectrically modulated biosensors emphasizing the design aspects.

## 15.2   GENERAL OVERVIEW OF BIOSENSORS

The International Union of Pure and Applied Chemistry (IUPAC) has defined biosensors as "any device that can use bio-molecular interactions at its surface to report this signal while rejecting unintended, nonspecific signal". Thus, biosensing principally involves the conversion of the biological signals (alteration in one or more physiochemical properties of the biological system) originated from bio-molecular interaction into an electrical signal that can be electronically processed and represented in a convenient form to the end user. The relative intensity of the biological signal represents the concentration of any specific biomolecule in the sample under question. There are two primary components of a biosensor, namely bio-receptor and transducer, as shown in Figure 15.1. The bio-receptor is a biological entity that has a specific affinity towards target biomolecules known as the analyte, i.e. from a collection of different biomolecules the bio-receptor biochemically interacts only analyte. The biomolecule detection in biosensors involves a well-defined sequence of generic steps. First, a specific bio-receptor is chemically immobilized to the sensor surface, known as bio-receptor functiona-lization. Next, in the incubation step, the biosensor is subjected to a biological sample containing the target biomolecules (analyte) for a sufficient time. The analyte diffuses towards the sensor surface and subsequently undergoes a selective binding with the functionalized bio-receptor, which is defined as ana-lyte/receptor conjugation. The quality of the generated biological signal strongly depends on the chemical properties of the bio-receptor and the specific functionali-zation process. The basic requirements of a good bio-receptor include simple and stable attachment with the sensor surface, high surface coverage, integrity of bio-logical activities after functionalization, and reasonable shelf-life. The transducer is a device capable of converting the biological signal from analyte-receptor interaction

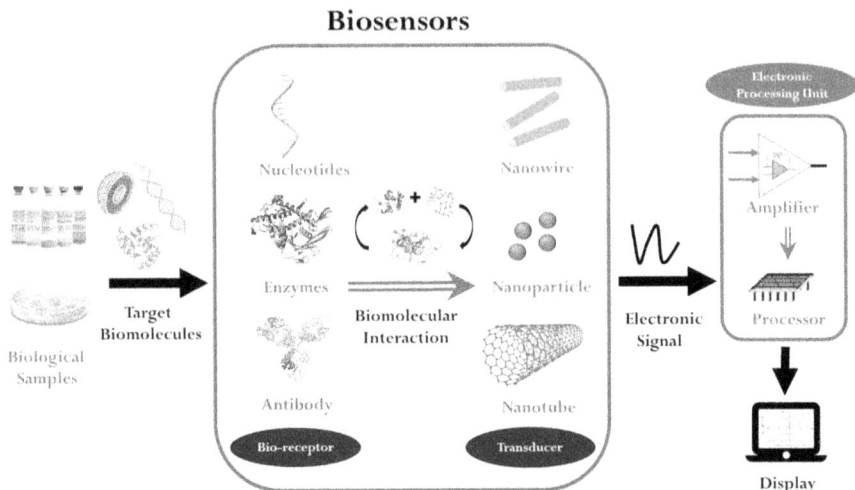

**FIGURE 15.1**  Generic representation of a biosensor and its primary components.

into a measurable and quantifiable electrical signal. The working principle and the efficiency of transducers widely vary depending upon the nature of biomolecular interaction and the transducing technology.

Owing to the gradual developments in biosensing technologies, a wide spectrum of biosensors are available, which can be classified on the basis of its bio-receptor element as well as its transduction mechanism. Therefore, the usefulness of any biosensing technology is defined in terms of its figures of merit (FOM), which offer a quantitative estimation of performance compared to that of other biosensing technologies. Both transient and steady-state responses of the biosensors are considered for evaluating the FOM. The principal FOM that is usually defined for biosensing are as follows.

## 15.2.1  SELECTIVITY

The selectivity indicates the ability of any biosensor to reliably report the presence of a specific analyte in a heterogeneous mixture. In other words, a highly selective biosensor can successfully distinguish between desired and undesirable biomolecular interactions in its response. The selectivity is often characterized by the signal-to-noise ratio (SNR) of the biosensor, where signal represents the biosensor response for desired bio-molecular interactions and noise represents the same for undesired bio-molecular interactions.

## 15.2.2  SENSITIVITY

The sensitivity (S) is the rate of change of transducer signal with analyte concentration as shown in Figure 15.2. A higher sensitivity implies a detectable change of transducer signal for a small change of analyte concentration.

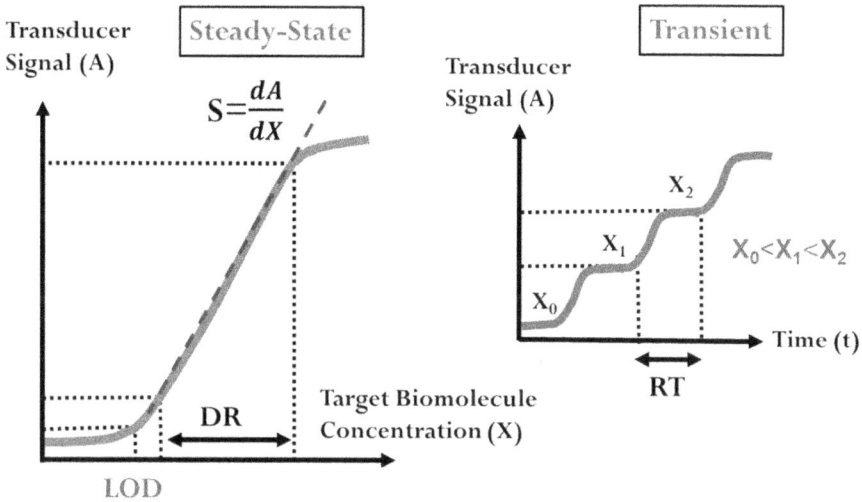

**FIGURE 15.2** Generic plots of steady-state and transient response of biosensors.

### 15.2.3 LIMIT OF DETECTION

The limit of detection (LOD) refers to the lowest analyte concentration at which a biosensor can produce a clearly distinguishable signal from noise in the environment, as shown in Figure 15.2. Therefore, the SNR of the biosensor will be less than one below the detection limit of a transducer.

### 15.2.4 DYNAMIC RANGE

The dynamic range (DR) of a biosensor describes the range of analyte concentrations over which the biosensor can reliably report a concentration variation, as shown in Figure 15.2. Generally, the response of a biosensor varies almost uniformly with analyte concentration within its dynamic range.

### 15.2.5 RESPONSE TIME

The response time (RT) of a biosensor is defined as the time needed for producing a detectable variation in its response, as shown in the inset of Figure 15.2. Hence, a smaller response time is needed to accomplish faster detection of the analyte.

## 15.3 BIOLOGICAL FIELD EFFECT TRANSISTOR (BIOFET)

The introduction of ion sensitive field effect transistor (ISFET) by Piet Bergveld in 1970 is widely acknowledged as the inception of FET-based electrochemical biosensing technology. Since then the field has rapidly evolved and has been an integral part of clinical medicine and diagnostic applications. The key advantages of such technology include the fast and label-free detection, scope of on-chip integration and miniaturization [1,2]. In the past four decades, the basic ISFET structure has undergone numerous modifications that either involve the receptor/ion-sensitive layer or the device structure/material. The latest developments in nano-fabrication and nano-material synthesis enable the exploration of various emerging device structures and materials for today's state-of-the-art FET-based biosensors [3,4].

Since the ISFETs represent the basic structural element for any BioFET, the working principle of ISFET is briefly discussed and consequent modifications for realizing the BioFET are indicated. The structure of ISFET is derived from that of MOSFET, where the metal/polysilicon gate electrode is replaced by a remote reference electrode, as shown in Figure 15.3a. Such a structure was initially introduced to measure the pH of an ionic solution (electrolyte) that is prepared by dissolving chemical compounds in any polar solvent (such as water). The ISFET offered significant advantages over the (then) existing glass-membrane electrode-based pH sensing technology in terms of operating temperature range, durability, and in-vivo monitoring [1,3,4].

The FET-based biosensors (BioFET) are realized by depositing a layer of bio-receptors above the gate-insulator region of ISFET, which is the general architecture for all the BioFETs, as shown in Figure 15.3b. After analyte incubation, the charged analytes are selectively attached with the bio-receptor at the surface of the BioFET, which alters charge densities near the gate insulator. The modulation in local charge densities in the gate insulator surface leads to a change in channel carrier concentration and thereby the channel conductivity of BioFET [2]. The principle transduction mechanism in such a BioFET is based on the electrical parameter (such as drain current, threshold voltage, trans-conductance) modulations due to the presence of the charged analyte, and hence, this class of BioFETs is commonly referred to as charge-modulated (CM) BioFET [5,6].

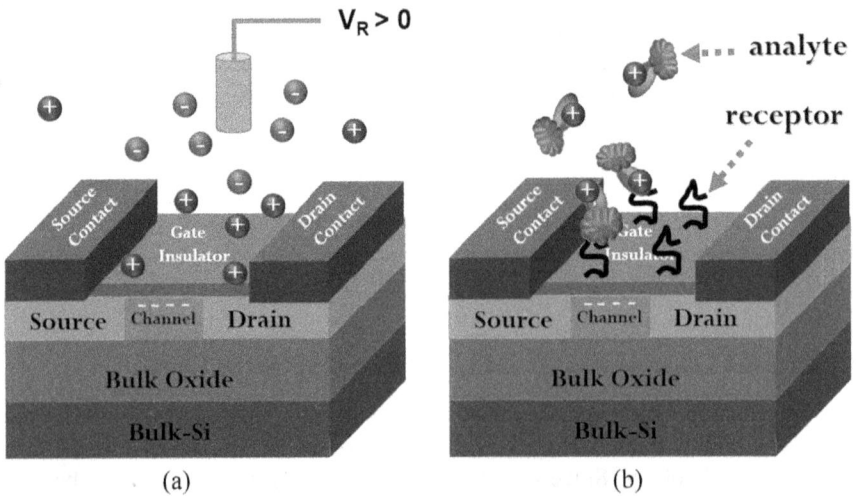

**FIGURE 15.3**   Schematic representation of (a) ISFET and (b) BioFET.

## 15.4 DIELECTRICALLY MODULATED BIOLOGICAL FIELD EFFECT TRANSISTOR (DM-BIOFET)

The inability of detecting charge-neutral biomolecules using CM-BioFET leads to the inception of the dielectric modulation-based transduction mechanism for BioFET. This class of biosensors is commonly known as the dielectrically modulated (DM) BioFETs. The DM-BioFET effectively integrates the advantages of nano-gap electrochemical biosensor with that of BioFET. A nano-gap electrochemical biosensor is a two-terminal device that can potentially serve as a site for biological signal generation due to the size compatibility of the nano-gap (5–100 nm). The contacts between the vertical nano-gap are usually realized by metal or heavily doped Si/Poly-Si that can successfully detect immune interaction and DNA hybridization events. The analyte/receptor binding alters the dielectric constant in the nano-gap that leads to a change in the capacitance between the terminals.

The DM-BioFET are realized by incorporating such nano-gap in the gate-stack region of a BioFET, as shown in Figure 15.4a. Specifically, a MOSFET structure is designed with a thicker gate insulator region, which is selectively etched out to form the nano-gap (~10 nm–20 nm) cavity regions. Next, the cavity regions are functionalized with appropriate bio-receptors, where after analyte incubation, the analyte selectively binds itself leading to an increase in the effective dielectric constant of these regions. For any applied gate bias, the increasing cavity dielectric constant reduces the electrostatic potential drop across the cavity region and thereby increases the channel surface potential, as depicted in Figure 15.4b. This effect is analogous to the overall gating effect modulation in DM-BioFET, and subsequently, a notable change in the terminal current and other electrical parameters is observed after biomolecule conjugation, as shown in Figure 15.4c. Furthermore, apart from dielectric modulation, the presence of an appreciable charge state in the biomolecules can alter the electrostatic potential profile across the cavity leading to further modulations in the overall gating effects in DM-BioFET. Therefore, in DM-BioFET, the biomolecular interactions in the cavity regions lead to the modulations in the electrical parameters, which are exploited for detection and quantifications of both charged and charge-neutral target biomolecules.

The presence of thick nano-gap cavities in the gate-stack region severely compromises the overall electrostatic integrity in the channel region of DM-BioFET. Hence, for an even larger channel length (~100–500 nm), the geometric scaling length appears comparable with the channel length, which results in prominent short channel effects and subsequent sensing performance degradation in DM-BioFET [6]. On the other hand, the invention of the transistor has propelled the progress of integrated circuit technology and the semiconductor industry has seen significant growth over the last few decades in terms of transistor count on a chip that led to an aggressive downscaling of the transistor size. In the nano-scale regime (sub-100 nm), the further miniaturization of bulk MOSFET encounters seemingly insurmountable barriers in terms of power dissipation and leakage. Consequently, a number of emerging nanostructures have been adopted to replace the conventional bulk-MOSFETs in the nanoscale regime of operation. The inherently superior electrostatic integrity in such nanostructures have been successfully exploited for effectively

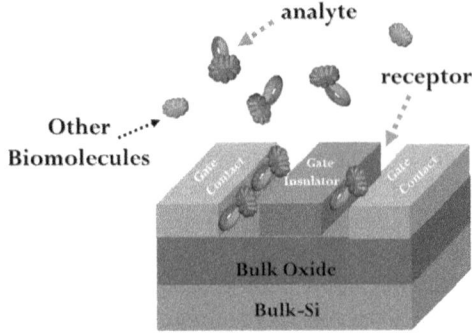

(a)

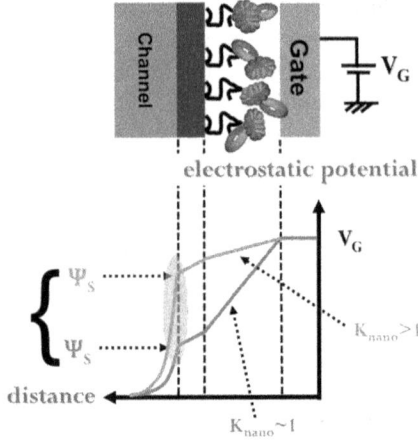

(b)

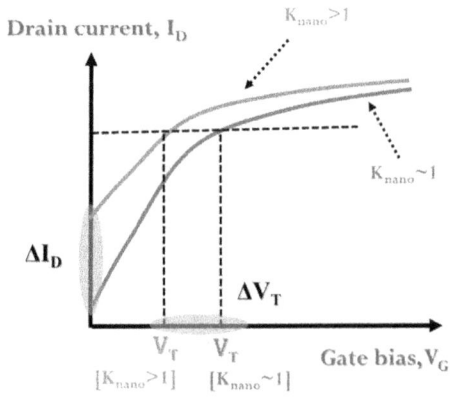

(c)

**FIGURE 15.4** Schematic representation of dielectrically modulated BioFET (DM-BioFET) biosensor.

reducing the detrimental short channel effects (SCE) and thereby improving the performance of nanoscale MOSFET.

Consequently, different nanostructures including nanowires (NW), nanosheets (NS), nanotubes (NT), and bio-tubes (BT) are being actively explored to realize robust and high-performance DM-BioFET designs. However, it is to be noted that to date, there is a relatively small number of experimental reports available on nanostructure-based DM-BioFET-based biosensors, whereas a rapidly increasing number of reports present numerical-simulation-based proofs-of-concept for different nanostructure transducer architectures for DM-BioFET design. Subsequently, in this section, both experimental and simulation-based development of nanostructure-based DM-BioFET are comprehensively summarized.

### 15.4.1 EXPERIMENTAL PROGRESS IN NANOSTRUCTURE DM-BioFET

The DM-BioFET has been experimentally demonstrated by Choi et al., and subsequently, the transduction mechanism has been validated through numerical device simulations. In this work, a tri-gated silicon on insulator (SOI) Fin-FET structure is realized with dual metal gate stacking with chromium (Cr) and Gold (Au) as the inner and outer gate materials, respectively. Subsequently, the Cr gate is selectively etched out from the sides which opens up nano-gap cavity regions between the gate insulator and the outer gate metal. The nano-gap is then functionalized with biotin, and finally the structure is introduced into a solution containing streptavidin. The steady-state electrical characteristics of the device have been monitored in each of these steps, which show a significant modulation in threshold voltage with biotin/streptavidin binding. The proposed architecture is also investigated through device-level simulation where the biotin/streptavidin binding is represented as an appropriate change in dielectric constant within the nano-gap cavity. The findings of experiment and simulation show qualitative agreement, which eventually establishes the dielectric constant modulation being the dominant physical mechanism for transduction, since the change in charge state is negligible in this case [7]. Similar DM-BioFinFET structure has been utilized by Kim et al. for investigating the relative effects of charge and dielectric constant variation for DNA hybridization in n- and p- channel DMFETs. Their results indicate that an n-channel DMFET has optimum transduction efficiency for the neutralized as well as positively charged DNA, whereas a p-channel DMTFET is the most suitable for negatively charged DNA. The results are also verified through device-level simulation [8]. Ahn et al. demonstrated a double-gated DM-BioNWFET structure, where nano-gaps are incorporated in the gate insulator, which offers relatively larger cavity occupancy due to the two-side opened walls cavity that ultimately facilitates a successful detection of the immune interaction between CRP antigen and its antibody. The experimental results are verified through device simulation that shows similar trends in drain current and threshold voltage modulation [9]. Kim et al. demonstrated a different DM-BioNWFET architecture, where two independent side gates are fabricated on the undoped channel region with a finite underlap from each other. The nano-gap cavity region between the gates is modified with (3-aminopropyl) triethoxysilane (APTES) and charged polyelectrolytes of poly-sodium 4-styrene sulfonate (PSS) is introduced followed by poly-allylamine hydrochloride (PAH). The negatively charged PSS binding with APTES results in a clear shift in I-V characteristics from the origin, which is then compensated by the binding of PSS with the positively charged PAH. This clearly exhibits the purely charge-modulated transduction in proposed DM-BioNWFET for biomolecule detection. In this work, the dielectric modulation–based transduction is further demonstrated by detecting the immune interaction of avian influenza antigen (AIa) fused with silica-binding proteins (SBP) and avian influenza antibody (anti-AI) [10]. Jae-Hyuk Ahn et al. systematically investigated the effects of charge and dielectric constant of the biomolecules and demonstrated that in DM-BioNWFET the dielectric modulation usually dominates over charge modulation, which implies a careful consideration of channel doping design in accordance with the charge states of the analyte can notably enhance the response of DM-BioNWFET [11].

However, the relatively smaller nano-gap dimensions in DM-BioFET can potentially impose difficulties for introducing solution containing analyte, as the fluid dynamics, in this case, is limited by an extremely low Reynolds's number. Gu et al. has attempted to address this problem by demonstrating a bulk DM-BioFET with a suspended chromium gate for detecting avian influenza antibody (ab-AI) in SBP-fused AI (SBP-AI) antigen modified gate-insulator surface. The simulation based verification and quantitative theoretical modeling of the dielectric modulation indicates effective dielectric constant variation in each step of detection [12]. On the other hand, the process level and environmental fluctuations in threshold voltage/drain current impose significant limitations in DMFET sensing performance. To address this problem, Kim et al. introduced a new sensing matrix based on the crucial gate voltage for maximum substrate current that shows reduced statistical fluctuations and SNR improvement during PNA detection. The simulation results corroborate the experimental finding and identify impact-ionization near the drain end of the channel to be the primary contributor to substrate current [13].

Although the above-mentioned performance improvement strategies are experimentally demonstrated for bulk DM-BioFET, the same can be adopted for nanostructure DM-BioFET, which is expected to further augment the overall sensing performance of this class of biosensors. Finally, the design of efficient read-out/detection circuitry for DM-BioFET, compatible with integrated circuit technology has been identified as another notable research challenge. In this context, Im et al. have experimentally demonstrated a low-noise read-out circuitry consisting of current-mirror, integration capacitor, and voltage follower circuit for a complete point-of-care diagnosis platform using an array of DM-BioFinFETs [14].

To summarize, the incorporation of nanostructure presents an attractive avenue for highly sensitive DM-BioFET design. However, the experimental exploration is still not sufficient for developing a detailed design-level understanding for realizing reliable and reproducible biosensor prototypes compatible with the lab-on-chip platform.

### 15.4.2 THEORETICAL PROGRESS IN NANOSTRUCTURE DM-BIOFET

The first experimental works on DM-BioFET have clearly indicated that the findings of experiment and numerical device simulation show qualitative agreement [7].

The subsequent work in this line exhibits that the numerical device simulation is an integral part of gaining insight into the transduction mechanism of DM-BioFET architectures.

In one of the initial works, Choi et al. proposed a surface potential-based analytical model under parabolic profile approximation that shows reasonable agreement between calculated and simulated threshold voltage shift for different biomolecule dielectric constants in bulk DM-BioFET. Their model also offers useful insight for optimizing different structural properties including such as nanogap length/thickness, gate length, gate-insulator thickness, channel thickness, and gate work function. This work demonstrates that the aggressive gate length scaling severely degrades the DM-BioFET sensing performance, which eventually becomes practically unacceptable [15]. Similar channel length scaling studies has been performed by Narang et al., where the gradual degradation in sensing performance of DM-BioFET is confirmed, and it has been observed that below 100 nm channel length the sensing performance is significantly compromised [16]. The sensing performance degradation with channel length scaling can be attributed to the increased short channel effects in presence of thick nano-gap cavities in the gate insulator regions which weakens the overall gate control over carrier injection component and enhances the drain control over the same [6]. Consequently, the nanostructure-based DM-BioFET design appears as a natural design-level remedy for this shortcoming, and a number of reports presented proof-of-concept for different nanostructure-based DM-BioFET designs using numerical device simulation, where the essential physics of the designs and their correlation with transduction mechanism is theoretically analyzed and analytically modeled.

To date, different nanostructures, namely nano-wire (NW), fin (F), nano-sheet (NS), nano-tube (NT), and bio-tube (BT) architectures are adopted for DM-BioFET design. It should be noted that in these nanostructure-based DM-BioFETs, the channel region is elevated on the substrate, and a thin isolation oxide usually wrapped around the outer surface of these nanostructures. Furthermore, the nano-gap cavity region is introduced between the gate electrode regions, which is also wrapped around the nanostructures, as depicted in Figure 15.5. Thus, the nanostructure channels form a quasi-one-dimensional channel, where typically the current conduction direction is considerably larger compared to other two dimensions. Such a nanostructure along with the wrapped gate architecture significantly augments the overall electrostatic integrity in the channel of these nanostructure DM-BioFET.

Consequently, the gradual development in each of these classes of DM-BioFETs is summarized in this section.

### 15.4.2.1  DM-BioNWFET

Similar to experimental research, to date, the NWFET structure is also the most theoretically explored nanostructure for dielectrically modulated biosensing applications. The NW is a cylindrical structure, where the electronic and transport properties of NW significantly depends on the diameter. In DM-BioNWFET, the one or multiple NW joins the source and drain regions, and the nano-gap cavity is introduced between the NW and the gate electrode that is deposited wrapping the NW, as depicted in Figure 15.5(a).

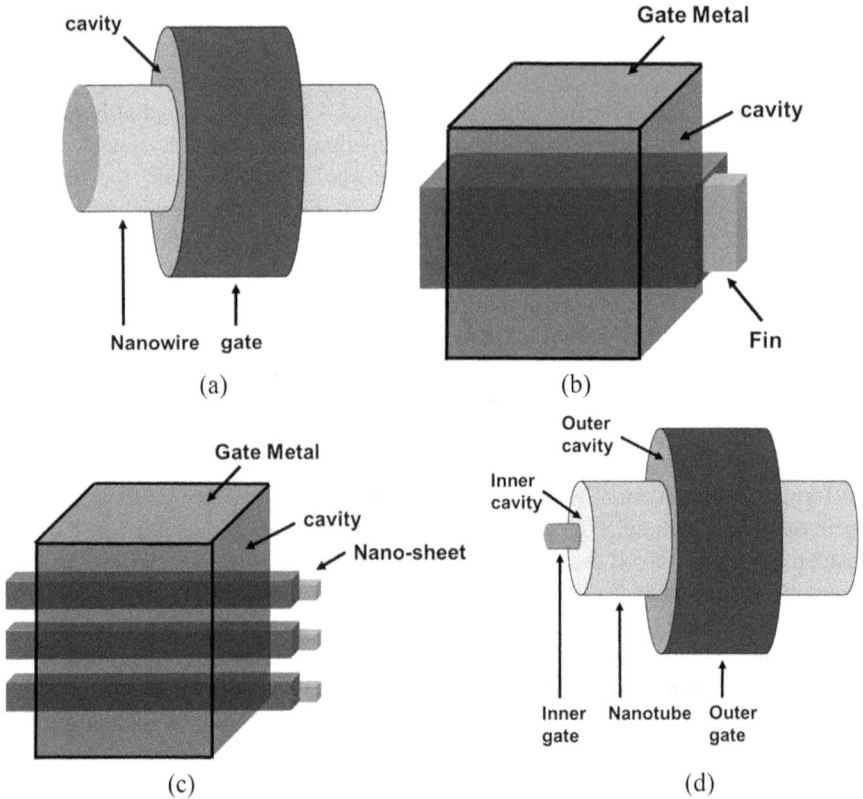

**FIGURE 15.5** Schematic representation of (a) nano-wire, (b) nano-sheet, (c) fin, and (d) nano-tube with nano-gap cavity region for dielectrically modulated biosensing applications.

In one of the early device simulation-based investigations on nanostructure-based DM-BioFET, Gautam et al. have introduced an analytical model for a gate all around (GAA) DM-BioNWFET that has successfully predicted the on-state and off-state drain current under different biomolecule dielectric constant and charge-density specifications. This work also exhibits a significant sensing performance improvement due to the incorporation of a vacuum gate in GAA-DM-BioNWFET architecture [17]. In this line, Kanungo et al. have demonstrated the concept of fringing field-based DM-BioNWFET, where a thicker gate insulator with underlaps is introduced near the source/channel and channel/drain region. The gate-induced fringing field in this structure primarily dominates the transduction mechanism and results in a significant performance improvement over the existing bulk DM-BioFET structures, which can be further augmented by the application of suitable backgate bias. Moreover, this work demonstrates that the sensing performance of DM-BioNWFET significantly depends on channel doping, and a large channel doping concentration is highly desirable for such a biosensor design [18]. Kanungo et al. have further extended this work by designing inverter and ring-oscillator-based detection circuits using such fringing field-based DM-BioNWFET [19]. This trend further suggests the suitability of incorporating the junction-less architecture for nanostructure-based DM-BioFET design, as the junction-less architecture is typically realized by incorporating high and uniform doping in the source, channel, and drain regions [20]. Consequently, Ahangari et al. have proposed a junction-less DM-BioNWFET architecture, which can simultaneously exploit the advantages of relatively higher channel doping in junction-less architecture and inherently strong immunity toward the short channel effects of nanowire architecture [21]. Finally, Saha et al. have demonstrated a reconfigurable DM-BioNWFET architecture, where the device electrostatics and thereby sensing performance can be tuned by using two independent gate terminals [22].

### 15.4.2.2  DM-BioFinFET

Apart from NW architecture, the FiNFET structure has also demonstrated significant potential in enhancing the channel electrostatic integrity, thereby realizing superior sensing performance in DM-BioFinFET. In FinFET, the channel resembles a fin-like cuboid nanostructure, where the height of the cuboid is usually larger than the width. Similar to DM-BioNWFET, in DM-BioFinFET, the one or multiple fin joins the source and drain regions, and the nano-gap cavity is introduced between the fin and the gate electrode, as depicted in Figure 15.5(b).

Very recently, Sehgal et al. have introduced a junction-less DM-BioFinFET design, where the sensing performance is found to be significantly influenced by the fin specifications. Furthermore, the work demonstrated that an optimized junction-less DM-BioFinFET can potentially outperform its junction-less DM-BioNWFET counterpart [23]. In this effect, Dixit et al. have incorporated the dielectric-ferroelectric stack in DM-BioFinFET design to exploit the negative capacitance induced enhanced on-sate/off-state current ratio and reduced sub-threshold swing in favor of dielectrically modulated biosensing applications. The work clearly exhibits that the incorporation of negative capacitance in the DM-BioFinFET design leads to significant improvements in drain current and threshold voltage sensitivities [24]. Dixit et al. have further introduced a useful material/device co-optimization strategy for DM-BioFinFET design, where a $GaAs_{1-x}Sb_x$ channel and raised source/drain architecture have been introduced. In this work, the higher mobility of $GaAs_{1-x}Sb_x$ and higher surface-to-volume ratio in raised source/drain architecture leads to notable sensing performance improvement in DM-BioFinFET compared to its conventional silicon-based design [25]. On the other hand, Kashyap et al. have demonstrated that the truncated fin architecture is particularly beneficial for DM-BioFinFET design, resulting in superior performance for both dielectric modulation and charge modulation [26].

### 15.4.2.3  DM-BioNSFET

Another emerging nanostructure that is drawing significant attention for designing high-performance transistor design is NSFET, where the channel is typically a cuboid-like nanostructure defined as NS. However, the NS can be perceived as a rotated fin structure, where the width of the cuboid is usually larger than the height. Consequently, in DM-BioNSFET, multiple NS connects the source and drain regions, and the nano-gap cavity is introduced around these NS and the common gate electrode region around the entire channel region, as shown in Figure 15.5(c).

Li et al. have proposed a DM-BioNSFET architecture for the first time, where multiple nanosheets are vertically stacked between source and drain region, and the nanogap cavity region is introduced between the nanosheets and gate electrodes. The results indicate that owing to the superior electrostics integrity of GAA NS architecture, the DM-BioNSFET biosensor outperforms its DM-BioNWFET counterpart [27].

### 15.4.2.4   DM-BioNTFET

Another emerging nanostructure for transistor design is the NT architecture, where the channel is wrapped around an inner gate terminal, and an outer gate terminal further surrounds the channel region. In DM-BioNTFET architecture, the nano-gap cavity regions are introduced in the insulator regions around the inner as well as outer gate terminals, as illustrated in Figure 15.5(d).

Sing et al. have reported the DM-BioNTFET design for the first time [28]. Tayal et al. have proposed and systematically studied the junction-less DM-BioNTFET structure. In this work, it has been demonstrated that the channel doping, channel thickness, and the difference in inner- and outer-gate work functions can significantly influence the sensing performance of junction-less DM-BioNTFET, where the outer cavity region essentially dominates the channel electrostatics compare to inner-cavity region. Furthermore, it has been observed that the advantages of superior electrostatics integrity in the channel region of NT, and the strong immunity towards the short channel effects in the junction-less structure together led to a significantly superior sensing performance compared to DM-BioNWFET [29].

### 15.4.2.5   DM-BioBTFET

Finally, another interesting nanostructured DM-BioFET design termed as DM-BioBTFET is recently introduced, which is structurally quite similar to DM-BioNTFET. In this structure the channel is sandwiched between the shell (outer) gate and the core (inner) gate structure, leading to superior gate control over channel electrostatics and thereby a superior sensing performance [30].

## 15.5   SUMMARY

To summarize the discussions, the sensing performances of different potential nanostructure-based DM-BioFETs are analyzed and are indicated in Figure 15.6. For this comparative study only the simulation-based reports are considered, and the sensing performance is assessed in terms of drain current sensitivity ($S_{ID}$) which is defined as given in Equation 15.1 [29]:

$$S_{ID} = \frac{I_{D\_no\_bio} - I_{D\_bio}}{I_{D\_bio}} \tag{15.1}$$

where $I_{D\_no\_bio}$ and $I_{D\_bio}$ are the drain current before and after biomolecule conjugation, respectively.

In general, it can be perceived that the introduction of nanostructure in DM-BioFET design leads to an acceptable drain current sensitivity ($S_{ID} > 10$) for scaled-down

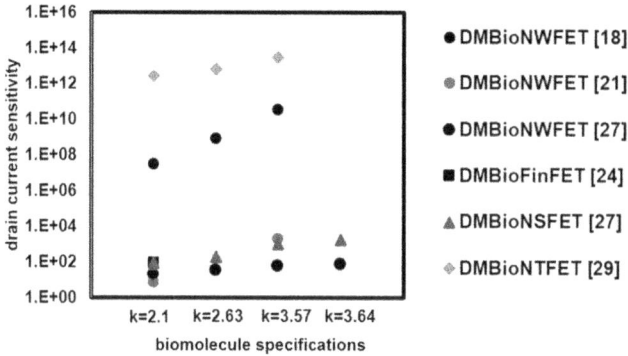

**FIGURE 15.6** Plots of reported drain current sensitivity as a function of biomolecule dielectric constants for different nanostructured DM-BioFET.

channel lengths. However, due to the presence of thick nano-gap cavity region, the gate coupling on channel electrostatic remains weaker even in the nanostructure in DM-BioFET. Furthermore, results also indicates, this limitation can be satisfactorily addressed by incorporating a relatively highly doped channel region, which drastically improve the sensing performance. Therefore, it can be surmised that the incorporation of junction-less architecture with nanostructure can pave the way for a superior dielectrically modulated biosensor design.

However, both theoretical and experimental research on nanostructured DM-BioFET design is still in its early stages of development, and substantial research efforts are still necessary to optimize the sensing performance for both dielectric-modulated and charge-modulated transduction mechanisms.

## ACKNOWLEDGMENT

This work is supported in part by an Outstanding Potential for Excellence in Research and Academics (OPERA) grant awarded to Sayan Kanungo by BITS-Pilani.

## REFERENCES

[1] Bergveld, Piet. "Thirty years of ISFETOLOGY: What happened in the past 30 years and what may happen in the next 30 years." *Sensors and Actuators B: Chemical* 88, no. 1 (2003): 1–20. DOI: 10.1016/S0925-4005(02)00301-5

[2] Grieshaber, Dorothee, Robert MacKenzie, Janos Vörös, and Erik Reimhult. "Electrochemical biosensors-sensor principles and architectures." *Sensors* 8, no. 3 (2008): 1400–1458. DOI: 10.3390/s80314000

[3] Shalev, Gil, Ariel Cohen, Amihood Doron, Andrew Machauf, Moran Horesh, Udi Virobnik, Daniela Ullien, and Ilan Levy. "Standard CMOS fabrication of a sensitive fully depleted electrolyte-insulator-semiconductor field effect transistor for biosensor applications." *Sensors* 9, no. 6 (2009): 4366–4379. DOI: 10.3390/s90604366

[4] Jimenez-Jorquera, Cecilia, Jahir Orozco, and Antoni Baldi. "ISFET-based microsensors for environmental monitoring." *Sensors* 10, no. 1 (2009): 61–83. DOI: 10.3390/s100100061

[5] Putzbach, William, and Niina J. Ronkainen. "Immobilization techniques in the fabrication of nanomaterial-based electrochemical biosensors: A review." *Sensors* 13, no. 4 (2013): 4811–4840. DOI: 10.3390/s130404811

[6] Kanungo, Sayan. "Introduction to dielectrically modulated biological field effect transistor." In 2018 International Symposium on Devices, Circuits and Systems (ISDCS), pp. 1–8. IEEE, 2018. DOI: 10.1109/ISDCS.2018.8379627

[7] Im, Hyungsoon, Xing-Jiu Huang, Bonsang Gu, and Yang-Kyu Choi. "A dielectric-modulated field-effect transistor for biosensing." *Nature Nanotechnology* 2, no. 7 (2007): 430–434. DOI: 10.1038/nnano.2007.180

[8] Kim, Chang-Hoon, Cheulhee Jung, Hyun Gyu Park, and Yang-Kyu Choi. "Novel dielectric modulated field-effect transistor for label-free DNA detection." *BioChip Journal* 2, no. 2 (2008): 127–134.

[9] Ahn, Jae-Hyuk, Jee-Yeon Kim, Maesoon Im, Jin-Woo Han, and Yang-Kyu Choi. "A nanogap-embedded nanowire field effect transistor for sensor applications: Immunosensor and humidity sensor." In Proceedings of the 14th International Conference on Miniaturized Systems for Chemistry and Life Sciences. 2010.

[10] Kim, Jee-Yeon, Jae-Hyuk Ahn, Dong-Il Moon, Sungho Kim, Tae Jung Park, Sang Yup Lee, and Yang-Kyu Choi. "A dual-gate field-effect transistor for label-free electrical detection of avian influenza." *BioNanoScience* 2, no. 1 (2012): 35–41. DOI: 10.1007/s12668-011-0035-0

[11] Ahn, Jae-Hyuk, Sung-Jin Choi, Maesoon Im, Sungho Kim, Chang-Hoon Kim, Jee-Yeon Kim, Tae Jung Park, Sang Yup Lee, and Yang-Kyu Choi. "Charge and dielectric effects of biomolecules on electrical characteristics of nanowire FET biosensors." *Applied Physics Letters* 111, no. 11 (2017): 113701. DOI: 10.1063/1.5003106

[12] Gu, Bonsang, Tae Jung Park, Jae-Hyuk Ahn, Xing-Jiu Huang, Sang Yup Lee, and Yang-Kyu Choi. "Nanogap field-effect transistor biosensors for electrical detection of avian influenza." *Small* 5, no. 21 (2009): 2407–2412. DOI: 10.1002/smll.200900450

[13] Kim, Chang-Hoon, Jae-Hyuk Ahn, Kyung-Bok Lee, Cheulhee Jung, Hyun Gyu Park, and Yang-Kyu Choi. "A new sensing metric to reduce data fluctuations in a nanogap-embedded field-effect transistor biosensor." *IEEE Transactions on Electron Devices* 59, no. 10 (2012): 2825–2831. DOI: 10.1109/TED.2012.2209650

[14] Im, Maesoon, Jae-Hyuk Ahn, Jin-Woo Han, Tae Jung Park, Sang Yup Lee, and Yang-Kyu Choi. "Development of a point-of-care testing platform with a nanogap-embedded separated double-gate field effect transistor array and its readout system for detection of avian influenza." *IEEE Sensors Journal* 11, no. 2 (2010): 351–360. DOI: 10.1109/JSEN.2010.2062502

[15] Choi, Ji-Min, Jin-Woo Han, Sung-Jin Choi, and Yang-Kyu Choi. "Analytical modeling of a nanogap-embedded FET for application as a biosensor." *IEEE Transactions on Electron Devices* 57, no. 12 (2010): 3477–3484. DOI: 10.1109/TED.2010.2076152

[16] Narang, Rakhi, Manoj Saxena, and Mridula Gupta. "Comparative analysis of dielectric-modulated FET and TFET-based biosensor." *IEEE Transactions on Nanotechnology* 14, no. 3 (2015): 427–435. DOI: 10.1109/TNANO.2015.2396899

[17] Gautam, Rajni, Manoj Saxena, R. S. Gupta, and Mridula Gupta. "Numerical model of gate-all-around MOSFET with vacuum gate dielectric for biomolecule detection." *IEEE Electron Device Letters* 33, no. 12 (2012): 1756–1758. DOI: 10.1109/LED.2012.2216247

[18] Kanungo, Sayan, Sanatan Chattopadhyay, Kunal Sinha, Partha Sarathi Gupta, and Hafizur Rahaman. "A device simulation-based investigation on dielectrically modulated fringing field-effect transistor for biosensing applications." *IEEE Sensors Journal* 17, no. 5 (2016): 1399–1406. DOI: 10.1109/JSEN.2016.2633621

[19] Kanungo, Sayan, Sabir Ali Mondal, Sanatan Chattopadhyay, and Hafizur Rahaman. "Design and investigation on bioinverter and bioring-oscillator for dielectrically modulated biosensing applications." *IEEE Transactions on Nanotechnology* 16, no. 6 (2017): 974–981. DOI: 10.1109/TNANO.2017.2736161

[20] Singh, Deepika, and Ganesh C. Patil. "Dielectric-Modulated Bulk-Planar Junctionless Field-Effect Transistor for Biosensing Applications." *IEEE Transactions on Electron Devices* 68, no. 7 (2021): 3545–3551. DOI: 10.1109/TED.2021.3083212

[21] Ahangari, Zahra. "Performance assessment of dual material gate dielectric modulated nanowire junctionless MOSFET for ultrasensitive detection of biomolecules." *RSC Advances* 6, no. 92 (2016): 89185–89191. DOI: 10.1039/C6RA17361F

[22] Saha, Priyanka, Dinesh Kumar Dash, and Subir Kumar Sarkar. "Nanowire reconfigurable FET as biosensor: Based on dielectric modulation approach." *Solid-State Electronics* 161 (2019): 107637. DOI: 10.1016/j.sse.2019.107637

[23] Sehgal, Himani Dua, Yogesh Pratap, Mridula Gupta, and Sneha Kabra. "Performance analysis and optimization of under-gate dielectric modulated Junctionless FinFET biosensor." *IEEE Sensors Journal* 21, no. 17 (2021): 18897–18904. DOI: 10.1109/JSEN.2021.3090263

[24] Dixit, Ankit, Dip Prakash Samajdar, and Vibhuti Chauhan. "Sensitivity analysis of a novel negative capacitance FinFET for label-free biosensing." *IEEE Transactions on Electron Devices* 68, no. 10 (2021): 5204–5210. DOI: 10.1109/TED.2021.3107368

[25] Dixit, Ankit, Dip Prakash Samajdar, and Navjeet Bagga. "Dielectric modulated GaAs1– x Sb X FinFET as a label-free biosensor: device proposal and investigation." *Semiconductor Science and Technology* 36, no. 9 (2021): 095033. DOI: 10.1088/1361-6641/ac0d97

[26] Kashyap, Mridul Prakash, Harshal Gudaghe, and Rishu Chaujar. "Compatibility of a Truncated Fin-FinFET as a k-modulated Biosensor with Optimum parameters for Pre-emptive Diagnosis of Diseases." *Computers and Electrical Engineering* 100 (2022): 107850. DOI: 10.1016/j.compeleceng.2022.107850

[27] Li, Cong, Feichen Liu, Ru Han, and Yiqi Zhuang. "A vertically stacked Nanosheet gate-all-around FET for biosensing application." *IEEE Access* 9 (2021): 63602–63610. DOI: 10.1109/ACCESS.2021.3074906

[28] Singh, Avtar, Saurabh Chaudhury, Manash Chanda, and Chandan Kumar Sarkar. "Split gated silicon nanotube FET for bio-sensing applications." *IET Circuits, Devices & Systems* 14, no. 8 (2020): 1289–1294. DOI: 10.1049/iet-cds.2020.0208

[29] Tayal, Shubham, Budhaditya Majumdar, Sandip Bhattacharya, and Sayan Kanungo. "Performance Analysis of the Dielectrically Modulated Junction-Less Nanotube Field Effect Transistor for Biomolecule Detection." *IEEE Transactions on NanoBioscience* (2022). DOI: 10.1109/TNB.2022.3172702

[30] Goel, Anubha, Sonam Rewari, Seema Verma, S. S. Deswal, and R. S. Gupta. "Dielectric modulated junctionless biotube FET (DM-JL-BT-FET) bio-sensor." *IEEE Sensors Journal* 21, no. 15 (2021): 16731–16743. DOI: 10.1109/JSEN.2021.3077540

# 16 Electrochemical Surface Plasmon Resonance for Efficient Sensing and Analysis

*Ashutosh Joshi, Manjuladevi V., and R. K. Gupta*
Department of Physics, Birla Institute of Technology and
Science, Pilani (BITS-Pilani), Rajasthan, India

## CONTENTS

## 16.1 INTRODUCTION TO SURFACE PLASMON RESONANCE

Surface plasmon resonance (SPR) is one of the popular optical phenomena that promises remarkable applications in the field of sensors. The SPR phenomenon is a classic example of light-matter interaction. Here, the surface plasmon polaritons (SPP) wave is generated at a metal-dielectric interface by the incidence of an electromagnetic (EM) wave under suitable optical conditions. The SPP wave is generated on the metal surface due to a quantized oscillation of free surface electrons induced by the incidence of EM waves [1]. The SPP waves are dependent on the dielectric properties of the metal and the dielectric medium forming the interface. The component of wave-vectors of the incident EM wave can be matched with that of the SPP wave leading to a maximum energy transfer from the EM wave to that of SPP waves (Equation 16.1). This is known as the resonance of surface plasmon [2]. At resonance,

$$k_x = k_p \qquad (16.1)$$

where $k_x$ and $k_p$ are the components of the wave vector of the incident EM wave and the SPP wave, respectively (Equation 16.2).

DOI: 10.1201/b23359-16

Here,

$$k_x = \frac{2\pi}{\lambda} n_{pri} \sin(\theta_i) \text{ and } k_p = \frac{\omega}{c} \sqrt{\frac{\varepsilon_1 \varepsilon_2}{\varepsilon_1 + \varepsilon_2}} \tag{16.2}$$

where $\lambda$ is the wavelength of the EM wave incident at an angle of $\theta i$ via a coupling medium of RI $n_{pri}$, $\omega$ is the angular frequency, $c$ is the speed of light in a vacuum, $\varepsilon_1$ and $\varepsilon_2$ are dielectric constants of the metal film and the dielectric medium either in form of an immobilized layer on the metal surface or medium with dispersed analytes.

For a given dielectric medium ($\varepsilon_2$), the wave vector of the incident EM wave can be matched with that of the SPP wave by altering either the wavelength of the incident light ($\lambda$) or the angle of incidence ($\theta i$) of the monochromatic light. The SPP waves are very sensitive to any change in the dielectrics of the medium which can be measured accurately and very precisely at a high resolution using the SPR phenomenon. This is the basis of a sensing mechanism and bio-analysis using the SPR. There is a plethora of literature available for sensing using the SPR phenomenon. Some of the gas, chemical and biosensing applications are shown in Table 16.1.

---

**TABLE 16.1**
**SPR-Based Sensing Applications in Different Fields**

| Nature of application | | Limit of detection (LOD) | References |
|---|---|---|---|
| Food quality and safety | • Domoic acid (DA) was detected in shellfish e.g. mussels, oysters and cockles in inhibition format of the SPR sensor. | 1 μg/g | [3] |
| | • Steroid hormone, progesterone was detected in milk using inhibition format. The progesterone derivatives were immobilized onto the gold surface of the SPR. | 3.6 ng/mL | [4] |
| | • Detection of contaminants, 4-nonylphenol in shellfish using the monoclonal specific antibodies and inhibition detection format in the SPR sensor. | 2 ng/g | [5] |
| | • Detection of *Escherichia coli* O157-H7 based on immunoassay sensing protocol using SPR sensor. | $10^4$ cells/mL | [6] |
| | • *Staphylococcal enterotoxin* B (SEB) was detected in milk and mushroom samples using the antibody of SEB immobilized onto the gold surface of the SPR sensor. | 1 ng/mL | [7,8] |
| | • Deoxynivalenol was detected in wheat using the SPR sensor. Deoxynivalenol conjugated with casein was immobilized onto the gold surface and the detection was done in inhibition format. | 2.5 ng/mL | [9] |

## TABLE 16.1 (Continued)
## SPR-Based Sensing Applications in Different Fields

| Nature of application | | Limit of detection (LOD) | References |
|---|---|---|---|
| | • Peanut allergens were detected using the immobilized specific antibody on the gold surface of a SPR sensor. | 700 ng/mL | [10] |
| | • The adulterants, non-milk proteins in dairy products, are detected employing polyclonal specific antibodies using a SPR sensor. | 200 ppb | [11] |
| Environmental monitoring | • The carcinogenic and environmental pollutant, benzo[a]pyrene was detected using the SPR sensor employing the indirect inhibition immunoassay format. | 10 ppt | [12] |
| | • Detection of dichlorodiphenyltrichloroethane (DDT) in water using monoclonal antibodies specific to DDT and its derivatives immobilized over the gold surface of the SPR sensor. | 15 pg/mL | [13] |
| | • Detection of $Cu^{2+}$ ions using a polymer composite of squarylium dye deposited on the gold surface of a SPR instrument. | 1pM | [14] |
| | • The explosive material and environmental pollutant, trinitrotoluene (TNT) was detected by monitoring the immunoreaction between trinitrophenol-bovine serum (TNP-BSA) conjugate and anti-TNP antibody using SPR sensor. | 60 ppt | [15] |
| | • Detection of pesticide, 2,4-dichlorophenoxyacetic acid (2,4-D) using the indirect inhibition immunoassay by monitoring the interaction between anti-2,4-D antibody and concanavalin A-2,4-D conjugate using SPR sensor. | 3 ppb | [16] |
| | • Detection of organophosphate pesticide chlorpyrifos and carbaryl using pesticide sensitive SPR sensor. | 1 ng/mL for carbaryl & 50 pg/mL for chlorpyrifos | [17–19] |
| | • Detection of heavy metals e.g. Cd, Zn and Ni using a protein metallothinein. | 100 ng/mL | [20] |
| Biomedical | • The markers for cardiac muscle injury, myoglobin and cardiac troponin I were detected using the antigen specific antibodies employing the SPR sensor. | 2.9 ng/mL for myoglobin & 1.4 ng/mL for troponin | [21] |

*(Continued)*

**TABLE 16.1 (Continued)**
**SPR-Based Sensing Applications in Different Fields**

| Nature of application | | Limit of detection (LOD) | References |
|---|---|---|---|
| | • The pancreatic cancer marker, carbohydrate antigen (CA 19-9) was detected by a specific antibody against CA 19-9 using a SPR sensor. | 410 U/mL | [22] |
| | • Detection of pregnancy markers, human chorionic gonadotropin hormone (hCG) was performed using a wavelength modulated SPR sensor with DNA-directed antibody immobilization method. | 0.5 ng/mL | [23] |
| | • Vascular endothelial growth factor protein (VEGF) can have a role in lung, breast and colorectal cancers. VEGF was detected using SPR imaging by forming a microarray of the RNA. | 1 pM | [24] |
| | • The prostate cancer marker, prostate specific antigen (PSA) was detected using the monoclonal antibodies against PSA employing the SPR sensor. | 0.15 ng/mL | [25] |

## 16.2   ADVANCEMENT IN SURFACE PLASMON RESONANCE

The sensing technology employing the SPR phenomenon has been improving gradually. The most commonly utilized SPR-based sensing technology relied on prism coupling configuration. Among the prism-based SPR instruments, the Kretschmann configuration is very popular due to its very large sensitivity, high resolution and ease of surface modification for specific molecular interaction [26].

A sensor based on the multiparameter measurement is highly reliable and practically impactful [27,28]. There are several attempts in different areas for the development of a multiparameter measurement system. One such example is an electronic nose. An electronic nose is essentially an electrochemical platform possessing a number of sensing elements that can be addressed independently [29]. There are several reports on sensing using an electronic nose [29]. The reports in the literature indicated the advantages of multichannel sensors for dealing with real samples. Here, the sensing area of different channels can be functionalized differently to address different analytes. Hence, a real sample consisting of a number of analytes can be analyzed using such devices directly without undergoing the complex sample preparation procedures. There is another way to analyze the real samples by measuring several physical parameters, simultaneously in a synchronous manner. Such measurement enhances the reliability of the sensors developed to work for real samples. Hence, there are few attempts to integrate several physical properties measurement units in a multiparameter measurement sensing unit which can be trained to work efficiently for real samples in wider ambient conditions [30,31].

The SPR sensors based on optical fibers are gaining attention due to several reasons. In an optical fiber–based SPR sensor, a simple modification of the optical fibers by deposition of gold layer and its suitable functionalization can make it possible to use them as a sensor. The device is very much cost effective and portable [32]. It has been demonstrated that an optical fiber–based SPR sensor can be developed using optical fiber and a smartphone [33]. It promises a point-of-care medical diagnosis at the doorstep. Optical fiber–based SPR sensors can be used for sensing multi-analytes by using multichannel fibers with different levels of functionalization of each fiber [34,35]. However, the optical fiber–based SPR sensors possesses low sensitivity and lower reproducibility.

The prism-based Kretschmann configuration of SPR setup offers very high sensitivity and resolution. The mutli-analytes can be addressed simultaneously using SPR imaging (SPRi) in the Kretschmann configuration [36]. In SPRi, the active area of the metal surface is functionalized with an array of spots of different specificity. On exposure to the real samples with multi-analytes, the spots can interact differently due to their specific binding with the corresponding analytes. These may change the RI of the individual spots differently. In the SPRi system, such changes in RI of the individual spots can be observed and measured simultaneously. In order to observe the change, the reflected intensity from each of the spots can be observed in real time using a charged coupled device (CCD) camera. The intensity data from each of the spots can be processed through a machine learning algorithm to yield a robust sensing performance from a SPRi device addressing real samples [2].

## 16.3   ELECTROCHEMICAL SPR AND APPLICATIONS

During sensing, in addition to the change in RI several other changes in physical parameters can take place on the active area of the transducer. Some of the major changes are like electrochemical, mass, photoactive changes due to electronic transition, vibrational bands, etc. The development of a multiparameter system may include a provision to measure these physical properties, simultaneously. Electrochemical SPR (eSPR) is getting large scientific attention as electrochemical change is the most commonly observed phenomenon in sensing platforms. Therefore, simultaneous and real-time measurement of electrochemical and optical (RI) properties can offer a robust sensing platform by revealing their dependencies for specific analytes and thereby a strong mathematical pattern for decision making.

There are two different modes of operation of SPR instruments configured in prism based Kretschmann methodology. They are wavelength interrogation and angular interrogation. The resolution of the wavelength interrogated SPR instrument is limited due to the resolution spectrometer e.g. 0.01 nm. However, the angular interrogated SPR instrument can offer a very high resolution as the angular scan can be achieved at a step of micro-degree. The instrumentation of angular interrogation is simple. Here, the angle of a monochromatic laser beam in a TM-mode incident at the metal-dielectric interface is changed and the corresponding reflected intensity is recorded. The SPR spectrum thus obtained reveals the reflected

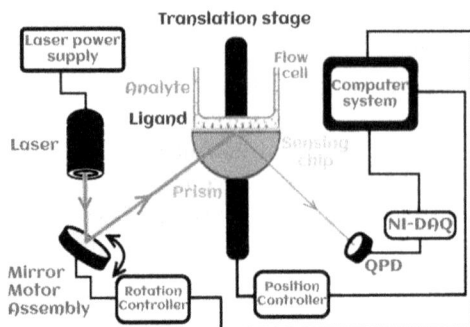

**FIGURE 16.1** Schematic of a Kretschmann-based SPR instrument based on a novel opto-mechanical scanning system (Indian Patent: 2644/DEL/2014).

intensity as a function of the angle of incidence of the EM wave. At resonance, the reflected intensity decreases to a minimum value. The corresponding angle is known as the resonance angle (RA). This can be easily observed by using a coupled arm goniometer equipped with lasers, optical components, sensing elements and detector. However, the goniometer can increase the cost of the equipment and limit the mobility for any field study. Therefore, a novel opto-mechanical scanning mechanism was adopted for the fabrication of the SPR instrument. The schematic of the equipment with the novel design is shown in Figure 16.1.

In the setup shown in Figure 16.1, a monochromatic laser beam in TM mode is allowed to incident on a scanning mirror. The angle of incidence at the metal-dielectric interface is changed by rotating the scanning mirror using a computer-controlled motor. As the angle of incidence is changed at the interface, the angle of reflection also changes accordingly. This may cause the reflected beam spot on the array detector to deflect from an initial position. The amount of spot displacement is fed back to the computer, which can compensate for this by adjusting the height of the prism assembly vertically. Once the original set point is achieved, the detector records the total intensity corresponding to the angle of incidence. This is a novel opto-mechanical scanning system [37]. This setup is very compact, low-weight, portable, low cost and yet very sensitive. The setup was calibrated using the standard glucose solution with the bare gold sensing chip. The result is shown in Figure 16.2.

The calibration curve is found to be linear (Figure 16.2). The sensitivity of the equipment is found to be 194.7°/RIU.

The setup can be used for sensing in different environments. It is therefore necessary to choose a suitable flow-cell for dispensing the analytes on the active area of the SPR sensor. The flow-cells for gas sensing and sensing in an aqueous medium and electrochemical cells can be fabricated. In 2010, Shan et al. reported that the local electrochemical current can be related to the optical signal arising from SPR. They indicated the role of the heterogeneous surface of the SPR chip in the electrochemical signal [38]. It is therefore essential to obtain a homogeneous immobilization over the gold sensing chip of SPR during electrochemical measurement. In an e-SPR, the electrochemical flow-cell needs to be integrated with the basic SPR setup (shown as a schematic Figure 16.3). The electrochemical flow-cell

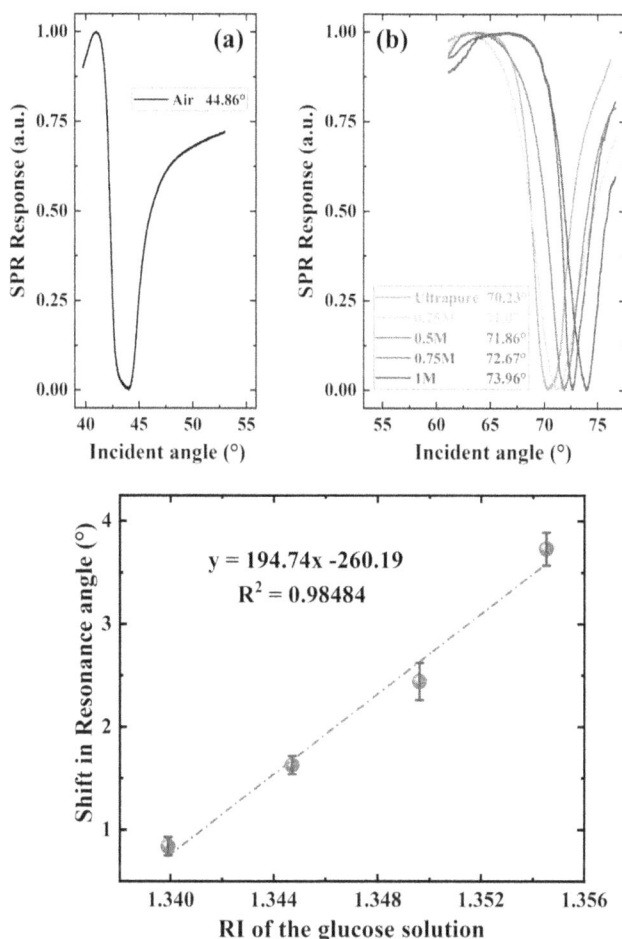

**FIGURE 16.2** (Left) The SPR spectra recorded for (a) gold-air interface and (b) gold-aqueous interface for different concentrations of glucose. (Right) A calibration curve showing the shift in resonance angle ($\Delta$RA) as a function of RI of the aqueous solution of glucose.

should be leak-proof, chemically inert and capable of housing reference and counter electrodes. In e-SPR, the working electrode should be the SPR sensing chip whereas Pt is a counter and Ag/AgCl are reference electrodes (Figure 16.3).

The flow-cell was fabricated in the laboratory and calibration was done using the electrolytes, polyvinyl (PVA) + KCl dissolved in 10 ml of ultrapure ion-free water medium. The electrochemical cyclic voltammogram (CV) is obtained by changing the voltage at a rate of 0.1 V/m (Figure 16.4). As the voltage sweeps, the change in SPR response is recorded as a shift in resonance angle, simultaneously. The CV curves show the usual trend of current as a function of applied voltage without any significant redox peaks. The trend is more like a charge storage behavior due to the presence of the electrolytes. Therefore, the surface capacitance values were calculated and its variation as a function of concentration of PVA in the aqueous

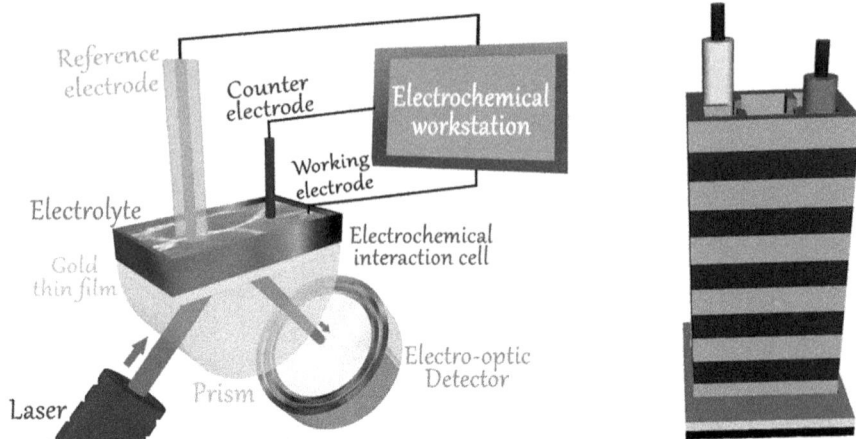

**FIGURE 16.3** (Left) Schematic of electrochemical SPR system. The three electrodes will be enclosed in an electrochemical flow-cell for dispensing analytes. (Right) Schematic of an electrochemical flow-cell.

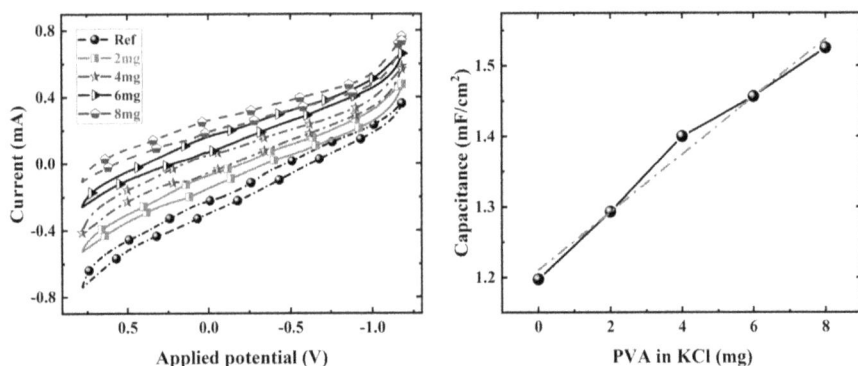

**FIGURE 16.4** (Left) Cyclic voltammogram curves for different concentrations of PVA dissolved in 10 mL of KCl (1M) solution of ultrapure ion-free water medium. (Right) Calibration curve showing the variation of surface capacitance as a function of concentration of PVA in the aqueous medium.

medium is shown in Figure 16.4. The slope of the linear trend of the capacitance is 40.9 mF/g-cm$^2$.

The change in capacitance can alter the dielectric properties (i.e. RI) of the medium adsorbed over the gold surface. This change was measured by recording the shift in RA measured, simultaneously (Figure 16.5), as a function of change in the concentration of PVA in the aqueous medium of 1M KCl. The calibration curve drawn from the SPR response is found to be linear with a slope of about 410.6 o/g (Figure 16.5). The slope of the calibration curve is an indicator of the sensitivity of the device [39]. Thus, the sensitivity for the developed eSPR is reasonably very high. These measurements clearly indicate a dependency of electrochemical properties with

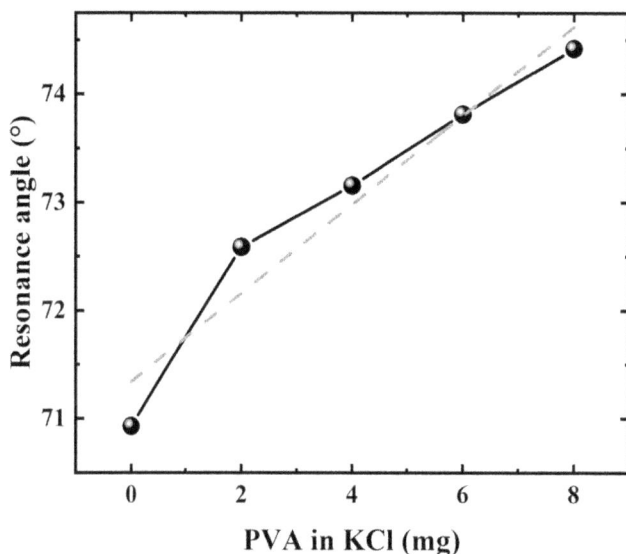

**FIGURE 16.5** The calibration curve shows the resonance angle as a function of the concentration of PVA in aqueous medium.

that of dielectric properties of the material at the metal-dielectric interface. These properties can change due to the interaction of the analytes with the ligands immobilized over the gold surface of an e-SPR system.

The time dependency of the current and the shift in the resonance angle can be calculated analytically [40]. The shift in resonance angle ($\Delta RA = \theta(t) - \theta_o$) due to electrochemical interaction is given by Equation 16.3:

$$\Delta RA(t) = B\left(\alpha_R D_R^{-\frac{1}{2}} - \alpha_o D_o^{-\frac{1}{2}}\right)(nF\pi^{1/2})^{-1}\int_0^t i(t')(t - t')^{-1/2}dt' \quad (16.3)$$

where B is some constant, $\alpha_R$ and $\alpha_o$ are changes in local refractive index per unit concentration of reduced and oxidized molecules, $n$ is a number of electrons transferred per reaction, $F$ is Faraday constant and $D_o$ and $D_R$ are diffusion constants for oxidized and reduced species, respectively. It is important to note that the temporal change in resonance angle is dependent on the convolution function of the current signal. Therefore, the Faradic current, as well as double layer current can shift the resonance angle during electrochemical SPR measurement. Heterogeneity of the electrode surface may arise due to non-uniform immobilization of the ligands. Therefore, sometimes the electrochemical double layer (EDL) formation on such e-SPR chip is expected to be non-homogeneous. This non-homogeneous feature can change $\Delta RA$ locally according to the above relation. Therefore, it is essential to fabricate a homogeneous sensing surface for a single-channel eSPR measurement. It can also be noted that the co-factor in the above equation is dependent on the $D_R$ and $D_o$.

The eSPR reveals valuable information related to the Faradic and ELD currents on the sensing chip of the SPR. Additionally in several cases, the states of redox reaction during interaction with the analytes yield a strong parameter for sensing and analysis. In 2013, Zhang et al. reported the fabrication of an e-SPR imaging instrument and demonstrated its capability for the measurement of cellular optical and electrochemical activities [41]. They analyzed A595 cells using the eSPR and related the change in morphology due to the death of the cells with the SPR response. The oxidation process due to cellular activity was related to the electrochemical response. The combination of these parameters yields valuable information on the cellular activities, which can lead to a path for bio-analysis. In 2015, Wu et al. reported real-time assessment of cancer cells treated with drug molecules daunorubicin (DNR) using eSPR technology [42]. Interestingly, on treatment with DNR, the HePG2 cancer cell residue is released. The change in morphology and mass leading to the mortality information of the cancer cell was related to the SPR signals. Similarly, the concentration of extracellular DNR residue was estimated from the electrochemical responses. This report indicated a potential application of eSPR for therapeutics and bio-analysis. In 2018, Golden et al. have employed eSPR for monitoring Faradaic processes optically, which occurs due to the change in redox state at the electrode surface [43].

Recently, Li et al. have demonstrated a forensic diagnostic application using eSPR. The synchronous signals of SPR and electrochemical were used to detect amphetamine-type stimulants (ATS) in human urine and serum samples. They claim the proposed methodology is very rapid, accurate and low cost [44].

Transferrin is a bio-marker for kidney disease. In 2010, Nakamoto et al. [45] reported eSPR measurements for the detection of transferring in urine samples. The SPR sensing chip was functionalized using anti-transferrin antibodies and was exposed to the urine sample. The binding of transferrin from the sample shifted the resonance angle, whereas the production of $H_2O_2$ as the by-product of the protein's interaction was detected from the electrochemical signal. The limit of detection (LOD) was reported to be around 20 ppb.

In 2021, Ribeiro et al. have used eSPR for the detection of breast cancer cell markers in the serum samples [46] using the suitable surface immobilized antibody. They reported that the LOD of the cancer cell increased from 21 U/mL to 0.0998 U/mL when the measurement was performed in the absence and presence of an electrochemical setup, respectively. The electrochemical process has given a great amplification for enhancing the detection limits of the cancer cell.

The neural activities at different locations of sciatic nerves from bullfrogs were studied using eSPR in imaging mode [47]. The combination of SPR and electrochemical responses revealed important information on the nervous system at molecular and cellular level.

In 2019, Qatamin and coworkers have functionalized the sensing chip of eSPR with a monoclonal antibody for a specific target antigen from the influenza virus. The SPR response was strategically modulated using the electrochemical potential sweep. This increases the sensitivity and the LOD of the influenza virus was reported to be 300 pM [48].

## 16.4   CONCLUSION

The optical phenomenon SPR has demonstrated immense potential for the next generation of sensors. The label-free sensing mechanism is extremely popular in the field of biosciences, pharmaceuticals and biomedical fields. The SPR-based developed technology is extremely sensitive for the detection of analytes in very low concentrations i.e. it is capable of detecting the interaction of a few molecules on the sensing chip. Such a high sensitivity and resolution can reduce the sensing/ bio-analysis time remarkably. The real samples from sputum/urine/serum do not require any further amplification for strengthening the signal-to-noise ratio. The next-generation SPR-based transducers have been conceptualized and adopted. An improved SPR sensing platform with multi-channels and/or multi-parameters measurement can improve the selectivity and sensitivity issues of a sensor. The analysis of multi-parameters obtained through such a sensing platform using the advanced mathematical algorithm (e.g. machine learning) can simplify the decision making during sensing. Such a system can yield reliable outcomes even from the real samples. The practical application of SPR-based sensors as point-of-care devices can be developed and commercialized. Enormous opportunities are available for the integration of measurement for several other physical parameters (viz. electrochemical, piezoelectric, thermal and optical spectroscopy) including the SPR responses. In this chapter, we have briefly reviewed the electrochemical SPR and some of its popular applications. It has been demonstrated by several researchers that synchronous analysis of both SPR and electrochemical responses can yield several valuable pieces of information about the system under investigation. It can be employed as a reliable sensing platform for biomolecules. The response of specific drugs on bio-active materials like cells and bacteria can be assessed using the eSPR. The dependencies of electrochemical responses on the dielectric properties at the metal-dielectric interface of the SPR chip can be analyzed using eSPR. In a nutshell, the immense potential of eSPR is evident and thus the field needs scientific and technological breakthroughs for the development of high-impact societal relevant devices.

## REFERENCES

[1] Raether, H. 1988. "Surface Plasmons on Smooth Surfaces." *Springer Tracts in Modern Physics*, 4–39. Berlin, Heidelberg: Springer Berlin Heidelberg. DOI: 10.1 007/BFb0048319 ISBN: 978-3-540-47441-8.

[2] Gupta, R.K., V.P. Devanarayanan, and V. Manjuladevi. 2018. *Microscopy Applied to Materials Sciences and Life Sciences*. Edited by A.V. Rane, S. Thomas, and N. Kalarikkal. Toronto; New Jersey: Apple Academic Press, 2019.: Apple Academic Press. 215–230.

[3] Traynor, I.M., L. Plumpton, T.L. Fodey, C. Higgins, and C.T. Elliott. 2006. "Immunobiosensor Detection of Domoic Acid as a Screening Test in Bivalve Molluscs: Comparison with Liquid Chromatography-Based Analysis." *Journal of AOAC International* 89 (3): 868–872. 10.1093/jaoac/89.3.868.

[4] Gillis, E.H., J.P. Gosling, J.M. Sreenan, and M. Kane. 2002. "Development and Validation of a Biosensor-Based Immunoassay for Progesterone in Bovine Milk." *Journal of Immunological Methods* 267 (2): 131–138. 10.1016/s0022-1759(02) 00166-7.

[5] Samsonova, J.V., N.A. Uskova, A.N. Andresyuk, M. Franek, and C.T. Elliott. 2004. "Biacore Biosensor Immunoassay for 4-Nonylphenols: Assay Optimization and Applicability for Shellfish Analysis." *Chemosphere* 57 (8): 975–985. 10.1016/j.chemosphere.2004.07.028

[6] Oh, Kim, Bae, Lee, and Choi. 2002. "Detection of Escherichia Coli $O_{157}$:$H_7$ Using Immunosensor Based on Surface Plasmon Resonance." *Journal of Microbiology and Biotechnology* 12 (5): 780–786. https://koreascience.kr:443/article/JAKO2002 11921405258.pdf

[7] Nedelkov, D., A. Rasooly, and R.W. Nelson. 2000. "Multitoxin Biosensor-Mass Spectrometry Analysis: A New Approach for Rapid, Real-Time, Sensitive Analysis of Staphylococcal Toxins in Food." *International Journal of Food Microbiology* 60 (1): 1–13. 10.1016/s0168-1605(00)00328-7

[8] Nedelkov, D., and R.W. Nelson. 2003. "Detection of Staphylococcal Enterotoxin B via Biomolecular Interaction Analysis Mass Spectrometry." *Applied and Environmental Microbiology* 69 (9): 5212–5215. 10.1128/AEM.69.9.5212-5215.2003

[9] Tüdös, A.J., E.R. Lucas-van den Bos, and E.C.A. Stigter. 2003. "Rapid Surface Plasmon Resonance-Based Inhibition Assay of Deoxynivalenol." *Journal of Agricultural and Food Chemistry* 51 (20): 5843–5848. 10.1021/jf030244d

[10] Mohammed, I., W.M. Mullett, E.P.C. Lai, and J.M. Yeung. 2001. "Is Biosensor a Viable Method for Food Allergen Detection?" *Analytica Chimica Acta* 444 (1): 97–102. 10.1016/s0003-2670(01)01166-7

[11] Haasnoot, W., K. Olieman, G. Cazemier, and R. Verheijen. 2001. "Direct Biosensor Immunoassays for the Detection of Nonmilk Proteins in Milk Powder." *Journal of Agricultural and Food Chemistry* 49 (11): 5201–5206. 10.1021/jf010440p

[12] Miura, N., M. Sasaki, K.V. Gobi, C. Kataoka, and Y. Shoyama. 2003. "Highly Sensitive and Selective Surface Plasmon Resonance Sensor for Detection of Sub-Ppb Levels of Benzo[a]Pyrene by Indirect Competitive Immunoreaction Method." *Biosensors & Bioelectronics* 18 (7): 953–959. 10.1016/s0956-5663(02)00242-7

[13] Mauriz, E., A. Calle, J.J. Manclús, A. Montoya, A. Hildebrandt, D. Barceló, and L.M. Lechuga. 2007. "Optical Immunosensor for Fast and Sensitive Detection of DDT and Related Compounds in River Water Samples." *Biosensors & Bioelectronics* 22 (7): 1410–1418. 10.1016/j.bios.2006.06.016

[14] Ock, Jang, Kim Roh, Kim, and Koh. 2001. "Optical Detection of Cu2+ Ion Using a SQ-Dye Containing Polymeric Thin-Film on Au Surface." *Microchemical Journal, Devoted to the Application of Microtechniques in All Branches of Science* 3 (70): 301–305. https://www.infona.pl//resource/bwmeta1.element.elsevier-33cb69f1-28ea-3d5b-8dbc-97559cb34709

[15] Shankaran, D.R., K.V. Gobi, T. Sakai, K. Matsumoto, T. Imato, K. Toko, and N. Miura. 2005. "A Novel Surface Plasmon Resonance Immunosensor for 2,4,6-Trinitrotoluene (TNT) Based on Indirect Competitive Immunoreaction: A Promising Approach for on-Site Landmine Detection." *IEEE Sensors Journal* 5 (4): 616–621. 10.1109/jsen.2005.848150

[16] Svitel, J., A. Dzgoev, K. Ramanathan, and B. Danielsson. 2000. "Surface Plasmon Resonance Based Pesticide Assay on a Renewable Biosensing Surface Using the Reversible Concanavalin A Monosaccharide Interaction." *Biosensors & Bioelectronics* 15 (7–8): 411–415. 10.1016/s0956-5663(00)00099-3

[17] Mauriz, E., A. Calle, J.J. Manclús, A. Montoya, A.M. Escuela, J.R. Sendra, and L.M. Lechuga. 2006. "Single and Multi-Analyte Surface Plasmon Resonance Assays for Simultaneous Detection of Cholinesterase Inhibiting Pesticides." *Sensors and Actuators. B, Chemical* 118 (1–2): 399–407. 10.1016/j.snb.2006.04.085

[18] Mauriz, E., A. Calle, A. Abad, A. Montoya, A. Hildebrandt, D. Barceló, and L.M. Lechuga. 2006. "Determination of Carbaryl in Natural Water Samples by a Surface

Plasmon Resonance Flow-through Immunosensor." *Biosensors & Bioelectronics* 21 (11): 2129–2136. 10.1016/j.bios.2005.10.013

[19] Mauriz, E., A. Calle, L.M. Lechuga, J. Quintana, A. Montoya, and J.J. Manclús. 2006. "Real-Time Detection of Chlorpyrifos at Part per Trillion Levels in Ground, Surface and Drinking Water Samples by a Portable Surface Plasmon Resonance Immunosensor." *Analytica Chimica Acta* 561 (1–2): 40–47. 10.1016/j.aca.2005.12.069

[20] Wu, L.P., Y.F. Li, C.Z. Huang, and Q. Zhang. 2006. "Visual Detection of Sudan Dyes Based on the Plasmon Resonance Light Scattering Signals of Silver Nanoparticles." *Analytical Chemistry* 78 (15): 5570–5577. 10.1021/ac0603577

[21] Oh, B.-K., W. Lee, Y.-K. Kim, W.H. Lee, and J.-W. Choi. 2004. "Surface Plasmon Resonance Immunosensor Using Self-Assembled Protein G for the Detection of Salmonella Paratyphi." *Journal of Biotechnology* 111 (1): 1–8. 10.1016/j.jbiotec. 2004.02.010

[22] Chung, J.W., R. Bernhardt, and J.C. Pyun (2006). Additive assay of cancer marker CA 19-9 by SPR biosensor. *Sensors and Actuators. B, Chemical*, 118(1–2), 28–32. 10.1016/j.snb.2006.04.015

[23] Ladd, J., C. Boozer, Q. Yu, S. Chen, J. Homola, and S. Jiang. 2004. "DNA-Directed Protein Immobilization on Mixed Self-Assembled Monolayers via a Streptavidin Bridge." *Langmuir: The ACS Journal of Surfaces and Colloids* 20 (19): 8090–8095. 10.1021/la049867r

[24] Li, Y., H.J. Lee, and R.M. Corn. 2007. "Detection of Protein Biomarkers Using RNA Aptamer Microarrays and Enzymatically Amplified Surface Plasmon Resonance Imaging." *Analytical Chemistry* 79 (3): 1082–1088. 10.1021/ac061 849m

[25] Besselink, G.A.J., R.P.H. Kooyman, P.J.H.J. van Os, G.H.M. Engbers, and R.B.M. Schasfoort. 2004. "Signal Amplification on Planar and Gel-Type Sensor Surfaces in Surface Plasmon Resonance-Based Detection of Prostate-Specific Antigen." *Analytical Biochemistry* 333 (1): 165–173. 10.1016/j.ab.2004.05.009

[26] Kretschmann, E. and H. Raether. 1968. "Radiative Decay of Nonradiative Surface Plasmons Excited by Light." *Zeitschrift Für Naturforschung A*, 23, 2135–2136.

[27] Torsi, L., A. Dodabalapur, L. Sabbatini, and P.G. Zambonin. 2000. "Multi-Parameter Gas Sensors Based on Organic Thin-Film-Transistors." *Sensors and Actuators. B, Chemical* 67 (3): 312–316. 10.1016/s0925-4005(00)00541-4

[28] Zhou, B., C. Bian, Tong, J., and Xia, S. (2017). Fabrication of a miniature multi-parameter sensor chip for water quality assessment. *Sensors (Basel, Switzerland)*, 17(12), 157. 10.3390/s17010157

[29] Karakaya, D., O. Ulucan, and M. Turkan. 2020. "Electronic Nose and Its Applications: A Survey." *International Journal of Automation and Computing* 17 (2): 179–209. 10.1007/s11633-019-1212-9

[30] Amer, M.-A., A. Turo, J. Salazar, L. Berlanga-Herrera, M.J. Garcia-Hernandez, and J.A. Chavez. 2020. "Multichannel QCM-Based System for Continuous Monitoring of Bacterial Biofilm Growth." *IEEE Transactions on Instrumentation and Measurement* 69 (6): 2982–2995. 10.1109/tim.2019.2929280

[31] Men, L., P. Lu, and Q. Chen (2008). Intelligent multiparameter sensing with fiber Bragg gratings. *Applied Physics Letters*, 93(7), 071110. 10.1063/1.2975186

[32] Jorgenson, R.C., and S.S. Yee. 1993. "A Fiber-Optic Chemical Sensor Based on Surface Plasmon Resonance." *Sensors and Actuators. B, Chemical* 12 (3): 213–220. 10.1016/0925-4005(93)80021-3

[33] Liu, Y., Q. Liu, S. Chen, F. Cheng, H. Wang, and W. Peng. 2015. "Surface Plasmon Resonance Biosensor Based on Smart Phone Platforms." *Scientific Reports* 5 (1): 12864. 10.1038/srep12864

[34] Peng, W., S. Banerji, Y.-C. Kim, and K.S. Booksh. 2005. "Investigation of Dual-Channel Fiber-Optic Surface Plasmon Resonance Sensing for Biological Applications." *Optics Letters* 30 (22): 2988–2990. 10.1364/ol.30.002988

[35] Zhang, S., B. Han, Y.-N. Zhang, Y. Liu, W. Zheng, and Y. Zhao. 2022. "Multichannel Fiber Optic SPR Sensors: Realization Methods, Application Status, and Future Prospects." *Laser & Photonics Reviews*, 2200009. 10.1002/lpor.202200009

[36] Peng, W., Y. Liu, P. Fang, X. Liu, Z. Gong, H. Wang, and F. Cheng. 2014. "Compact Surface Plasmon Resonance Imaging Sensing System Based on General Optoelectronic Components." *Optics Express* 22 (5): 6174–6185. doi:10.1364/OE.22.006174

[37] Devanarayanan, V.P., V. Manjuladevi, and R.K. Gupta. 2016. "Surface Plasmon Resonance Sensor Based on a New Opto-Mechanical Scanning Mechanism." *Sensors and Actuators. B, Chemical* 227: 643–648. 10.1016/j.snb.2016.01.027

[38] Hickel, W., D. Kamp, and W. Knoll. 1989. "Surface-Plasmon Microscopy." *Nature* 339 (6221): 186–186. 10.1038/339186a0

[39] Devanarayanan, V.P., V. Manjuladevi, M. Poonia, R.K. Gupta, S.K. Gupta, and J. Akhtar. 2016. "Measurement of Optical Anisotropy in Ultrathin Films Using Surface Plasmon Resonance." *Journal of Molecular Structure* 1103: 281–285. 10.1016/j.molstruc.2015.09.018

[40] Wang, S., X. Huang, X. Shan, K.J. Foley, and N. Tao. 2010. "Electrochemical Surface Plasmon Resonance: Basic Formalism and Experimental Validation." *Analytical Chemistry* 82 (3): 935–941. 10.1021/ac902178f

[41] Zhang, L.L., X. Chen, H.T. Wei, H. Li, J.H. Sun, H.Y. Cai, J.L. Chen, and D.F. Cui (2013). An electrochemical surface plasmon resonance imaging system targeting cell analysis. *The Review of Scientific Instruments*, 84(8), 085005. 10.1063/1.4819027

[42] Wu, C., F.U. Rehman, J. Li, J. Ye, Y. Zhang, M. Su, H. Jiang, and X. Wang. 2015. "Real-Time Evaluation of Live Cancer Cells by an in Situ Surface Plasmon Resonance and Electrochemical Study." *ACS Applied Materials & Interfaces* 7 (44): 24848–24854. 10.1021/acsami.5b08066

[43] Golden, J., M.D. Yates, M. Halsted, and L. Tender. 2018. "Application of Electrochemical Surface Plasmon Resonance (ESPR) to the Study of Electroactive Microbial Biofilms." *Physical Chemistry Chemical Physics: PCCP* 20 (40): 25648–25656. 10.1039/c8cp03898h

[44] Li, C., D. Han, Z. Liang, F. Han, W. Fu, W. Wang, D. Han, Y. Wang, and L. Niu. 2022. "Novel Electrochemical-Surface Plasmon Resonance (EC-SPR) Sensor for Amphetamine-Type Stimulants Detection Based on Molecularly Imprinted Strategy." *Sensors and Actuators. B, Chemical* 369 (132258): 132258. 10.1016/j.snb.2022.132258

[45] Nakamoto, K., R. Kurita, and O. Niwa. 2010. "One-Chip Biosensor for Simultaneous Disease Marker/Calibration Substance Measurement in Human Urine by Electrochemical Surface Plasmon Resonance Method." *Biosensors & Bioelectronics* 26 (4): 1536–1542. 10.1016/j.bios.2010.07.107

[46] Ribeiro, J.A., M.G.F. Sales, and C.M. Pereira. 2021. "Electrochemistry-Assisted Surface Plasmon Resonance Biosensor for Detection of CA 15-3." *Analytical Chemistry* 93 (22): 7815–7824. 10.1021/acs.analchem.0c05367

[47] Li, H., L. Zhang, X. Chen, J. Sun, and D. Cui. 2014. "SPR Imaging Combined with Cyclic Voltammetry for the Detection of Neural Activity." *AIP Advances* 4 (3): 031342. 10.1063/1.4869848

[48] Qatamin, A.H., J.H. Ghithan, M. Moreno, B.M. Nunn, K.B. Jones, F.P. Zamborini, R.S. Keynton, M.G. O'Toole, and S.B. Mendes. 2019. "Detection of Influenza Virus by Electrochemical Surface Plasmon Resonance under Potential Modulation." *Applied Optics* 58 (11): 2839–2844. 10.1364/AO.58.002839

# 17 Carbon-Based Electrochemical Devices for Energy Storage

*Saroj Sundar Baral*

## CONTENTS

## 17.1 INTRODUCTION

Energy storage has gained intense interest from researchers and commercial companies due to the world's economy's rapid growth and the rise in the use of portable electronics and electronic vehicles. Different types of energy, including thermal, chemical, electrical, and electrochemical energy, can be stored (Iqbal et al. 2019). Using electricity, thermal energy produces heating (geezers) or cooling (air conditioners). The most sources of renewable energy that produce electricity are wind and solar power. EESDs combines electrical and chemical energy within a short period and store in chemical form (Iqbal et al. 2019). Super-capacitors and rechargeable batteries are the EESDs that store and deliver electricity effectively and therefore used in a large number of industries (S. Chen et al. 2020; Gogotsi and Simon 2011; Liu et al. 2010). Due to these industries' rapid development, the needs for related EESDs are expanding in variety and specificity for various applications. For high mileage and quick charging, the increasingly popular EVs need EESDs having large energy/power densities or a load-bearing capability to act as structural body elements to reduce overall weight/volume (Bruce et al. 2012; Kang and Ceder 2009; Asp and Greenhalgh 2014). These new features require innovative EESDs, which are compact, flexible, wearable, large energy/power density, and have high physical strength. Traditional EESs, on the other hand, are heavy, inflexible, and not intended for load-bearing. They also have low energy/power demands; therefore, they are vital to creating essential materials, which rely on novel designs for the devices' structural components. However, the progress in energy conversion and storage relies heavily on nanostructured materials. They have distinctive dimensions with novel qualities and potential.

In light of this, the current review offers an outline of the development of EESDs and their fundamental purposes. Additional discussion of the numerous nanomaterials employed in EESDs during the last few decade has been included. Sustainable energy storage materials developments also have been emphasized as a future trend.

## 17.2  HISTORY OF EESDS AND MATERIALS

Numerous energy storage technologies, including fuel cells, batteries, capacitors, solar cells, etc., have been produced to date. Among them, a British scientist named William Grove invented the fuel cell in 1839, which was the first energy storage technology capable of producing a significant quantity of energy (Zackay and Ofek 2017). Grove's method was applied by NASA to create the first fuel cell utilized commercially in 1960 (Zackay and Ofek 2017). After that, several changes were made to enhance its capacity to produce electricity. Following that, scientists concentrated on the energy storage technologies like the battery and supercapacitor. Gaston Plante created a battery in the year 1859. Following the development of sticky plates in 1880, Emile Alphonse Faure was commercially launch the battery (Whittingham 2012). During development, he improved the battery's storage capacity by coating Pb plates with a paste made of Pb powder and $H_2SO_4$. Other researchers have employed this procedure extensively, and an increase in storage capacity was observed. Sellon has patented a method for creating batteries utilizing Faure's technology, which is having a larger capacity than Faure's battery (Iqbal et al. 2019). Further, Standard Oil of Ohio and American Oil Company developed and patented the first electronic double-layer supercapacitors in year 1961 (SOHIO). Following that, these devices greatly caught the researchers' attention, and a notable advancement in their development was noted. The invention of solar cell technology was another huge advancement in energy storage. Becquerel created the first solar cell in 1839 and introduced the idea of a solar cell (Iqbal et al. 2019). Scientist have investigated variety of resources to further improve the storage capacity and stability of these devices that are discussed in the following sections.

## 17.3  FUNDAMENTALS AND TYPES OF EESDS

Supercapacitors and batteries are the two basic categories of EESDs. Figure 17.1a depicts the standard components of planar supercapacitors and batteries. In the charge and discharge methods, supercapacitors quickly transfer charge (either ions or electrons) to store and release electrical energy. The CVC (Figure 17.1b) and linear galvanostatic charge-discharge curves demonstrate this quick charge-discharge characteristic (Figure 17.1c). Supercapacitors can operate for a significant number of cycles at close to 100% Coulombic efficiency. As a result, supercapacitors are

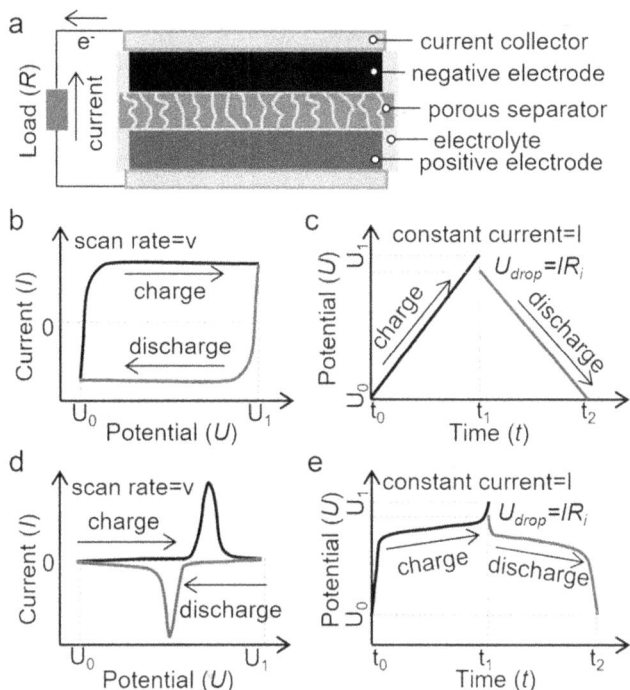

**FIGURE 17.1** Structure and electrochemical behavior of supercapacitors and batteries. Reprinted with permission from (S. Chen et al. 2020). Copyright 2022 American Chemical Society.

utilize in devices that need a high power output, including electric drills, cranes, and EV acceleration systems. The electrodes of supercapacitors are facilitated by two different types of capacitance. One type of capacitance is known as an electric double layer (EDL), and it occurs when electrolyte ions quickly absorb or desorb from the surface of an active material (Nishino 1996). The other is pseudo-capacitance, which is brought on by quick redox processes that take place during charge and discharge (Simon and Gogotsi 2008). Metal oxides, hydroxides, sulphides, nitrides, and conductive polymers are examples of pseudo-capacitive active materials. They all have a significantly large specific capacitance (SC) compared to CBMs (Simon and Gogotsi 2008). Keep in mind that due to their substantial SSA, nanostructured metal oxides exhibit considerable EDL capacitance.

The existence of redox surface groups and doped heteroatoms in the carbon materials make it a pseudo-capacitance. Examples include oxidized CNTs, RGO, B/N/P/S-doped RGO, and reduced graphene oxide (RGO)(S. H. Lee et al. 2010; Wu et al. 2012). Due to the latter's improved conductivity and stability, hybridizing carbon-pseudo-capacitors is a common technique to boost the former's usable capacitance and cycle performance (Frackowiak and Béguin 2001; D.-W. Wang et al. 2009). A supercapacitor's anode and cathode materials may be the same or different depending on whether the construction is symmetric or asymmetric. The working voltage of asymmetric capacitors is often higher, which leads to a substantially large ED.

Due to the materials' limited capacity for ion diffusion and electron transport, batteries use electrochemical redox processes to accumulate and discharge electrical energy. Batteries, therefore, have a significantly better ED. However, it has lower power density compared to supercapacitors. Rechargeable batteries are the subject of this review because they exhibit reversible redox processes while charge and discharge are shown by the existence of anodic and cathodic peaks in the CV curve (Figure 17.1d) and one plateau in the GCD curves (Figure 17.1e). Materials used as anodes include metals, metal oxides/sulfides, metal-organic complexes carbon Si, P, and organics, that are reduced while charging and oxidized after discharge (S. Chen et al. 2020). Typical cathode materials include sulphur, metal oxides/sulfides, metal hydroxides, metal phosphates, sulphur, oxygen (also air), iodine, and organics (S. Chen et al. 2020). These materials are oxidized during the charging process but reduced during discharge. Batteries with a large voltage output range and energy densities are employed to a variety of applications under varying working conditions thanks to versatile anode and cathode material combinations.

Hybrid cells, which use a capacitive electrode and battery electrode on the same component, have been created for the pursuit of EESDs having high energy and power densities. $NiOOH/Ni(OH)_2$ and $PbO_2$ are examples of capacitive carbon anodes that can be used in combination with the cathodes for NiCd and lead acid batteries to create hybrid cells with higher output voltages that operate in aqueous electrolytes.

## 17.4 CBMS FOR SUPERCAPACITORS AND BATTERIES

Many materials have been created that have demonstrated their potential for use in energy storage devices, including graphene, polymers, metal oxides, and carbon nanotubes.

### 17.4.1 CNT-Based Materials

CNTs are the most popular and effective one-dimensional nanostructure material for use in energy storage. They have good storage capacity; large surface-to-weight ratio; and outstanding electrical, thermal, and mechanical qualities (H. Zhang et al. 2008). A review of the literature showed that many papers have discussed the usage of CNT-based materials for EESDs.

He et al. (Yang He et al. 2010) created a $Fe_3O_4$/CNT composite and examined its electrochemical characteristics, for instance. This material was found to be stable up to 145 cycles and to have a great discharge capacity (656 mAh g1). The creation of a conductive network within the composite may be responsible for these materials' excellent electrochemical performance. MWCNT/S composite materials have been created by Wei and colleagues for use in batteries. They demonstrated the good retention capacity of the synthesized materials, which was 96.5% for 100 cycles, and determined that it may be used as cathode for Li-ion batteries. The uniform dispersion of MWCNT, that improves ion movement inside the nanocomposite, may be responsible for the composite's enhanced electrochemical characteristics (Wei et al. 2011). Then, utilizing a solution mineralization technique, Kim et al.

(Kim et al. 2010) created a core-shell wire of nano-size using CNT and $FePO_4$ and examined their electrochemical characteristics. In comparison to previous composites that have been published, they revealed that nanocomposite shows higher SC (1,000 mA g1) and significantly large rate. The core-shell assembly of the nanocomposite, which improved Li-ion transport and enhanced electrochemical characteristics, may be employed as environmentally acceptable cathode for batteries, according to researchers. $CNT/RuO_2$ core-shell composite was made by Jian et al. (Jian et al. 2014) using a sol-gel method, and the electrochemical activity was examined. While the composite was charged and discharged at 100 mA g1, they saw that it demonstrated outstanding efficiency (79%) over-potentials of 0.51 V and 0.21 V. Chemical vapor deposition was used by Xie and his associates (Xie et al. 2015) to create an effective electrode material utilizing tin (Sn) and CNT. The surface of carbon paper was used to generate the Sn/CNT nanopillar, which served as a free-standing anode material for a Na-ion battery application. The synthesized electrode displayed good cyclability up to 100 cycles and an exceptional reversible capacity (RC) (887 mAh cm2). Hong et al. (Hong et al. 2019) formulation of c-$Fe_2O_3$/CNT composite is intended for use in the field of energy storage. It demonstrated outstanding RC (518.5 mAh g1) and high cycling capacity (1186.8 mAh g1) after 400 cycles at 200 mA g1. These characteristics of the c-$Fe_2O_3$/CNT composite give it an advantage over other materials for Li-ion battery applications.

As a result, outstanding chemical and thermal resilience, CNT is also frequently utilized in the manufacture of supercapacitors (SC). Numerous publications on the production of CNT-based composite for use in SCs are currently available. As an illustration, Chen et al. (Y. Chen et al. 2015) described the manufacture of PPy electrodes based on CNT and examined their capacitive performance. This material outperformed pristine conducting polymers in terms of electrochemical performance and overcame the shortcoming of CP-based SC, namely quick capacitance degradation. The manufactured electrode demonstrated exceptional stability, flexibility, long life, and the capacity to preserve 95% capacitance even after 10,000 cycles, according to the authors. Later, Sun et al. (J. Sun et al. 2016) created a composite material for stretchable yarn SC applications utilizing PPy, CNT, and urethane. The PPy@CNTs@urethane hybrid has outstanding capacitive performance and can tolerate around 80% of the applied strain. This demonstrated their ability to become a top SC. $MnO_2$/CNT and $Fe_2O_3$/CNT macro-films have been created by Gu et al. (Gu and Wei 2016) for use in SCs. In comparison to pure CNT, $MnO_2$, and $Fe_2O_3$, the scientists discovered that the stated film is stretchy and has strong capacitive performance. The synergistic combination of each component is responsible for the exceptional behavior. A stretchable film with good electrochemical performance was created by Yu et al. (J. Yu et al. 2016) and has a SC of 1,147.12 mF $cm^2$ at 10 mV s1. These materials can withstand a 200% omni-directional strain, which is twice as much as what biaxial stretchy SCs made of CNT can withstand. This is explained by the interfacial interaction between the substrate molecule and the CNT. Then, using electrospinning techniques, Simotwo and colleagues (Simotwo et al. 2016) created polyaniline (PANI) and polyaniline-based CNT (PANI/CNT) nanofibers. They demonstrated how the synthetic materials might be a candidate for use in SC electrodes. When compared to pure PANI

(308 F g1), the PANI/CNT hybrid electrode performed better (SC of 385 F g1). The PANI/CNT hybrid electrode performs at its peak due to the interconnected network's improved electron transport to the active site.

## 17.4.2 GRAPHENE-BASED MATERIALS

The graphene-based materials used in various EESDs were covered in the following sections. Because there is a high van der Waals force, graphene is a more durable electrode material than CNTs. Additionally, they have a lot of surface area, are microporous, and have strong electrical conductivity, which makes them a good option for energy storage use. Reddy and colleagues (Iqbal et al. 2019) have created nitrogen-doped graphene films using a CVD process and studied its electrochemical characteristics in this regard. The capacitive behavior of the composite film was noticeably better than that of the virgin graphene film. The inclusion of numerous flaws on the surface of graphene, which increased the contact between the N and graphene, may be the cause of the composite film's enhanced reversible discharge capabilities. Due to the composite film's intercalation capabilities, Li batteries could potentially benefit from it.

Later, Wu et al. (Wu et al. 2011) demonstrated a comparison study by synthesizing graphene materials with nitrogen and boron dopants for use in LIB. It used a heat treatment process to create the aforementioned materials. The researcher found that the composite had outstanding rate capability and cyclability in addition to outstanding specific capacities of 199 mAh g1 and 235 mAh g1 at 25 A g1. For heavy-duty LIBs, these composite qualities may be advantageous. Manganese oxide ($Mn_3O_4$) and rGO were used to create an anode material for the LiB used by Wang et al. (Xing et al. 2017). The surface of the rGO sheets has been grown with $Mn_3O_4$ nanoparticles. Due to the outstanding interaction between these moieties, several electrochemical parameters, including SC and cyclability have been improved. It is demonstrated an exceptional 900 mAh g1 SC and are good for battery use. Following that, Sun and his team (Y. Sun et al. 2011) used a solution phase technique to create composite made of $MoO_2$ and graphene. The authors examined the synthesized nano-hybrid for use in LIBs and discovered that it has a good reversible SC (848.6 mAh g1) and cyclability (100% up to 70 cycles) at a current density (CD) of 500 mA g1. These two moieties' better specific capacities and capacity for charge storage result from their synergistic features. This nano-hybrid can also be used on an industrial scale because it was created without the use of seed crystals, templates, or additives.

For usage in LiB, Abe et al. (L. Wang et al. 2013) looked into performance of a $LiFePO_4$/graphite composite. The authors discovered that the material in question had good performance after more than 5,000 charge-discharge cycles. The performance is higher and the overall capacity will be lower as the N/P ratio decreases. Aging by cycling may be to blame for the decline in capability. A $Na_3V_2(PO_4)_3$-integrated graphene microsphere was created by Zhang et al. (J. Zhang et al. 2017) using a spray drying technique (Figure 17.2). The creation of a highly conductive route, which improves electron transport and leads to an increase in electrochemical properties, may be responsible for the composite's enhanced electrochemical capabilities. These materials might have used as cathode in Na-ion batteries.

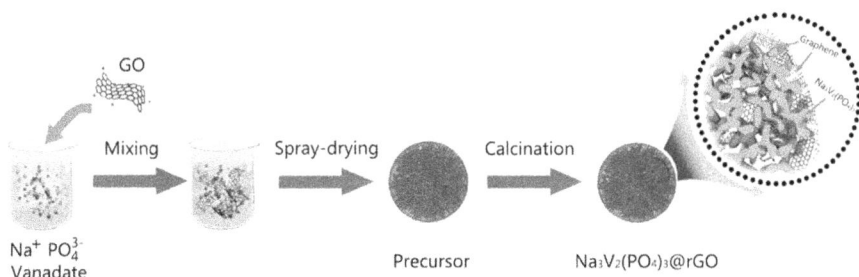

**FIGURE 17.2** Synthesis schematic of NVP@rGO microspheres. Reprinted with permission from (J. Zhang et al. 2017). Copyright 2022 American Chemical Society.

After that, He et al. (Yanyan He et al. 2017) created a composite material employing rGO, nickel oxide (NiO), and tin oxide ($SnO_2$) that served as an anode for Li-ion and Na-ion battery applications. The $NiO/SnO_2/rGO$ composite, according to the scientists, demonstrated an outstanding SC of 800 mAh g1 at a CD of 1,000 mA g1, as well as enhanced cyclability (400 cycles). The homogenous combination of these materials shortens the Li-ion diffusion path and boosts electron flow, is likely responsible for the composite's superior electrochemical properties. These improvements were made possible by the improvement in electrochemical properties.

Investigations into the electrochemical characteristics of the aforementioned foam revealed that it has a very high SC (123 mA/g at 5,000 mA/g of CD) and a very high cyclability (up to 10,000 cycles). Additionally, the synthesized foam's exceptional stability and capacity to charge and discharge at quick rates (80 s and 3,100 s, respectively) may expand their application in Al-ion batteries.

A $TiO_2/GO$ composite for a Na-ion capacitor was made by Le et al. (Le et al. 2017) using a microwave-assisted solvothermal method. The Na-ion capacitor's significant flaw, according to the literature, is its high capacitive performance. At a CD of 0.2C, the scientists discovered that the manufactured materials had better capacity (268 mAh g1). The SC of the aforementioned materials was discovered to be 126 mAh g1 after the cyclability through 18,000 cycles. As a result, it could be a contender for the Na-ion capacitor. Zuo and colleagues developed a high-performance 3D composite material utilizing graphene and $SnO_2$ (Figure 17.3). The material showed a remarkable RC of 720.8 mAh g1, according to the authors' tests of their electrochemical properties. The 3D structure and large surface area of the composite are likely responsible for its exceptional electrochemical characteristics. The increased surface area gives the molecules more room to interact, which leads to faster kinetics and improved electron transport, which increases their potential for use in batteries and super capacitors (Zuo et al. 2017).

### 17.4.3 ACTIVATED CBMs

Activated carbon (AC), a superior CBM, shares similarities with graphene and has application potential in energy storage devices (Z. Fan et al. 2011). The AC also should have small pores to enhance its electrochemical characteristics. By adding additional components, like conductive polymer, metal oxides, CBMs, these characteristics of the

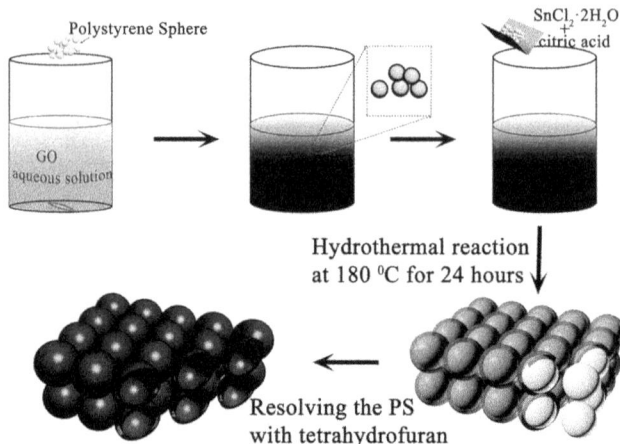

**FIGURE 17.3**   A schematic representation of the synthesis process. Reproduced from Ref. (Zuo et al. 2017) from the Royal Society of Chemistry.

AC have been further enhanced. For this report, Fan et al. used graphene, MnO2, and activated carbon nanofiber to create an effective energy storage material (ACN). The hybrid materials that were created have been evaluated for use in supercapacitors. They discovered a possible candidate for aelectrochemical supercapacitor use and discovered that the composite materials had spectacular SC (97%) after 1000 cycles (Z. Fan et al. 2011). Later, activated carbon cloth (ACC) was created by Wang et al. (Gongming Wang et al. 2014) for use in solid-state supercapacitors. With a SC of 8.8 mFg1 at a scan rate of 10 mV/s, they demonstrated excellent electrochemical characteristics and were discovered to be significantly greater than that of regular carbon cloth. The authors also showed that the manufactured ACC has a rapid charging rate and retains 50% of its capacitance. As a result, it was determined that this material would make the best flexible and high-performance capacitor. Zhang and his team have created a porous, ultrahigh-activated CBM for use in Li-S batteries. With a SC of 1,105 mA h g1, the synthesized materials displayed significant surface area (3,164 m2 g1) and good electrochemical characteristics. After 800 cycles, the manufactured material had outstanding retention capacity (up to 51%) and was identified as a viable cathode material for Li-S battery applications. Due to the high surface area and wide pore volume that aid in the homogeneous dispersion of sulphur into the AC, the stated composite has outstanding characteristics, which is why an improvement in electrical behavior is shown (S. Zhang et al. 2014). For an effective anode material for Na-ion batteries, Luo et al. (Luo et al. 2015) have also looked into the electrochemical characteristics of CNT, graphene, graphite, and activated CBMs. After that, Li et al. (Li et al. 2016) added nitrogen (N) to the AC and investigated its electrochemical functionality. They discovered that the addition of N improves the electrochemical characteristics, particularly a rise in energy density (ED) (230 Wh kg[1]). Even after 8000 cycles, the nitrogen-doped activated carbon showed significant specific retention (76.3%). Due to these qualities, the synthetic composite material is used in hybrid supercapacitors. Additionally, employing activated pinecone carbon, Barzegar et al.

(Barzegar et al. 2017) developed an asymmetric capacitor (APC). They have been used in low-cost, environmentally friendly carbon-based capacitors because of their huge surface area (808 m$^2$ g$^1$), the bright SC of 69 F g$^1$ at a CD of 0.5 A g$^1$, and exceptional ED (24.6 Wh kg$^1$). Lee and coworkers have created sulfur-incorporated activated carbon composite to further enhance the electrochemical characteristics of AC. They demonstrated that adding sulphur to the AC improved the aforementioned composite's electrochemical performance and allowed for usage in Li-S battery applications. The interaction between sulphur and AC intensifies, shortening the distance between layers and enhancing the composite's capacity for Li-ion transit and storage. The authors discovered that the composite had a high SC (1,351 mA h g$^1$) and that the SC only decreased after 300 cycles by a certain amount (920 mA h g$^1$) (J. S. Lee et al. 2017). By utilizing potassium hydroxide as a surface-etching agent during a high-temperature annealing process, Tai et al. (Tai et al. 2017) have created AC from graphite. Yu et al. (L. Yu et al. 2018) recently created a flexible electrode by a one-step chemical method using MXene (Ti$_3$-C$_2$Tx) and AC. The synthesized material's electrochemical capacitance performance was evaluated, and it was found that the composite electrode displayed outstanding capacitive behavior. According to the authors, the synthetic material has outstanding charge retention properties and superior capacitance (126 F g$^1$) at a CD of 0.1 A g$^1$ (57.9%). MXene was used as a binder, conductive additive, and to give the electrode a flexible backbone in this situation.

### 17.4.4 CONDUCTING POLYMER-BASED MATERIALS

In addition to having high mechanical strength and flexibility, carbon-based fibers (CFs) like carbon nanotube (CNT) and graphene fibres also have strong electrical conductivity and electrochemical characteristics. They can be put together to create textile and wire-shaped EESDs, which can serve as unique power sources for wearable, tiny, flexible electronics and intelligent apparel (Figure 17.4). High-strength CF textiles can also be used to assemble structural EESDs as active electrodes and current collectors.

Wang and colleagues (Gang Wang et al. 2015) developed a SnS$_2$/PANI hybrid and demonstrated its promise in LIB application. Due to the lamellar sandwich structure of the SnS$_2$/PANI nano-plates, they showed that the hybrid material has improved RC and rate capability. Due to the sandwich structure's ability to prevent SnS$_2$ nano-plate stacking and promote quick electron transport between the two moieties, an increase in capacity and cyclability is seen. Additionally, the interaction of PANI with exfoliated SnS$_2$ provides a short path length for Li+ ion transport, further enhancing their capacity for charge-discharge. This results in increased cyclability, coulombic efficiency, and energy storage capacity. The SnO$_2$/PPy hybrid composite was subsequently made by Cao et al. (Cao et al. 2016) who also investigated its electrochemical performance for energy storage. They showed that the composite material had a remarkable SC (646 mA h g$^1$) and an exceptional ability to retain charge (98%) even after 150 cycles.

According to the authors, PPy prevents materials from directly contacting the electrolyte, causing a stable solid electrolyte interface (SEI) layer to form. These layers are crucial to improve both stability and rate capability. Then, Parveen et al.

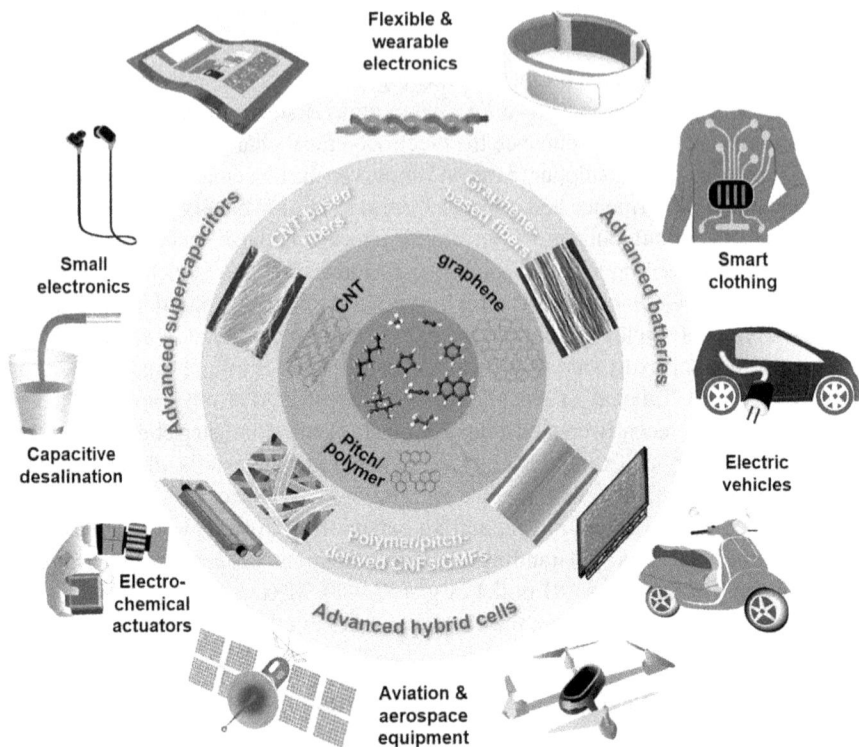

**FIGURE 17.4** Overview of CFs for advanced EESDs and their potential applications in various fields. Reprinted with permission from (S. Chen et al. 2020). Copyright 2022 American Chemical Society.

(Parveen et al. 2017) created a composite of polyaniline, $TiO_2$, and graphene, and discovered that it had significantly better cyclic and stability properties than PANI/GN and pure PANI (80% retention after 1,000 cycles). The ternary composite's improved electrochemical characteristics were primarily responsible for quicker contact due to the porous structure involving the electrolyte and the electrode. The substantial surface area and greater SC were achieved due to faster electron transportation.

The use of CP is also found in the supercapacitor industry. The CP-based supercapacitor is a matter of widespread research. One such example is the GO nano-sheet, nano-wafer hybrid material that Gao and colleagues (Gao et al. 2013) produced and investigated. At 5 $mVs^1$ CD, aforementioned composite material displayed a remarkable capacitance of 329.5 F $g^1$. They demonstrated the improved life cycle of the aforementioned hybrid material as well. These characteristics of the hybrid make them a possible contender for use in supercapacitors. The hierarchical structure of the hybrid composite, which allows for quick electron transport between the two moieties, may be the cause of its remarkable electrochemical capabilities. By employing a chemical polymerization method with polyaniline (PANI) and rGO, Fan et al. (W. Fan et al. 2013) created a hybrid hollow sphere.

They demonstrated that the hybrid electrode has exceptional cyclability and stability as well as a huge capacitance of 614 F g1 at 1 A g$^1$. The creation of a conductive network between moieties is primarily accountable for composite's exceptional electrochemical characteristics.

A rise in electrochemical properties is observed due to wrapping of rGO on the PANI surface. A hybrid graphene-PPy aerogel was created by Ye and colleagues (Ye and Feng 2014), and its capacitive performance has been studied. The synthetic aerogel demonstrated exceptional stability, rate capability, and capacitance (253 F g$^1$). The vast space provided by PPy's high surface area also prevents aggregation and restacking of graphene sheets, which speeds up the electron transmission. Due to these characteristics, hybrid materials are effective for high-performance supercapacitors. Tao et al. (Tao et al. 2013) created a ternary hybrid material based on polypyrrole (PPy), $MnO_2$, and carbon fiber (CF) for use in supercapacitors. The authors showed that the material might serve as an electrode for EESD because it had an outstanding capacitance (69.3 F cm$^3$) and high ED (6.16 103 Wh cm$^3$). Mesoporous PANI/G hybrid was created by Wang et al. (Q. Wang et al. 2014) and its electrochemical characteristics were investigated. The mesoporous material had a large capacitance of 749 F g$^1$ when the CD was 0.5 A g$^1$, which is significantly greater than that of an electrode based on nascent PANI (315 F g$^1$). This might be controlled by the interaction between PANI and graphene. A flexible PANI/RGO/MWCNT ternary hybrid film has been created, as reported by Fan and colleagues (H. Fan et al. 2014) in a different investigation. The composite film demonstrated improved capacitance and stability, making it a potential top choice as a component in supercapacitor applications. The MWCNT served as a connecting element, interconnecting to the RGO flaws, forming a conductive channel, and facilitating quick electron transport. So, employing graphene, $SnO_2$, and PANI, Jin et al. (Jin and Jia 2015) created a ternary hybrid material and assessed its electrochemical performance. They demonstrated that the artificially created composite materials have a remarkable capacitance of 913.4 F g$^1$, as well as strong cyclic stability (1,000 cycles), and rate capability. The hybrid composite's great stability and enhanced mechanical strength are both a result of the homogenous dispersion of graphene nano-sheets. Through in-situ chemical oxidative polymerization, Liu et al. (Lu et al. 2018) have created hollow composite materials based on polypyrrole (PPy) and carbon microspheres. The hollow carbon microsphere's SC of 508 F g$^1$ at 1 A g$^1$ is increased by the 15 nm PPy coating. This improvement is primarily controlled by the development of a thin PPy coating on the carbon microsphere, which leads to an increase in electron transport.

## 17.4.5 FULLERENE MATERIALS

Another significant class of carbon is called fullerene (C60), in which the carbon atoms are linked together by single and double bonds to form a closed mesh-like structure (Kroto et al. 1985). Due to their chemical and thermal stability, they have drawn a lot of attention since their discovery in 1985. These materials are used in energy storage devices because of their superior conductivity. For usage in secondary batteries, Arie et al. (Arie et al. 2009) have created a fullerene-coated silicon thin film anode material. They used a plasma coating method to coat the fullerene, and they discovered that the material has an excellent SC of >2,000 mAh g$^1$. According to the author, the fullerene coating improves the materials' conductivity, which makes them more useful for battery applications. Later, Troshin and colleagues (Troshin et al. 2011) developed fullerene (C70)-based materials for solar cells, including (W. Fan et al. 2013) fullerene, PCPP, and [phenyl-C71-propionic acid butyl ester ((W. Fan et al. 2013) PCPB). The C70 molecule was discovered to function as a better acceptor than photovoltaic materials composed of the C60 molecule and has a 10% greater short circuit CD. These characteristics indicated that these materials might be used in bulk heterojunction solar cells. The efficiency of photovoltaic systems is also increased by grafting the C60 molecule onto graphene, as demonstrated by Yu et al. (D. Yu et al. 2011). These materials were employed in solar cells as electron acceptors, improving the efficiency and performance with a 1.22% power conversion. Non-transition metal (cluster (C60)) cathode have been produced by Zhang et al. (R. Zhang et al. 2015) for use in Mg-battery applications. This cathode's precise capacity was discovered to be 50 mAh g$^1$. The additional electrons across C60 cluster and increase in cathode's efficacy were primarily responsible for the improved performance of these materials. According to Noh et al. (Noh et al. 2017), nitrogen-doped fullerene (NC60) was created and used as a catalyst in fuel cell and batteries.

## 17.5 PROSPECTS AND CONCLUSION

Numerous studies on the creation of energy storage materials have been published, but much more work has to be done. A review of the literature showed that several author reported on three-dimensional composite materials and that two-dimensional nanostructures have been created in vast quantities. Therefore, future work should concentrate on creating these kinds of materials. Literature also demonstrates that minor structural changes can enhance electrochemical characteristics. Therefore, it is necessary to select materials with good electrochemical properties while making batteries. Additionally, new composite materials such as metal-organic/inorganic frameworks have recently drawn a lot of interest from scientists and technologists, yet there aren't many articles that discuss them. Additionally, according to the literature, different materials have been created to take the role of silicon in solar cell devices. However, these materials still perform poorly. So, to improve power conversion efficiency, synthesized materials and hybrid materials must be developed soon. Similar to this, the materials demonstrated good SC in supercapacitor applications, but maintaining their stability is a significant difficulty. Therefore, future research should concentrate on creating such materials with great stability and superior electrochemical characteristics.

Further improvement of these materials can undoubtedly increase their ability to store energy and their stability, which is also crucial for EESDs. In conclusion, the current review paper emphasizes the development of EESDs as well as their fundamental purposes across a range of EESD kinds. Additionally, a thorough discussion of the numerous nanomaterials employed in EESDs during the last few years has also been included.

## REFERENCES

Arie, Arenst Andreas, Jin O. Song, and Joong Kee Lee. 2009. "Structural and Electrochemical Properties of Fullerene-Coated Silicon Thin Film as Anode Materials for Lithium Secondary Batteries." *Materials Chemistry and Physics* 113 (1): 249–254. doi:10.1016/j.matchemphys.2008.07.082

Asp, Leif E., and Emile S. Greenhalgh. 2014. "Structural Power Composites." *Composites Science and Technology* 101 (September): 41–61. doi:10.1016/j.compscitech.2014.06.020

Barzegar, Farshad, Abdulhakeem Bello, Julien K. Dangbegnon, Ncholu Manyala, and Xiaohua Xia. 2017. "Asymmetric Supercapacitor Based on Activated Expanded Graphite and Pinecone Tree Activated Carbon with Excellent Stability." *Applied Energy* 207 (December): 417–426. doi:10.1016/j.apenergy.2017.05.110

Bruce, Peter G., Stefan A. Freunberger, Laurence J. Hardwick, and Jean-Marie Tarascon. 2012. "Li–O2 and Li–S Batteries with High Energy Storage." *Nature Materials* 11 (1): 19–29. doi:10.1038/nmat3191

Cao, Zhenzhen, Hongyan Yang, Peng Dou, Chao Wang, Jiao Zheng, and Xinhua Xu. 2016. "Synthesis of Three-Dimensional Hollow SnO 2 @PPy Nanotube Arrays via Template-Assisted Method and Chemical Vapor-Phase Polymerization as High Performance Anodes for Lithium-Ion Batteries." *Electrochimica Acta* 209 (August): 700–708. doi:10.1016/j.electacta.2016.01.158

Chen, Shaohua, Ling Qiu, and Hui-Ming Cheng. 2020. "Carbon-Based Fibers for Advanced Electrochemical Energy Storage Devices." *Chemical Reviews* 120 (5): 2811–2878. doi:10.1021/acs.chemrev.9b00466

Chen, Yanli, Lianhuan Du, Peihua Yang, Peng Sun, Xiang Yu, and Wenjie Mai. 2015. "Significantly Enhanced Robustness and Electrochemical Performance of Flexible Carbon Nanotube-Based Supercapacitors by Electrodepositing Polypyrrole." *Journal of Power Sources* 287 (August): 68–74. doi:10.1016/j.jpowsour.2015.04.026

Fan, Haosen, Ning Zhao, Hao Wang, Jian Xu, and Feng Pan. 2014. "3D Conductive Network-Based Free-Standing PANI–RGO–MWNTs Hybrid Film for High-Performance Flexible Supercapacitor." *Journal of Materials Chemistry A* 2 (31): 12340–12347. doi:10.1039/C4TA02118E

Fan, Wei, Chao Zhang, Weng Weei Tjiu, Kumari Pallathadka Pramoda, Chaobin He, and Tianxi Liu. 2013. "Graphene-Wrapped Polyaniline Hollow Spheres As Novel Hybrid Electrode Materials for Supercapacitor Applications." *ACS Applied Materials & Interfaces* 5 (8): 3382–3391. doi:10.1021/am4003827

Fan, Zhuangjun, Jun Yan, Tong Wei, Linjie Zhi, Guoqing Ning, Tianyou Li, and Fei Wei. 2011. "Asymmetric Supercapacitors Based on Graphene/MnO2 and Activated Carbon Nanofiber Electrodes with High Power and Energy Density." *Advanced Functional Materials* 21 (12): 2366–2375. doi:10.1002/adfm.201100058

Frackowiak, Elzbieta, and François Béguin. 2001. "Carbon Materials for the Electrochemical Storage of Energy in Capacitors." *Carbon* 39 (6): 937–950. doi:10.1016/S0008-6223 (00)00183-4

Gao, Zan, Wanlu Yang, Jun Wang, Bin Wang, Zhanshuang Li, Qi Liu, Milin Zhang, and Lianhe Liu. 2013. "A New Partially Reduced Graphene Oxide Nanosheet/Polyaniline Nanowafer Hybrid as Supercapacitor Electrode Material." *Energy & Fuels* 27 (1): 568–575. doi:10.1021/ef301795g

Gogotsi, Y., and P. Simon. 2011. "True Performance Metrics in Electrochemical Energy Storage." *Science* 334 (6058): 917–918. doi:10.1126/science.1213003

Gu, Taoli, and Bingqing Wei. 2016. "High-Performance All-Solid-State Asymmetric Stretchable Supercapacitors Based on Wrinkled MnO 2 /CNT and Fe 2 O 3 /CNT Macrofilms." *Journal of Materials Chemistry A* 4 (31): 12289–12295. doi:10.1039/C6TA04712B

He, Yang, Ling Huang, Jin-Shu Cai, Xiao-Mei Zheng, and Shi-Gang Sun. 2010. "Structure and Electrochemical Performance of Nanostructured Fe3O4/Carbon Nanotube Composites as Anodes for Lithium Ion Batteries." *Electrochimica Acta* 55 (3): 1140–1144. doi:10.1016/j.electacta.2009.10.014

He, Yanyan, Aihua Li, Caifu Dong, Chuanchuan Li, and Liqiang Xu. 2017. "Mesoporous Tin-Based Oxide Nanospheres/Reduced Graphene Composites as Advanced Anodes for Lithium-Ion Half/Full Cells and Sodium-Ion Batteries." *Chemistry - A European Journal* 23 (55): 13724–13733. doi:10.1002/chem.201702225

Hong, Min, Yanjie Su, Chao Zhou, Lu Yao, Jing Hu, Zhi Yang, Liying Zhang, Zhihua Zhou, Nantao Hu, and Yafei Zhang. 2019. "Scalable Synthesis of γ-Fe2O3/CNT Composite as High-Performance Anode Material for Lithium-Ion Batteries." *Journal of Alloys and Compounds* 770 (January): 116–124. doi:10.1016/j.jallcom.2018.08.118

Iqbal, Sajid, Halima Khatoon, Ashiq Hussain Pandit, and Sharif Ahmad. 2019. "Recent Development of Carbon Based Materials for Energy Storage Devices." *Materials Science for Energy Technologies* 2 (3): 417–428. doi:10.1016/j.mset.2019.04.006

Jian, Zelang, Pan Liu, Fujun Li, Ping He, Xianwei Guo, Mingwei Chen, and Haoshen Zhou. 2014. "Core-Shell-Structured CNT@RuO 2 Composite as a High-Performance Cathode Catalyst for Rechargeable Li-O 2 Batteries." *Angewandte Chemie International Edition* 53 (2): 442–446. doi:10.1002/anie.201307976

Jin, Yuhong, and Mengqiu Jia. 2015. "Design and Synthesis of Nanostructured Graphene-SnO2-Polyaniline Ternary Composite and Their Excellent Supercapacitor Performance." *Colloids and Surfaces A: Physicochemical and Engineering Aspects* 464 (January): 17–25. doi:10.1016/j.colsurfa.2014.09.032

Kang, Byoungwoo, and Gerbrand Ceder. 2009. "Battery Materials for Ultrafast Charging and Discharging." *Nature* 458 (7235): 190–193. doi:10.1038/nature07853

Kim, Sung-Wook, Jungki Ryu, Chan Beum Park, and Kisuk Kang. 2010. "Carbon Nanotube-Amorphous FePO4 Core–Shell Nanowires as Cathode Material for Li Ion Batteries." *Chemical Communications* 46 (39): 7409. doi:10.1039/c0cc02524k

Kroto, H. W., J. R. Heath, S. C. O'Brien, R. F. Curl, and R. E. Smalley. 1985. "C60: Buckminsterfullerene." *Nature* 318 (6042): 162–163. doi:10.1038/318162a0

Le, Zaiyuan, Fang Liu, Ping Nie, Xinru Li, Xiaoyan Liu, Zhenfeng Bian, Gen Chen, Hao Bin Wu, and Yunfeng Lu. 2017. "Pseudocapacitive Sodium Storage in Mesoporous Single-Crystal-like TiO 2 –Graphene Nanocomposite Enables High-Performance Sodium-Ion Capacitors." *ACS Nano* 11 (3): 2952–2960. doi:10.1021/acsnano.6b08332

Lee, Jun Seop, Wooyoung Kim, Jyongsik Jang, and Arumugam Manthiram. 2017. "Sulfur-Embedded Activated Multichannel Carbon Nanofiber Composites for Long-Life, High-Rate Lithium-Sulfur Batteries." *Advanced Energy Materials* 7 (5): 1601943. doi:10.1002/aenm.201601943

Lee, Sun Hwa, Hyun Wook Kim, Jin Ok Hwang, Won Jun Lee, Joon Kwon, Christopher W. Bielawski, Rodney S. Ruoff, and Sang Ouk Kim. 2010. "Three-Dimensional Self-Assembly of Graphene Oxide Platelets into Mechanically Flexible Macroporous Carbon Films." *Angewandte Chemie International Edition* 49 (52): 10084–10088. doi:10.1002/anie.201006240

Li, Bing, Fang Dai, Qiangfeng Xiao, Li Yang, Jingmei Shen, Cunman Zhang, and Mei Cai. 2016. "Nitrogen-Doped Activated Carbon for a High Energy Hybrid Supercapacitor." *Energy & Environmental Science* 9 (1): 102–106. doi:10.1039/C5EE03149D

Liu, Chang, Feng Li, Lai-Peng Ma, and Hui-Ming Cheng. 2010. "Advanced Materials for Energy Storage." *Advanced Materials* 22 (8): E28–E62. doi:10.1002/adma.200903328

Lu, Qun, Jia Liu, Xianyou Wang, Bing Lu, Manfang Chen, and Meihong Liu. 2018. "Construction and Characterizations of Hollow Carbon Microsphere@polypyrrole Composite for the High Performance Supercapacitor." *Journal of Energy Storage* 18 (August): 62–71. doi:10.1016/j.est.2018.04.022

Luo, Xu-Feng, Cheng-Hsien Yang, You-Yu Peng, Nen-Wen Pu, Ming-Der Ger, Chien-Te Hsieh, and Jeng-Kuei Chang. 2015. "Graphene Nanosheets, Carbon Nanotubes, Graphite, and Activated Carbon as Anode Materials for Sodium-Ion Batteries." *Journal of Materials Chemistry A* 3 (19): 10320–10326. doi:10.1039/C5TA00727E

Nishino, Atsushi. 1996. "Capacitors: Operating Principles, Current Market and Technical Trends." *Journal of Power Sources* 60 (2): 137–147. doi:10.1016/S0378-7753(96)80003-6

Noh, Seung Hyo, Choah Kwon, Jeemin Hwang, Takeo Ohsaka, Beom-Jun Kim, Tae-Young Kim, Young-Gi Yoon, Zhongwei Chen, Min Ho Seo, and Byungchan Han. 2017. "Self-Assembled Nitrogen-Doped Fullerenes and Their Catalysis for Fuel Cell and Rechargeable Metal–Air Battery Applications." *Nanoscale* 9 (22): 7373–7379. doi:1 0.1039/C7NR00930E

Parveen, Nazish, Mohammad Omaish Ansari, Thi Hiep Han, and Moo Hwan Cho. 2017. "Simple and Rapid Synthesis of Ternary Polyaniline/Titanium Oxide/Graphene by Simultaneous TiO2 Generation and Aniline Oxidation as Hybrid Materials for Supercapacitor Applications." *Journal of Solid State Electrochemistry* 21 (1): 57–68. doi:10.1007/s10008-016-3310-8

Simon, Patrice, and Yury Gogotsi. 2008. "Materials for Electrochemical Capacitors." *Nature Materials* 7 (11): 845–854. doi:10.1038/nmat2297

Simotwo, Silas K., Christopher DelRe, and Vibha Kalra. 2016. "Supercapacitor Electrodes Based on High-Purity Electrospun Polyaniline and Polyaniline–Carbon Nanotube Nanofibers." *ACS Applied Materials & Interfaces* 8 (33): 21261–21269. doi:10.1021/acsami.6b03463

Sun, Jinfeng, Yan Huang, Chenxi Fu, Zhengyue Wang, Yang Huang, Minshen Zhu, Chunyi Zhi, and Hong Hu. 2016. "High-Performance Stretchable Yarn Supercapacitor Based on PPy@CNTs@urethane Elastic Fiber Core Spun Yarn." *Nano Energy* 27 (September): 230–237. doi:10.1016/j.nanoen.2016.07.008

Sun, Yongming, Xianluo Hu, Wei Luo, and Yunhui Huang. 2011. "Self-Assembled Hierarchical MoO 2 /Graphene Nanoarchitectures and Their Application as a High-Performance Anode Material for Lithium-Ion Batteries." *ACS Nano* 5 (9): 7100–7107. doi:10.1021/nn201802c

Tai, Zhixin, Qing Zhang, Yajie Liu, Huakun Liu, and Shixue Dou. 2017. "Activated Carbon from the Graphite with Increased Rate Capability for the Potassium Ion Battery." *Carbon* 123 (October): 54–61. doi:10.1016/j.carbon.2017.07.041

Tao, Jiayou, Nishuang Liu, Wenzhen Ma, Longwei Ding, Luying Li, Jun Su, and Yihua Gao. 2013. "Solid-State High Performance Flexible Supercapacitors Based on Polypyrrole-MnO2-Carbon Fiber Hybrid Structure." *Scientific Reports* 3 (1): 2286. doi:10.1038/srep02286

Troshin, Pavel A., Harald Hoppe, Alexander S. Peregudov, Martin Egginger, Sviatoslav Shokhovets, Gerhard Gobsch, N. Serdar Sariciftci, and Vladimir F. Razumov. 2011. "[70]Fullerene-Based Materials for Organic Solar Cells." *ChemSusChem* 4 (1): 119–124. doi:10.1002/cssc.201000246

Wang, Da-Wei, Feng Li, Jinping Zhao, Wencai Ren, Zhi-Gang Chen, Jun Tan, Zhong-Shuai Wu, Ian Gentle, Gao Qing Lu, and Hui-Ming Cheng. 2009. "Fabrication of Graphene/Polyaniline Composite Paper via In Situ Anodic Electropolymerization for High-Performance Flexible Electrode." *ACS Nano* 3 (7): 1745–1752. doi:10.1021/nn900297m

Wang, Gang, Jun Peng, Lili Zhang, Jun Zhang, Bin Dai, Mingyuan Zhu, Lili Xia, and Feng Yu. 2015. "Two-Dimensional SnS 2 @PANI Nanoplates with High Capacity and Excellent Stability for Lithium-Ion Batteries." *Journal of Materials Chemistry A* 3 (7): 3659–3666. doi:10.1039/C4TA06384H

Wang, Gongming, Hanyu Wang, Xihong Lu, Yichuan Ling, Minghao Yu, Teng Zhai, Yexiang Tong, and Yat Li. 2014. "Solid-State Supercapacitor Based on Activated

Carbon Cloths Exhibits Excellent Rate Capability." *Advanced Materials* 26 (17): 2676–2682. doi:10.1002/adma.201304756

Wang, Li, Xiangming He, Jianjun Li, Jian Gao, Mou Fang, Guangyu Tian, Jianlong Wang, and Shoushan Fan. 2013. "Graphene-Coated Plastic Film as Current Collector for Lithium/Sulfur Batteries." *Journal of Power Sources* 239 (October): 623–627. doi:10.1 016/j.jpowsour.2013.02.008

Wang, Qian, Jun Yan, Zhuangjun Fan, Tong Wei, Milin Zhang, and Xiaoyan Jing. 2014. "Mesoporous Polyaniline Film on Ultra-Thin Graphene Sheets for High Performance Supercapacitors." *Journal of Power Sources* 247 (February): 197–203. doi:10.1016/ j.jpowsour.2013.08.076

Wei, Wei, Jiulin Wang, Longjie Zhou, Jun Yang, Bernd Schumann, and Yanna NuLi. 2011. "CNT Enhanced Sulfur Composite Cathode Material for High Rate Lithium Battery." *Electrochemistry Communications* 13 (5): 399–402. doi:10.1016/j.elecom.2011.02.001

Whittingham, M. S. 2012. "History, Evolution, and Future Status of Energy Storage." *Proceedings of the IEEE* 100 (Special Centennial Issue): 1518–1534. doi:10.1109/ JPROC.2012.2190170

Wu, Zhong-Shuai, Wencai Ren, Li Xu, Feng Li, and Hui-Ming Cheng. 2011. "Doped Graphene Sheets As Anode Materials with Superhigh Rate and Large Capacity for Lithium Ion Batteries." *ACS Nano* 5 (7): 5463–5471. doi:10.1021/nn2006249

Wu, Zhong-Shuai, Andreas Winter, Long Chen, Yi Sun, Andrey Turchanin, Xinliang Feng, and Klaus Müllen. 2012. "Three-Dimensional Nitrogen and Boron Co-Doped Graphene for High-Performance All-Solid-State Supercapacitors." *Advanced Materials* 24 (37): 5130–5135. doi:10.1002/adma.201201948

Xie, Xiuqiang, Katja Kretschmer, Jinqiang Zhang, Bing Sun, Dawei Su, and Guoxiu Wang. 2015. "Sn@CNT Nanopillars Grown Perpendicularly on Carbon Paper: A Novel Free-Standing Anode for Sodium Ion Batteries." *Nano Energy* 13 (April): 208–217. doi:1 0.1016/j.nanoen.2015.02.022

Xing, Yue-Ming, Xiao-Hua Zhang, Dai-Huo Liu, Wen-Hao Li, Ling-Na Sun, Hong-Bo Geng, Jing-Ping Zhang, Hong-Yu Guan, and Xing-Long Wu. 2017. "Porous Amorphous Co 2 P/N,B-Co-Doped Carbon Composite as an Improved Anode Material for Sodium-Ion Batteries." *ChemElectroChem* 4 (6): 1395–1401. doi:10.1002/celc.201700093

Ye, Shibing, and Jiachun Feng. 2014. "Self-Assembled Three-Dimensional Hierarchical Graphene/Polypyrrole Nanotube Hybrid Aerogel and Its Application for Supercapacitors." *ACS Applied Materials & Interfaces* 6 (12): 9671–9679. doi:10.1 021/am502077p

Yu, Dingshan, Kyusoon Park, Michael Durstock, and Liming Dai. 2011. "Fullerene-Grafted Graphene for Efficient Bulk Heterojunction Polymer Photovoltaic Devices." *The Journal of Physical Chemistry Letters* 2 (10): 1113–1118. doi:10.1021/jz200428y

Yu, Jiali, Weibang Lu, Shaopeng Pei, Ke Gong, Liyun Wang, Linghui Meng, Yudong Huang, et al. 2016. "Omnidirectionally Stretchable High-Performance Supercapacitor Based on Isotropic Buckled Carbon Nanotube Films." *ACS Nano* 10 (5): 5204–5211. doi:10.1021/acsnano.6b00752

Yu, Lanyong, Longfeng Hu, Babak Anasori, Yi-Tao Liu, Qizhen Zhu, Peng Zhang, Yury Gogotsi, and Bin Xu. 2018. "MXene-Bonded Activated Carbon as a Flexible Electrode for High-Performance Supercapacitors." *ACS Energy Letters* 3 (7): 1597–1603. doi: 10.1021/acsenergylett.8b00718

Zackay, Barak, and Eran O. Ofek. 2017. "How to COAAD Images. I. Optimal Source Detection and Photometry of Point Sources Using Ensembles of Images." *The Astrophysical Journal* 836 (2): 187. doi:10.3847/1538-4357/836/2/187

Zhang, Hao, Gaoping Cao, Zhiyong Wang, Yusheng Yang, Zujin Shi, and Zhennan Gu. 2008. "Growth of Manganese Oxide Nanoflowers on Vertically-Aligned Carbon

Nanotube Arrays for High-Rate Electrochemical Capacitive Energy Storage." *Nano Letters* 8 (9): 2664–2668. doi:10.1021/nl800925j

Zhang, Jiexin, Yongjin Fang, Lifen Xiao, Jiangfeng Qian, Yuliang Cao, Xinping Ai, and Hanxi Yang. 2017. "Graphene-Scaffolded Na 3 V 2 (PO 4) 3 Microsphere Cathode with High Rate Capability and Cycling Stability for Sodium Ion Batteries." *ACS Applied Materials & Interfaces* 9 (8): 7177–7184. doi:10.1021/acsami.6b16000

Zhang, Ruigang, Fuminori Mizuno, and Chen Ling. 2015. "Fullerenes: Non-Transition Metal Clusters as Rechargeable Magnesium Battery Cathodes." *Chemical Communications* 51 (6): 1108–1111. doi:10.1039/C4CC08139K

Zhang, Songtao, Mingbo Zheng, Zixia Lin, Nianwu Li, Yijie Liu, Bin Zhao, Huan Pang, Jieming Cao, Ping He, and Yi Shi. 2014. "Activated Carbon with Ultrahigh Specific Surface Area Synthesized from Natural Plant Material for Lithium–Sulfur Batteries." *Journal of Materials Chemistry A* 2 (38): 15889–15896. doi:10.1039/C4TA03503H

Zuo, Shi-yong, Zhi-guo Wu, Shuan-kui Li, De Yan, Yan-hua Liu, Feng-yi Wang, Ren-fu Zhuo, Bai-song Geng, Jun Wang, and Peng-xun Yan. 2017. "High Rate Performance SnO 2 Based Three-Dimensional Graphene Composite Electrode for Lithium-Ion Battery Applications." *RSC Advances* 7 (29): 18054–18059. doi:10.1039/C6RA28258J

# 18 Electrochemical Biosensors for Rare Cell Isolation

*Sarthak Singh, Dev Choudhary, and Jegatha Nambi Krishnan*
Department of Chemical Engineering, BITS Pilani, K K Birla Goa Campus, Goa, India

## CONTENTS

## 18.1 INTRODUCTION

Rare cells are low abundance cells with concentrations less than 1,000 per millilitre volume of a sample [1]. Their efficient selection, enumeration, and isolation are essential in a various applications such as cancer research, discovering circulating tumour cells (CTCs) in blood for early diagnosis and prognosis, non-invasive isolation, and characterization of fetal cells from maternal blood, environmental monitoring, detection of T lymphocytes or autoimmune diseases, as well as isolation and identification of pathogens and protozoan parasites in numerous food and water samples. However, their isolation from various complex heterogeneous samples is fraught with challenges. Common types of rare cells include CTC, fetal, stem, somatic, T cell, bacteria, virus, and protozoa/helminths [1]. According to the World Health Organisation (WHO), cancer, in 2019, ranked as the first- or second-leading cause of premature mortality in 112 of 183 countries and ranked third or fourth in a further 23 countries, including India. This worldwide burden of cancer is estimated to reach 28.4 million in 2040 [2]. Cancer can manifest itself in over 200 different forms, such as lung, bladder, prostate, bone, breast, colon, neuron, hematologic, skin, leukaemia, and ovary [3,4]. Therefore, CTCs early detection through screening has gained significant interest and attention to reduce this future burden and suffering.

Current conventional methods for the detection of rare cells and treatment methods are based on high-end complex clinical settings and molecular tools. These include computed tomography imaging (CT-scan), X-ray, positron emission tomography (PET), magnetic resonance imaging (MRI), sonography, biopsy, cytology, endoscopy, radioimmunoassay (RIA), enzyme-linked immunosorbent assay (ELISA), polymerase chain reaction (PCR), immune-histochemistry (IHC), thermography, and flow-cytometry [4–8]. Although the present methods and technologies are effective, they are cumbersome, intrusive, non-robust, costly, time-consuming, and are limited to large hospital laboratories as it requires skilled personnel for authenticated results potentiates the demand for the development of specific, novel, and ultra-sensitive micro devices. Microtechnology and nanotechnology breakthroughs have enabled miniaturized, on-site, and cost-effective early cancer diagnosis [4]. Therefore, scientists and engineers are adopting a wide variety of microfluidic-based devices for low abundance rare cell manipulation, evaluation, and collection enabling high sensitivity, throughput, and non-invasive real-time monitoring with electric and electrochemical systems needed for point-of-care (POC) diagnosis [9].

Cancer is abnormal and uncontrolled cell growth caused due to a sequence of cellular occurrences such as genomic changes in normal cells followed by protein and transcriptome alteration, tumour generation, cancer-cell migration, and tumour metastasis [10]. This proliferation of cancer cells is marked by release of different biomarkers. This morphological or biochemical attribute of the cancerous cells are different than normal cells, which help in their spotting [8]. Biomarkers are chemical or biological molecules detected in bodily tissues and fluids (e.g. blood, urine, and saliva) under various clinical circumstances and act as an accurate real-time indication of cancer in the body. They are either products or by-products of numerous intracellular events such as genomic mutations and transcriptional or post-transcriptional aberrations [4]. Biomarkers are generally classified based on characteristics as imaging biomarker and molecular biomarker and based on application as diagnostics biomarkers, prognostic biomarkers, and predictive biomarkers. The diagnostic biomarkers assist in the detection of a specific disease, for example, prognostic biomarkers assist in assessment of disease recurrence and the predictive biomarkers assist in the prediction of disease prior to symptoms appearing [6]. Although several types of biomarkers are known, such as protein/surface antigens, microRNA, circulating tumour DNA (ctDNA), cytokines, MMPs, exosomes, CTCs, the detection of protein and CTCs are commonly accepted for diagnosis by medical and clinical teams [4,11,12]. These biomarkers/biomolecules are variably demonstrated in or on tumour cells and are denotative of cancer advancement [8]. Figure 18.1B represents the biomarkers' origin, migration, and colonization of cancer cells. The aforementioned biomarkers are also used in monitoring the response to the treatment allowing patient follow-up. In diagnostic assemblies, biomarkers should be an element that can simply

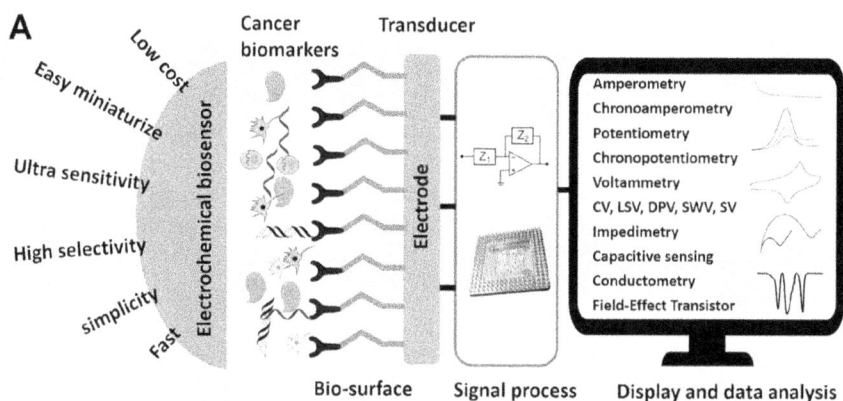

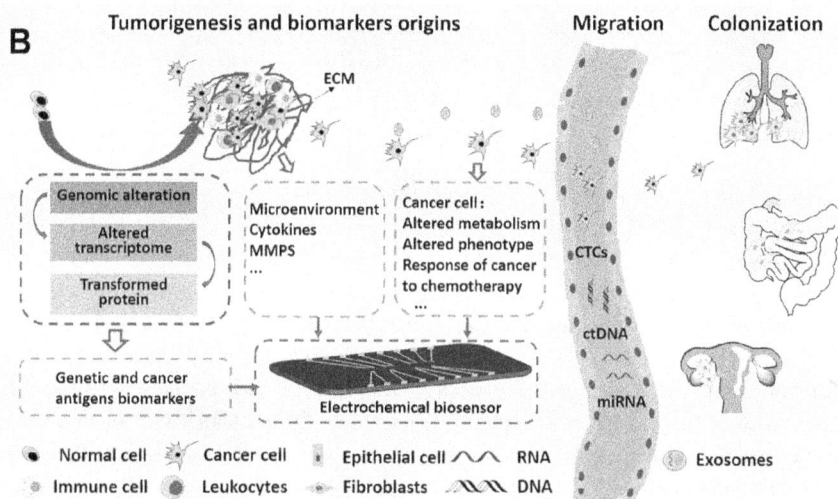

**FIGURE 18.1** Schematic diagram of advantages, component parts, and various measurement methods of the electrochemical biosensor (A) and the cancer biomarker originating from tumorigenesis, cancer cell migration, and colonization to a second site (B) [10]. Adapted with permission from [10]. Copyright (2020) Journal of Electrochmeical Society.

be removable out of the physiological fluids of rehabilitants in a non-invasive method; it shouldn't be present in a healthy body [8].

For cancer biomarker analysis, a biosensor is emerging as an attractive alternative to aforementioned standard methods such as IHC and ELISA. The electrochemical techniques have widespread use in numerous biosensors owing to their high sensitivity, portability, robustness, wider economic feasibility, specificity, superior analytical performance, customizability, and advantage of rapid response because of direct appraisement of physiological fluids (e.g. blood, saliva, serum, milk, urine) using a non-invasive method [8]. Therefore, electrochemical biosensors are widely accepted not just for clinical diagnosis but also for daily health monitoring [4]. This development of non-invasive detection of cancer biomarkers using biosensors is imperative for detection and isolation of rare cells thereby increasing the success rate in diagnosis [11]. Also, the ambiguity surrounding the false positive prostate specific antigen (PSA) test can be eliminated using biosensors [3].

In this chapter, the emphasis is placed on the emerging biosensor-based electrochemical techniques for rare cell detection. As per latest research, the incorporation of microfluidic technology and cancer-on-chip is discussed that has downsized the biosensors further and enabled a more user-friendly detection process. Finally, the challenges, possible solutions, and future scope are elucidated.

## 18.2 BIOSENSORS

A biosensor is a device that translates chemical signals into an electrical reaction by detecting and quantifying biomarkers [6]. They are built based on the type of the molecules to be detected, consisting of binding sites for the biomarkers of interest. Since the invention of measuring blood and tissue oxygen tensions continuously using oxygen electrodes by Leland Charles Clark Jr., the father of biosensors, in the 1950s [13] and commercial launch of the first glucose biosensor by Yellow Spring Instrument Company in 1975 [14], significant improvements have been achieved in both bioreceptor and transducer areas.

Biosensor integration with miniaturized electrochemical systems/transducers are potential alternative tools to conventional assays for rare cell early diagnosis, staging, and prognosis. The use of biosensors as a diagnostic tool holds vast potential. Due to simple, rapid, and cost-effective detection, electrochemical biosensors are also used for environmental monitoring and food quality control [6]. Over the last ten years, biosensor research has rapidly grown in a wide range of illness diagnoses with researchers working on cardiovascular diseases, cancer, and diabetes all around the world. The primary goal is to increase the specificity and sensitivity of biomarker-based quantification methodologies resulting in more dependable, stable, and sensitive biosensing platforms. This wasaccomplished by improving sensor fabrication and manufaturing quality, developing advanced surface chemistries to increase the affinity between target biomarkers and surface ligands, and signal amplification studies using nanomaterials such as magnetic gold particles or quantum dots [7].

Choosing the conductive surface on which the bioreceptors or bio-recognition components (e.g. enzymes, proteins, receptors, antigens, nucleic acids, or antibodies) are immobilized is the first step in the design of a biosensor that is responsible for selectivity. For visual examination of results or response, the conductive surface on which the bio-recognition element is immobilized must be lucent [6]. Aiming to acknowledge a biomarker, recognition elements also known as bioreceptors are used, they interconnect with the biomarkers available on cell surface or the shucked off extracellular domains (ECDs) and thereby produce a relative dose-dependent response. These biochemical motion generated by the interlinkage of the bioreceptor with the biomarker requires further selection of a specific type of bio-transducer. In particular, changes in the proton concentration are determined by the potentiometric bio-transducer, discharge or uptake of the electrons are measured by amperometry bio-transducer, absorption/fluorescence/light emission or reflectance measured using the optical detection, and changes in the mass due to piezoelectric biosensors [8].

Figure 18.1 represents the advantages of electrochemical biosensors, biomarker detections using biosensors equipped with transducers, and various electrochemical measurement methods. The biosensor machine mainly comprises three primary components: the biological recognition surface; the signal transducer; the electronic signal processor that relays, amplifies, and presents the information [3,6]. The transducer, which could be electrochemical, piezoelectric, optical, thermometric, magnetic, or mechanical, transforms the biological signal to an electrical output when the molecular identification constituent recognizes a signal from the environment in the pattern of an analyte [7].

The analyte detected by a biosensor in the case of a rare cell is a tumour biomarker [3]. Therefore, the biosensors can judge whether a tumour is present and whether it is benign or cancerous, and whether a treatment effectively reduces or eliminates cancerous cells by tracking the extent of specific proteins expressed and/or exuded by tumour cells. As most cancers involve multiple biomarkers, biosensors capable of detecting numerous analytes could be particularly beneficial for cancer diagnosis and monitoring. The capacity of a biosensor is to screen for numerous markers simultaneously and aids diagnosis with reduced consumption of both money and time [3]. Biosensors are less invasive and provide more accurate results, which aid in diagnosis.

## 18.3   RARE CELL TYPES AND THEIR ISOLATION

There are many rare cell types currently known such as circulating tumour cells (CTCs), cancer stem cells, circulating endothelial cells, antigen-specific T cells, invariant natural killer T cells, endothelial progenitor cells, circulating rare cells, and many more. Currently, the rare cell research has progressed expeditiously providing much more information on type of biosensor to be used for their isolation. Therefore, the rare cell isolation methods are also discussed.

### 18.3.1   Circulating Tumour Cells (CTCs)

These cells are discharged from the primary tumour into the bloodstream, which makes them the principal supporter of metastasis, which is the main cause of death from cancer. Many scientists targeted these cells to understand the tumour biology and also to improve the clinical management of the disease. The absolute system for the analysis of CTCs must comprise of the maximum efficiency of discovery in real time [15] (Figures 18.2 and 18.3).

## 18.4   ELECTROCHEMICAL BIOSENSORS

The compatibility of the bio-transducer with the bioreceptor is critical to the biosensor's success [8]. The biosensor's recognition element is a crucial component. According to the detection mechanism, biosensors are classified as electrochemical, mass, calorimetric, optical, and aptasensors [6]. Because of great sensitivity, portability, ease of use, specificity, and quick response, electrochemical bio-transducers are widely used in biosensing. The majority of electrochemical research takes

| CTCs Isolation | | | CTCs Detection |
| --- | --- | --- | --- |
| | CELLSEARCH | anti-EpCAM | CD45; CK |
| | IMAGESTREAM | anti-EpCAM; flow cytometry | CD45; CK |
| | MULTICOLOUR FLOW CYTOMETRY | anti-CD45-; flow cytometry | CD45; EpCAM; CK |
| IMMUNOAFFYNITY | MAGNETIC SIFTER | anti-EpCAM | CD45; EpCAM; CK; EGFR mutation |
| | GILUPI | anti-EpCAM | CD45; EpCAM; CK |
| | ISOLATION OF EPITHELIAL TUMOUR CELLS: ISET | Size | CK; PSA |
| | SCREENCELL | Size | CD45; CK |
| | MEMS | Size | not established |
| | ONCOQUICK | Density | CK |
| | CTC-CHIP | Microfluidics | CK;CD45; PSA |
| PHYSICAL PROPERTIES | HB-CHIP | Microfluidics | CK;CD45; PSA |
| | MICROFLUIDIC PLATFORM | Microfluidics | CD45; CK |
| | DIELECTROPHORETIC FIELD FLOW FRACTIONATION CHAMBER | Microfluidics | not established |
| | MULTI-ORIFICE FLOW FRACTIONATION DEVICE-MOFF | Microfluidics | CD45; EpCAM; CK |

**FIGURE 18.2** Showcasing different methods of CTC isolation and detection in a tabular form [15]. Adapted with permission from [15]. Copyright (2014) Sensors.

| Biosensor Principle | Subtype of Transductor | Limit of Detection |
| --- | --- | --- |
| APTASENSOR | Quantum dot label | 5 fM of CEA fragments |
| APTASENSOR | Resonance frequency shifts | 4 LNCap cells/10 mL of blood |
| ELECTROCHEMICAL | Amperometry | DU145 cells concentration of 125 cells per sensor |
| ELECTROCHEMICAL | Impedance | - |
| ELECTROCHEMICAL | Conductometry | 10 MDA-MB-468 cells/mL of blood |
| ELECTROCHEMICAL | Conductometry | 10 ± 1 MCF-7 cells/mL of blood |
| ELECTROCHEMICAL | Conductometry | - |
| MASS CHANGE | Ultrasound | - |
| OPTICAL | Reflectometric interference spectroscopy | 1,000 PANC-1 cells/mL |

**FIGURE 18.3** Showcasing different biosensors for CTC detection and their limit in a tabular form [15]. Adapted with permission from [15]. Copyright (2014) Sensors.

advantage of this characteristic attempting to find the ideal combination of the conductivity enhancer (nanoparticles) and the target bioreceptor, which provide an enhanced readout [8]. Electrochemical sensors provide a useful platform for immobilization of bioreceptors of various types that can generate analyte concentration signals. Receptor proteins, enzymes, antibodies, antigens, and nucleic acids are the examples of recognition elements [3].

Early biosensors relied on purified recognition elements extracted from biological or environmental systems. Many of the biosensor-recognition elements are at present getting manufactured in the laboratory because of improvements in technology and synthetic chemistry, authorizing for increased biosensor function stability and repeatability [3].

Since all reactions are usually detected only near the electrode surface, the electrodes themselves are critical to the effectiveness of the electrochemical biosensor. Electrode material, surface variation, and size all have a significant influence on the sensing capabilities of a particular electrode, depending on the function selected [14]. A standard electrochemical workstation comprises of an auxiliary electrode, a working electrode, and a reference electrode system. The biochemical events relating the targeted bioreceptor interaction occur on the working electrode's surface [8] and the counter electrode connects the electrolytic solution, allowing current to flow to the working electrode [14]. Generally, the electrochemical biosensors involve plane/modified glassy carbon electrodes (GCE), Ag/AgCl (in saturated KCl) and Pt wire as working electrode, reference electrode, and counter/ auxiliary electrode, respectively [12,16–18].

The different types of electrochemical biosensing techniques such as amperometry, impedance spectroscopy, conductometry, and various field effect transistors are discussed below. Currently, the sensing platforms have progressed from electrochemical electrodes to capacitor sensors providing high sensitivity, label-free, and cost-effective detection using a capacitor surface [8]. Therefore, the capacitive sensing methods are also discussed.

## 18.4.1 Amperometry

Amperometry is the most widely used technique in the detection of rare cell biomarkers, where current developing from the reduction or the oxidation of electroactive species in the biochemical reaction is measured continuously at a steady applied voltage. Clark's oxygen electrode is the first amperometry biosensor for detecting glucose, where current isresulted in proportion to the oxygen concentration [19]. Kim and co-workers applied an amperometry immunosensor with AuNP/Den for an early identification of lung cancer biomarkers Annexin II and MUC5AC, which has shown limit of the detection (LOD) of 280 ± 8.0 pg/ml [16]. Chandra et al. demonstrated that an amperometry nano-biosensor is capable of detecting multi-drug resistant cancer cells ($MDR_{CC}$) by sensing permeability glycoprotein (P-gp) in serum and mixed cell samples. It offers linear range (LR) of 50 to $10^5$ cells/ml and LOD of 23 ± 2 cells/ml [12]. In another work, Zhang and co-workers applied an amperometry technique for microRNA-21 (miR-21) detection that's based on 2'-O-methyl modified DNAzyme and DSN-assisted target- recycling without miRNA labeling, enrichment, or PCR amplification [20]. Furthermore, the detection of 5-methylated cytosines (5-mC) in DNA or DNA methylation, one of the most frequent molecular phenomena in human cancer, using immunosensor and DNA sensor–based amperometry is reported [21]. This utilization of electrochemical biosensor avoids the bisulphite and PCR quantification making it suitable for point-of-care applications. The immunosensor detects rare cells with LR of 23 to 24,000 pM and LOD of 6.8 pM, whereas a DNA sensor detects rare cells with a LR of 139 to 5,000 pM and LOD of 42 pM [21].

### 18.4.2 IMPEDIMETRIC

Electrochemical impedance spectroscopy (EIS) is a commonly accustomed impedance method, because of its low-excitation voltage, high sensitivity, and fast speed, impedance biosensors are a reassuring approach for cancer-biomarker detection. The low-excitation voltage of 5 mV or 10 mV is lower than that needed in voltametric methods i.e., 200 mV to 600 mV [10]. This limits the problem of electrode heating, making it appropriate for the long-term and real-time sensing. More importantly, EIS-based biosensors might work on both Faradaic and non-Faradaic modes allowing label-free detection by measuring changes in form of capacitance and/or resistance [22]. The use of interdigitated microelectrodes (IDμE) with EIS offers additional advantages [22].

Chan et al. reported an electrochemical impedance spectroscopy (EIS) biosensor to detect MCF-7 human breast cancer cell lines in whole blood with LOD of 5 cells/ml. In this work, three electrode systems with surface 3D microarray coated with a gold layer have been used, which enhances the capture efficiency of rare cells [18]. In a study conducted by Chen and colleagues, an impedance-based high-density EIS biosensor is developed for breast tumour cell (MCF-7) detection using Au electrode and CMOS substrate [23]. Furthermore, the cancer cells have also been isolated using self-propelled nano/micromotors functionalized with aptamers as bioreceptors, since the oxygen bubbles generated on the nanomotors accelerates movement as well as chemical and biochemical reaction. Based on this concept of electrochemical aptamer-based biosensor, Tabrizi and co-workers targeted human promyelocytic leukaemia cells (HL-60) from human serum using EIS with LOD and LR of 250 cells/ml and 25 to $5 \times 10^5$ cells/ml, respectively [24].

Other impedimetric technique is called capacitive sensing, where capacitive alteration are occurring due to revamping in the obstruction induced by the surface dielectric, charge distribution, or local conductance are measured [22].

Capacitive sensing depends on the capacitive coupling phenomenon. The input to the system in this method is human body capacitance. These sensors can be used to detect anything as far as that object is conductive or has dielectric dissimilar from that of air [25]. An et al. put forward a dielectrophoretic (DEP) microfluidic enrichment platform which has inbred differential contactless capacitive sensor for detection of rare tumour cells. In this work, DEP forces were resulted using sinusoidal function signal, with 16 $V_{pp}$ amplitude and a frequency of 1 MHz, are applied to target cells which are dangled in chamber to pivot them towards sensing electrodes. Furthermore, capacitive measurements were carried out at 300 kHz frequency to locate cells [26].

### 18.4.3  CONDUCTOMETRIC SENSING

Conductometric sensing is a rapidly emerging research field with thrilling new prospects for the advancement of a new analytical tools for biotechnological and bioanalytical applications. They have many distinctive advantages over other sensor such as acceptable for miniaturization and a large-scale manufacturing. Moreover, they don't need a reference electrode but do require a low operatable voltage [27]. These sensors typically contain a planar glass mounting which are integrated with pairs of gold electrodes on one plane of the planar configuration, and the signal is driven due to electrical resistance in the middle of two parallel electrodes by multiple biochemical reactions which changes the ionic compositions in the solution [28]. Tang et al. drafted a design of a conductometric immunosensor in order to get AFP sensing on an anti-AFP capture antibody modified inter-digitated gold transducer utilizing horseradish peroxidase (HRP)-labeled carbon-nanoparticles in order to generate a $H_2O_2$–KI system [29].

### 18.4.4  FIELD-EFFECT TRANSISTOR (FET)

Field-effect transistors (FETs) are appealing for point-of-care operations because of their low detection time credentials. There are some recent studies which shows that incorporation of nanoparticles in FET could help in amplify their sensitivity as a biosensor. Lately, graphene- based materials are also in study for a breast cancer FET biosensor because of its outstanding electrical and mechanical properties [30]. Sun et al. proposed a novel organic field effect transistor (OFET) formed biosensor using 2,6-bis(4-formylphenyl) anthracene (BFPA) substance as a preventive and viable layer for the ultrasensitive and diagnosis of alpha-fetoprotein (AFP)-biomarkers up to femtomolar level (53 fM for threshold voltage (Vth) as a sensing signal and 45 fM for $I_{ds}$ as sensing-signal) in the human serum. In addition, this device shows an improved reliability in monitor targeting the AFP-biomarkers in human serums and has the ability to distinguish between the liver cancer patients from the healthy individuals. This OFET-based biosensor could be used in the future as a definitive tool in the prior diagnosis of liver cancer [31].

## 18.5  FUTURE OF ELECTROCHEMICAL BIOSENSORS

The future for electrochemical biosensors is promising though there are many unanswered questions. Although significant progress has been made in this field, there are still many obstacles to overcome, particularly those requiring the development of novel materials to increase the specificity and sensitivity of biorecognition events and biosensor stability. Apart from that, miniaturization plays a vital role in point-of-care (POC) diagnosis. Therefore, the success depends on developing both miniaturized systems (including electrodes) and biorecognition elements. The integration of multi-channeled microfluidics-based diagnostic automation could lead to down-sized and smart point-of-care devices with public appeal. During a diagnostic experiment, microfluidics enables for regulated laminar flow of participating reactants throughout microchannels, resulting in lower material consumption, smaller sample sizes, real-time analysis, and high throughput screening [32]. On the other hand, this necessitates the requirement of pumping components with valves that adds complication to POC devices for observing numerous forms of cancer at preliminary stages.

The absence of specific, sensitive, low cost, quick and easy detection, and isolation of rare cells has brought more in the growth of electrochemical biosensors research. The emerging technologies and discoveries in lab-on-chip microdevices and the nano-sensors fields offer chances for the growth of completely new and better biosensors [33]. In addition, electrochemical biosensors are easy to integrate into innumerable platforms (i.e., microfluidics) where different assay steps can be involved, making them a good alternative for detection of DNA in actual samples. The synergy between electrochemical detection and the use of NPs, either as labels or as carriers, is given to the devices with evolved sensitivity in addition to evolved stability, if compared to other technologies using enzymes or other labels. With a boost in nanotechnology research in general and in NMs in particular, the price of NP-based DNA biosensors is anticipated to go hand in hand with mass production, making them an ideal tool for easy-to-use devices and future point-of-care for diagnostics, security, and environmental applications.

A growth in electrochemical-DNA biosensors predicated on microchip devices can be observed. Microfluidics platforms (lateral-flow devices or lab-on-chip devices) are the marvellous pathway for integrated DNA technology. These mediums have thus far integrated into a single device, with all the steps necessary for detecting DNA such as: sample pre-treatment, amplification reaction, labellng, pre-concentration, and lastly detection. Majorly, such a level of integration is expected for preferable path for the DNA technology to point-of-care operations in many settings where there are fewer resources and unavailability of specialized personnel to carry out such level of analysis. To escalate more the sensitivity of DNA detection, the integration of isothermal amplification techniques with electrochemical detection can be expected to bring noteworthy advantages. This integration may streamline the design and functioning of DNA biosensors. The possibility of multi-detecting of different DNA sequences or SNPs in the same run (sample) can be reached using various NPs (with various catalytic/electrochemical properties). Studying different DNA sequences in tandem with a fast, inexpensive, and easy-to-use biosensor will greatly simplify the screening of mutations, besides other applications.

## 18.6   CONCLUSION

Currently, different types of cancer are diagnosed only after they have spread throughout the body; at that point, the cancer is essentially untreatable. Therefore, there is an urgent need for accurate, early detection and diagnosis. The utilization of these emerging biosensor technology has been contributory in the prior detection of rare cancer cells, improving a patient's overall chance of survival. The epicentre of advancement in cancer-diagnostic techniques has been recently repositioned to biomarkers because of several advantages, such as efficient detection at a very low concentrations of the biomarkers, well-defined endpoints, numerous biomarkers that can be utilized for parallel detection, and the procedure is relatively speedy and economical.

Biosensors can also be used in medicine to detect glucose levels in diabetic blood, identify infections, and diagnose and monitor cancer; and in the environment to detect dangerous pesticides or bacteria in the air, water, or food. Recently, the military has become very much interested in the advancements of the biosensors as counter-bioterrorism devices, which can detect biological and chemical warfare materials in order to prevent illness or exposure. Biosensors might potentially be implanted on the human body to monitor vital signs, correct irregularities, or even signal a cry for help in an emergency, according to the futuristic vision. Biosensors have an almost infinite number of uses in theory.

## REFERENCES

[1] Dharmasiri, Udara, Małgorzata A. Witek, Andre A. Adams, and Steven A. Soper. "Microsystems for the capture of low-abundance cells." *Annual Review of Analytical Chemistry* 3 (2010): 409–431.

[2] Sung, Hyuna, Jacques Ferlay, Rebecca L. Siegel, Mathieu Laversanne, Isabelle Soerjomataram, Ahmedin Jemal, and Freddie Bray. "Global cancer statistics 2020: GLOBOCAN estimates of incidence and mortality worldwide for 36 cancers in 185 countries." *CA: A Cancer Journal for Clinicians* 71, no. 3 (2021): 209–249.

[3] Bohunicky, Brian, and Shaker A. Mousa. "Biosensors: the new wave in cancer diagnosis." *Nanotechnology, Science and Applications* 4 (2011): 1.

[4] Mahato, Kuldeep, Ashutosh Kumar, Pawan Kumar Maurya, and Pranjal Chandra. "Shifting paradigm of cancer diagnoses in clinically relevant samples based on miniaturized electrochemical nanobiosensors and microfluidic devices." *Biosensors and Bioelectronics* 100 (2018): 411–428.

[5] da Silva, Everson T.S.G., Dênio E.P. Souto, José T.C. Barragan, Juliana de F. Giarola, Ana C.M. de Moraes, and Lauro T. Kubota. "Electrochemical biosensors in point-of-care devices: recent advances and future trends." *ChemElectroChem* 4, no. 4 (2017): 778–794.

[6] Prabhakar, Bala, Pravin Shende, and Steffi Augustine. "Current trends and emerging diagnostic techniques for lung cancer." *Biomedicine & Pharmacotherapy* 106 (2018): 1586–1599.

[7] Altintas, Zeynep, and Ibtisam Tothill. "Biomarkers and biosensors for the early diagnosis of lung cancer." *Sensors and Actuators B: Chemical* 188 (2013): 988–998.

[8] Mittal, Sunil, Hardeep Kaur, Nandini Gautam, and Anil K. Mantha. "Biosensors for breast cancer diagnosis: A review of bioreceptors, biotransducers and signal amplification strategies." *Biosensors and Bioelectronics* 88 (2017): 217–231.

[9] Hiramoto, Kaoru, Kosuke Ino, Yuji Nashimoto, Kentaro Ito, and Hitoshi Shiku. "Electric and electrochemical microfluidic devices for cell analysis." *Frontiers in Chemistry* 7 (2019): 396.

[10] Cui, Feiyun, Zhiru Zhou, and H. Susan Zhou. "Measurement and analysis of cancer biomarkers based on electrochemical biosensors." *Journal of the Electrochemical Society* 167, no. 3 (2019): 037525.

[11] Freitas, Maria, Henri P.A. Nouws, and Cristina Delerue-Matos. "Electrochemical Biosensing in Cancer Diagnostics and Follow-up." *Electroanalysis* 30, no. 8 (2018): 1584–1603.

[12] Chandra, Pranjal, Hui-Bog Noh, Ramjee Pallela, and Yoon-Bo Shim. "Ultrasensitive detection of drug resistant cancer cells in biological matrixes using an amperometric nanobiosensor." *Biosensors and Bioelectronics* 70 (2015): 418–425.

[13] Qlark Jr, L.C. "Monitor and control of blood and tissue oxygen tensions." *Asaio Journal* 2, no. 1 (1956): 41–48.

[14] Grieshaber, Dorothee, Robert MacKenzie, Janos Vörös, and Erik Reimhult. "Electrochemical biosensors-sensor principles and architectures." *Sensors* 8, no. 3 (2008): 1400–1458.

[15] Costa, Clotilde, Miguel Abal, Rafael López-López, and Laura Muinelo-Romay. "Biosensors for the detection of circulating tumour cells." *Sensors* 14, no. 3 (2014): 4856–4875.

[16] Kim, Dong-Min, Hui-Bog Noh, Deog Su Park, Seung-Hee Ryu, Ja Seok Koo, and Yoon-Bo Shim. "Immunosensors for detection of Annexin II and MUC5AC for early diagnosis of lung cancer." *Biosensors and Bioelectronics* 25, no. 2 (2009): 456–462.

[17] Moscovici, Mario, Alyajahan Bhimji, and Shana O. Kelley. "Rapid and specific electrochemical detection of prostate cancer cells using an aperture sensor array." *Lab on a Chip* 13, no. 5 (2013): 940–946.

[18] An, Li, Guangtong Wang, Yu Han, Tianchan Li, Peng Jin, and Shaoqin Liu. "Electrochemical biosensor for cancer cell detection based on a surface 3D microarray." *Lab on a Chip* 18, no. 2 (2018): 335–342.

[19] Clark Jr, Leland C., and Champ Lyons. "Electrode systems for continuous monitoring in cardiovascular surgery." *Annals of the New York Academy of Sciences* 102, no. 1 (1962): 29–45.

[20] Zhang, Xi, Dongzhi Wu, Zhijing Liu, Shuxian Cai, Yanping Zhao, Mei Chen, Yaokun Xia, Chunyan Li, Jing Zhang, and Jinghua Chen. "An ultrasensitive label-free electrochemical biosensor for microRNA-21 detection based on a 2′-O-methyl modified DNAzyme and duplex-specific nuclease assisted target recycling." *Chemical Communications* 50, no. 82 (2014): 12375–12377.

[21] Povedano, Eloy, Eva Vargas, Víctor Ruiz-Valdepeñas Montiel, Rebeca M. Torrente-Rodríguez, María Pedrero, Rodrigo Barderas, Pablo San Segundo-Acosta et al. "Electrochemical affinity biosensors for fast detection of gene-specific methylations with no need for bisulfite and amplification treatments." *Scientific Reports* 8, no. 1 (2018): 1–11.

[22] Arya, Sunil K., Pavel Zhurauski, Pawan Jolly, Marina R. Batistuti, Marcelo Mulato, and Pedro Estrela. "Capacitive aptasensor based on interdigitated electrode for breast cancer detection in undiluted human serum." *Biosensors and Bioelectronics* 102 (2018): 106–112.

[23] Chen, Yu, Chee Chung Wong, Tze Sian Pui, Revanth Nadipalli, Roshan Weerasekera, Jegatha Chandran, Hao Yu, and Abdur R.A. Rahman. "CMOS high density electrical impedance biosensor array for tumor cell detection." *Sensors and Actuators B: Chemical* 173 (2012): 903–907.

[24] Tabrizi, Mahmoud Amouzadeh, Mojtaba Shamsipur, Reza Saber, and Saeed Sarkar. "Isolation of HL-60 cancer cells from the human serum sample using MnO2-PEI/Ni/Au/aptamer as a novel nanomotor and electrochemical determination of thereof by aptamer/gold nanoparticles-poly (3, 4-ethylene dioxythiophene) modified GC electrode." *Biosensors and Bioelectronics* 110 (2018): 141–146.

[25] Shadwani, Mayank, Shivani Sachan, and Palak Sachan. "Capacitive sensing & its applications." *International Journal of Engineering Research and General Science* 4, no. 3 (2016): 265–269.

[26] Do, Loc Quang, Ha Tran Thi Thuy, Tung Thanh Bui, Van Thanh Dau, Ngoc-Viet Nguyen, Trinh Chu Duc, and Chun-Ping Jen. "Dielectrophoresis microfluidic enrichment platform with built-in capacitive sensor for rare tumor cell detection." *BioChip Journal* 12, no. 2 (2018): 114–122.

[27] Liang, Jiaming, Jing Wang, Luwei Zhang, Sijia Wang, Cuiping Yao, and Zhenxi Zhang. "Conductometric immunoassay of alpha-fetoprotein in sera of liver cancer patients using bienzyme-functionalized nanometer-sized silica beads." *Analyst* 144, no. 1 (2019): 265–273.

[28] Christy, Francis A., and Pranav S. Shrivastav. "Conductometric studies on cation-crown ether complexes: a review." *Critical Reviews in Analytical Chemistry* 41, no. 3 (2011): 236–269.

[29] Tang, Juan, Jianxin Huang, Biling Su, Huafeng Chen, and Dianping Tang. "Sandwich-type conductometric immunoassay of alpha-fetoprotein in human serum using carbon nanoparticles as labels." *Biochemical Engineering Journal* 53, no. 2 (2011): 223–228.

[30] Novodchuk, I., M. Bajcsy, and M. Yavuz. "Graphene-based field effect transistor biosensors for breast cancer detection: A review on biosensing strategies." *Carbon* 172 (2021): 431–453.

[31] Sun, Chenfang, Rui Li, Yaru Song, Xiaoqian Jiang, Congcong Zhang, Shanshan Cheng, and Wenping Hu. "Ultrasensitive and reliable organic field-effect transistor-based biosensors in early liver cancer diagnosis." *Analytical Chemistry* 93, no. 15 (2021): 6188–6194.

[32] Ranjan, Rajeev, Elena N. Esimbekova, and Valentina A. Kratasyuk. "Rapid biosensing tools for cancer biomarkers." *Biosensors and Bioelectronics* 87 (2017): 918–930.

[33] Bahadır, Elif Burcu, and Mustafa Kemal Sezgintürk. "Applications of commercial biosensors in clinical, food, environmental, and biothreat/biowarfare analyses." *Analytical Biochemistry* 478 (2015): 107–120.

# Index

For Product Safety Concerns and Information please contact our EU
representative  GPSR@taylorandfrancis.com
Taylor & Francis Verlag GmbH, Kaufingerstraße 24, 80331 München, Germany

www.ingramcontent.com/pod-product-compliance
Lightning Source LLC
Chambersburg PA
CBHW060335220326
41598CB00023B/2714